SUB-SAHARAN
AFRICA

ABC-CLIO'S
NATURE AND HUMAN SOCIETIES SERIES

Australia, New Zealand, and the Pacific: An Environmental History
Donald S. Garden

Canada and Arctic North America: An Environmental History
Graeme Wynn

Mediterranean, The: An Environmental History
J. Donald Hughes

Northeast and Midwest United States: An Environmental History
John T. Cumbler

Northern Europe: An Environmental History
Tamara L. Whited, Jens I. Engels, Richard C. Hoffmann,
Hilde Ibsen, and Wybren Verstegen

Russia: An Environmental History
Alexei Karimov and Irina Merzliakova

South Asia: An Environmental History
Richard Hugh Grove

Southeast Asia: An Environmental History
Peter Boomgaard

Southern United States: An Environmental History
Donald E. Davis

Sub-Saharan Africa: An Environmental History
Gregory H. Maddox

United States Great Plains and Intermountain West: An Environmental History
James E. Sherow

United States West Coast: An Environmental History
Adam M. Sowards

NATURE AND HUMAN SOCIETIES

SUB-SAHARAN AFRICA
An Environmental History

Gregory H. Maddox

A B C ⬥ C L I O

Santa Barbara, California   *   Denver, Colorado   *   Oxford, England

Library of Congress Cataloging-in-Publication Data
Maddox, Gregory.
    Sub-Saharan Africa : an environmental history / Gregory H. Maddox.
        p. cm. — (Nature and human societies)
    Includes bibliographical references and index.
    ISBN 1-85109-555-1 (hard cover : alk. paper) — ISBN 1-85109-560-8
(ebook) 1. Africa, Sub-Saharan--Environmental conditions.
2. Human ecology — Africa, Sub-Saharan. 3. Agriculture--Environmental aspects — Africa, Sub-Saharan.
I. Title. II. Series.
    GE160.A357M33 2006
    304.20967—dc22

                                                                 2005034782

09   08   07   06   10 9 8 7 6 5 4 3 2 1

This book is also available on the World Wide Web as an eBook.
Visit abc-clio.com for details.

ABC-CLIO, Inc.
130 Cremona Drive, P.O. Box 1911
Santa Barbara, California 93116–1911

The Acquisitions Editor for this title was Steven Danver, the Project Editor was Carla Roberts, the Media Editor was Sharon Daugherty, the Media Manager was Caroline Price, the Assistant Production Editor was Cisca Schreefel, the Production Manager was Don Schmidt, and the Manufacturing Coordinator was George Smyser.

This book is printed on acid-free paper.♾
Manufactured in the United States of America

CONTENTS

Series Foreword   *vii*

Acknowledgments   *xi*

CHAPTER 1: AFRICAN ENVIRONMENTS
AND THE ORIGINS OF HUMANITY   1

CHAPTER 2: AFRICAN ENVIRONMENTS AND THE
DEVELOPMENT OF FOOD PRODUCTION SYSTEMS   23

CHAPTER 3: AFRICAN ENVIRONMENTS AND
THE DEVELOPMENT OF COMPLEX SOCIETIES   49

CHAPTER 4: AFRICAN ENVIRONMENTS AND
THE ERA OF THE COLUMBIAN EXCHANGE   75

CHAPTER 5: AFRICAN ENVIRONMENTS AND THE
REORGANIZATION OF SPACE UNDER COLONIAL RULE   103

CHAPTER 6: AFRICAN ENVIRONMENTS IN THE AGE OF
CONSERVATION AND DEVELOPMENT   137

CHAPTER 7: CASE STUDIES   169

    CASE STUDY 1: THE FIRST FRONTIER: THE SAHARA DESERT   169

    CASE STUDY 2: DEATH AND LIFE ON THE SERENGETI:
    WILDERNESS IN AFRICA AS MANAGED LANDSCAPE   181

    CASE STUDY 3: TRANSFORMING THE LANDSCAPE:
    FOOD PRODUCTION AND AGRICULTURE IN EASTERN,
    CENTRAL, AND SOUTHERN AFRICA   197

Documents   207

Chronology   257

Glossary   267

Bibliography   297

Index   325

SERIES FOREWORD

Long ago, only time and the elements shaped the face of the earth, the black abysses of the oceans, and the winds and blue welkin of heaven. As continents floated on the mantle, they collided and threw up mountains or drifted apart and made seas. Volcanoes built mountains out of fiery material from deep within the earth. Mountains and rivers of ice ground and gorged. Winds and waters sculpted and razed. Erosion buffered and salted the seas. The concert of living things created and balanced the gases of the air and moderated the earth's temperature.

The world is very different now. From the moment our ancestors emerged from the southern forests and grasslands to follow the melting glaciers or to cross the seas, all has changed. Today the universal force transforming the earth, the seas, and the air is for the first time a single form of life: we humans. We shape the world, sometimes for our purposes and often by accident. Where forests once towered, fertile fields or barren deserts or crowded cities now lie. Where the sun once warmed the heather, forests now shade the land. One creature we exterminate only to bring another from across the globe to take its place. We pull down mountains and excavate craters and caverns, drain swamps and make lakes, divert, straighten, and stop rivers. From the highest winds to the deepest currents, the world teems with chemical concoctions only we can brew. Even the very climate warms from our activity.

And as we work our will upon the land, as we grasp the things around us to fashion into instruments of our survival, our social relations, and our creativity, we find in turn our lives and even our individual and collective destinies shaped and given direction by natural forces, some controlled, some uncontrolled, and some unleashed. What is more, uniquely among the creatures, we come to know and love the places where we live. For us, the world has always abounded with unseen life and manifest meaning. Invisible beings have hidden in springs, in mountains, in groves, in the quiet sky and in the thunder of the clouds, and in the deep waters. Places of beauty from magnificent mountains to small, winding brooks captured our imaginations and our affection. We have perceived a mind like our own, but greater, designing, creating, and guiding the universe around us.

The authors of the books in this series endeavor to tell the remarkable epic of the intertwined fates of humanity and the natural world. It is a story only now coming to be fully known. Although traditional historians told the drama of men and women of the past, for more than three decades now many have added the natural world as a third actor. Environmental history by that name emerged in the 1970s in the United States. Historians quickly took an interest and created a professional society, the American Society for Environmental History, and a professional journal, now called *Environmental History.* American environmental history flourished and attracted foreign scholars. By 1990 the international dimensions were clear; European scholars joined together to create the European Society for Environmental History in 2001, with its journal, *Environment and History.* A Latin American and Caribbean Society for Environmental History should not be far behind. With an abundant and growing literature of world environmental history now available, a true world environmental history can appear.

This series is organized geographically into regions determined as much as possible by environmental and ecological factors, and secondarily by historical and historiographical boundaries. Befitting the vast environmental historical literature on the United States, four volumes tell the stories of the North, the South, the Plains and Mountain West, and the Pacific Coast. Other volumes trace the environmental histories of Canada and Alaska, Latin America and the Caribbean, Northern Europe, the Mediterranean region, sub-Saharan Africa, Russia and the former Soviet Union, South Asia, Southeast Asia, East Asia, and Australia and Oceania. Authors from around the globe, experts in the various regions, have written the volumes, almost all of which are the first to convey the complete environmental history of their subjects. Each author has, as much as possible, written the twin stories of the human influence on the land and of the land's manifold influence on its human occupants. Every volume contains a narrative analysis of a region along with a body of reference material. This series constitutes the most complete environmental history of the globe ever assembled, chronicling the astonishing tragedies and triumphs of the human transformation of the earth.

The process of creating the series, recruiting the authors from around the world, and editing their manuscripts has been an immensely rewarding experience for me. I cannot thank the authors enough for all of their effort in realizing these volumes. I owe a great debt too to my editors at ABC-CLIO: Kevin Downing (now with Greenwood Publishing Group), who first approached me about the series; and Steven Danver, who has shepherded the volumes through delays and crises to publication. Their unfaltering support for and belief in the series were essential to its successful completion.

—Mark Stoll
Department of History
Texas Tech University
Lubbock, Texas

ACKNOWLEDGMENTS

I would like to thank Mark Stoll for giving me the opportunity to write this volume. His comments on the manuscript strengthened it greatly. Steve Danver and Carla Roberts proved very patient editors in seeing this through. My home institution, Texas Southern University, provided a variety of support for the project. All errors and omissions remain my own.

Finally, I'd like to thank my family. My wife Sheryl McCurdy provided inspiration and support. My stepson Anthony lost his soccer partner for a little while. My daughter Kate, now a young woman, started taking the first steps on her own journey as I finished this one.

AFRICAN ENVIRONMENTS AND THE ORIGINS OF HUMANITY

I f we define environmental history to be the history of the interaction of human communities with the natural world, then no environment can be said to have a longer history than that of Africa. On the plains, mountains, and valleys of eastern Africa some 6 million years ago the ancestors of modern humans split from the ancestors of chimpanzees and began the long and still not completely understood evolution to modern humanity. Despite the longest history of human action within and upon a set of landscapes, African environments still arouse images of primeval nature, of beautiful and dangerous "nature run riot" as the novelist Joseph Conrad put it in the *The Heart of Darkness* (1926). Such "myths of wild Africa" (the phrase comes from an excellent study of conservation by Jonathan Adams and Thomas McShane [1996]) not only color popular perceptions (and not just outside Africa but within its growing urban areas as well), but creep into the specialist literature, both technical and historical, both academic and development-oriented. One of the pioneering scholars, Philip Curtin, in the field of African environmental history has commented waggishly to the effect that while environmental history generally studies the nasty things humans do to the environment, in Africa it studies the nasty things the environment does to humans.

Yet it was from such a mother that humanity emerged. African environments nurtured the tiny earliest human communities. She taught them well how to survive and even thrive. She threw at them extremes of variability and change in both space and time. It was from nature that humans learned to struggle, to compete, to wring from a place unforgiving of error sustenance sufficient to survive. At the same time, within this parent's harsh embrace, these humans learned to cooperate, to remember, to teach and learn, to love, to contemplate the sublime in the world around them. In waves, groups of prodigals left the mother that birthed them; hardened by their experiences, they found new worlds that they sought to master. But they took with them, implanted in their genes and remembered in their brains, the lessons taught by their parent. Those that remained sought to gain more from their mother. They learned to nurture animals and plants, to transform rocks into tools and weapons and ore into metal.

Cape Town and Table Mountain, South Africa, 2004. (Photo courtesy of Sheryl McCurdy)

They transformed their communities as they transformed their landscapes and began to compete with each other, often violently. But the mother remained harsh, and as if to warn of the sin of hubris, often brought crashing down the edifices her children built. In the effort to defend themselves against the vagaries of her wrath and to wring from what they deemed wealth, the children, both those who stayed near home and those who wandered, wrought great changes on the face of the mother, threatening to destroy her ability to provide sustenance.

Hence, any attempt to review the history of Africans and their environments must emphasize two central themes. First, African environments are specific to their contexts. Models of human environmental interaction developed for different landscapes and in different historical contexts will not apply in the same way in African history. Whether we are speaking of theories about the invention of agriculture in the distant past or models of conservation developed in the twentieth century, all must bend to African realities.

Second, African environmental history must account for the extreme variability of African environments across time and in space. As with the other continental landmasses, Africa contains a diversity of landscapes. However, in Africa the extremes of variability are often found within a short distance. Similarly, Africa has witnessed extremes of climate variability, along any time scale one wishes to choose. The reality of the threat of drought or flood confronts African communities every year. The El Niño–Southern Oscillation effect has for the last four millennia caused significant variation in rainfall in eastern and southern Africa at periodic intervals, while the North Atlantic Oscillation has generated a similar system in western Africa. The amounts and timing of rainfall in various localities have proven quite different over decades and centuries. The long-term climate shifts, such as the Little Ice Age defined for the Northern Hemisphere, which lasted from about 1560 to 1850, had their counterparts in Africa. All directly affected the way human communities used the environments in which they lived.

These two elements then provide the frame for examining the effects of human agency on African landscapes. Debates over anthropogenic (caused by humans) change in Africa have mirrored those debates elsewhere in the world. Many observers have argued that the twentieth century saw significant degradation in African environments. Forests have been cleared for timber and agricultural land. Africa's wildlife, though still containing the largest number of large animals in the wild in the world, have suffered from both subsistence and commercial hunting as well as the loss of habitat. Erosion has made once productive grazing and agricultural land unproductive. By the end of the twentieth century, pollutants began to contaminate some of Africa's rivers, lakes, and air. Some observers have seen the cause in the rapid rise in African populations in the twentieth century coupled with the development of relatively technically simple production techniques. Others have suggested the cause lies in the incorporation of African societies in a global economic system in which the African commodities, like resources and labor, are produced for the benefit of outsiders. Any general explanation of anthropogenic change must take into account the specificity of African environments and their variability.

This volume then will serve as a brief overview of African environmental history. In summary, two major themes will run through its chapters. First, it will emphasize how the extreme variability of African environments shaped human responses to the landscapes in which they lived, and then the volume will demonstrate how Africans developed the social and technical means of coping with this variability. Second, it will emphasize the ingenuity and tenacity with which African societies sought to control their landscapes. Rather than living at the mercy of the environment, it will show how Africans developed to ensure

The African continent.

survival, sometimes at the expense of long-term sustainability. It will conclude with an analysis of the dramatic changes in the environment over roughly the last century as exploitation of the resources in Africa's environments increased.

This volume will focus on Africa south of the Sahara (or sub-Saharan Africa as it is alternatively called). Such a division of the continent has both geographic and historical justifications but is also fundamentally an external construction. Since about 2500 B.C.E., when the Sahara entered its current extremely dry phase, it has served as a barrier, or more accurately, a filter for biological and human contact. The regions of Africa bordering the Mediterranean have since that time had a fundamentally Mediterranean climate and have historically been part of the Mediterranean world. Yet before that time, for long periods the Sahara received enough rainfall to support more animals and hence humans. In fact, as this volume will show, its drying marks one of the major turning points in the environmental history of the continent. Even after its final drying, the Sahara remained more a filter than an absolute barrier. Nomads, herding camels after about the beginning of the current era, lived in the desert and transported goods and people across it.

Given the fluidness of the barrier posed by the Sahara, Africa south of the Sahara, despite its diversity, can conveniently be considered a unit. Straddling the equator, its climates and environments are tropical, except at its very southern tip. Its seasons depend more on the variation of the rains. Its temperatures vary largely by altitude, not latitude. Despite its great linguistic diversity—African peoples speak some 1,200 different languages—there has been a long debate on the degree of cultural unity among the peoples of the continent. This volume will try to maintain a balance between understanding the singularity of specific ways that Africans have operated in and upon their particular environments and the commonality created across African environments.

Africa, then, is humanity's first home. Its environments shaped human beings as biological animals. Indeed, its variability provides the key to the evolution of humanity as a species. It can be argued that humans developed speech and memory, the key elements of what one might call consciousness, precisely to cope with a variable environment. These elements allowed humans to adapt their behavior, their ways of life, rather than continuing to be subject to the iron laws of biological adaptation. Eventually, human actions brought great changes to African environments, to the point where their ability to survive in them is now threatened.

GEOGRAPHY, GEOLOGY, AND CLIMATE

Efforts to describe the geography of Africa have eased considerably since the nineteenth century when Europeans had little knowledge of the interior of the continent. Today we have the technical means to draw maps based on satellite and aerial imagery. More importantly, interconnectivity has increased to the point that the communities in every part of the continent are linked in some way or another with the rest of the globe. These links are not always benign or easy, but "globalization" is a real phenomenon. Globalization can perhaps best be measured in terms of knowledge and interaction, and in both cases knowledge about African geography and climates, both inside and outside the continent, and the interaction between African communities and the rest of the world, have increased dramatically over the last two centuries. Today, we can know what Africa looks like at many different levels, but the trick is to know how it came to look that way. For the environmental historian this task is as difficult, if not more so, than for the historian of politics or culture since much of what we study does not keep its own records. We shall begin our exploration of African environmental history with a general description of the continent's geography and climate; this approach will allow us to begin to explore the process of change.

Africa's geography forms the basis of both its environmental and human histories. Today over 800 million people live in Africa, and the continent is divided into fifty-three separate countries. Africa is in essence a giant island, connected to the larger Eurasian landmass only at Suez (which is now cut by a canal). It is primarily tropical, split almost in half along its north/south axis by the equator, with most of its landmass falling between the Tropics of Cancer and Capricorn. The movement then of the Intertropical Convergence Zone across the equator creates the monsoon variation in prevailing wind and hence rainfall that determines the climate of the continent. Altitude plays the most important role in determining local climate.

By convention, we refer to several specific regions when we discuss Africa. North Africa borders the Mediterranean and is bounded by the Sahara Desert. West Africa includes Africa south of the Sahara from the Atlantic to Cameroon. Central Africa stretches from Chad to Zambia. East Africa refers specifically to the nations of Tanzania, Kenya, and Uganda, whereas eastern Africa includes all the territory from Ethiopia to Mozambique east of the Great Lakes. South Africa refers to the nation of that name, whereas southern Africa refers to all the territory south of the Zambezi River.

Geographers divide the African continent into nine types of eco-physiognomic zones based on general topography, climate, and resulting vegetation. These zones can be further subdivided into thirty-two physiographic regions with several of those having smaller subregions where topography or waterways create smaller environmental zones in the midst of larger ones. As climate has changed, the extent of many of these zones of climate has varied. Populations of plants and animals have gone through periods of isolation that have led to endemism, especially in some of the highland regions.

Most of eastern and southern Africa consists of high plateaus over 1,000 meters above sea level. The western bulge of the continent in the north lies lower. Two significant features emerge from this situation. First, the higher altitude regions tend to have milder temperature regimes, and indeed some fairly large regions in eastern and southern Africa have very pleasant climates. Second, the rivers that cut through eastern and southern Africa often pass over waterfalls and rapids as they descend to the coast that have in the past limited their usefulness for navigation. Rivers in western Africa such as the Senegal and Niger flow rather more unobstructed to the sea. The two largest rivers of Africa, the Nile

Aerial view of the Niger River. (Yann Arthus-Bertrand/Corbis)

and the Congo, are so long that despite obstructions, they both carried large amounts of human traffic for thousands of years. The lower Nile, from the first cataract at the modern site of Aswan north, developed a large floodplain that served as a travel route through the arid regions of North Africa and became the home of one of the earlier sedentary civilizations, ancient Egypt. South of the first cataract, the Nile was not particularly navigable because of the waterfalls and the great swamps of the Sud, but it still served as a well-watered corridor for human contact, even if at a slower and more difficult pace.

The Congo (formerly called the Zaire) and its tributaries flow through the Central African rainforests. The falls at Malebo Pool in the western Democratic Republic of the Congo (DRC) blocked easy water transport all the way to the Atlantic Ocean, but above the Pool, great expanses of the Congo and its major tributaries such as the Ubangi and the Lualaba eventually allowed the development of extensive riverine transport.

In the case of the Nile and the Niger, the rivers also served as sites of the early development of agricultural and urban communities. The Upper Nile and the Niger Bend both flow through arid environments. Both also have extensive floodplains. As such, they proved crucibles for agricultural and urban development. The Inner Delta of the Niger located on the western arm of the great bend in the river also probably served as the site of the domestication of African rice. Many other African rivers, however, did not create the conditions necessary for very early urban development. Although many had moderately extensive floodplains, such as the Rufiji Delta in modern Tanzania, most also harbored a number of deadly diseases and their vectors such as malaria and river blindness. While the great rivers of the south, the Zambezi, the Limpopo, and others, provided permanent sources of water, they proved rather less important as means of transport and communication except in very localized ways.

The African Great Lakes also provided important resources for humans. The faulting of the Great Rift Valley in eastern Africa created large basins that captured much of the water brought by the monsoons in that area. These are some of the oldest and deepest lakes in the world. Lake Tanganyika dates to 10 million years ago. Lakes Malawi, Tanganyika, Victoria Nyanza, and several smaller ones eventually find a partial outlet in the Nile. Indeed, the level of the Nile flood has been used to determine the level of rainfall in eastern and central Africa. Similarly, Lake Chad drains much of west-central Africa. The levels of those lakes also serve as historical markers of the rainfall level. At various points in the past, especially Lake Chad has covered a much larger area than currently, and all have seen much lower levels.

Altitude plays a critical role in determining climate in Africa. Generally, the higher in elevation an area, the cooler and wetter it is. The plains of western, central, eastern, and southern Africa all face the threat of great variability in rainfall. In addition, their soils are often of great age, sandy, and generally of poor fertility. Many are laterite, meaning the soils contain relatively high concentrations of iron that cause them, if exposed to water and then drying, to turn into a rock-hard surface that can take years to break up. Even the areas covered in the past by rainforest are extremely fragile when cleared. Highland regions have in turn been of critical importance both because they generally receive more and more consistent rainfall than surrounding lowlands and because many of the most important are volcanic in origin. Volcanic rock weathers into highly fertile soil (McCann 1999b).

In West and Central Africa, the rainforest results from an area of high and regular rainfall. Much of the area had at one time a botanical climax with a high canopy and a relatively open area underneath, because not enough sunlight penetrated to permit the growth of plants. This climax was always incomplete because events such as storms, fires, and even grazing by large animals such as elephants created breaks in the canopy. In many areas little of the high canopy remains. Agriculture over the last 2,000 years has resulted in a mosaic of dense bush, some high trees, and open farmland. The West African forest was farmed more intensively before the twentieth century than the Central African forest. Logging activities greatly accelerated the clearance of the forest after the beginning of the twentieth century.

The tectonic activity associated with the creation of the Rift Valley over the last 30 million years in eastern Africa has created an extremely diverse set of environments in close proximity. The area from the Ethiopian highlands through Kenya, Tanzania, Uganda, Burundi, Rwanda, the DRC, Malawi, and down to the Drakensberg Mountains of southern Africa served as the original home of humanity and continued to create the conditions for developing diverse human societies that exploited the resulting ecological variation. In addition, the volcanic areas of the Cameroon highlands and the Ahaggar and Tibesti Mountains in the Sahara have served as havens for animal and human populations.

The final geographic feature that merits mention here are the deserts of Africa. The Sahara, the largest desert in the world, and its smaller counterpart in southern Africa, the Kalahari, host some of the most hostile environments in the world for human habitation. As a result, they remain extremely sparsely populated. The Kalahari has served as the last refuge of speakers of the Khoisan languages practicing foraging lifeways, which until the beginning of the current era dominated much of eastern and southern Africa. The Sahara has loomed even larger in the history of Africa. During the historic era, it has served as a filter between sub-Saharan Africa and the Mediterranean world of which North Africa was a part. Yet the Sahara has not always been either as dry as it is now or as much of a barrier to contact as it became. Indeed, the drying of the Sahara that climaxed in about 2500 B.C.E. had a momentous impact on the civilizations of the Nile Valley and North Africa as well as sub-Saharan Africa. And even since the final drying of the Sahara, its boundaries, as measured by both vegetation and rainfall, have shifted often in response to both climatic variation and human use (Grove 1978).

These variations in environment created the conditions for a diverse floral and faunal population on the continent. The close-packed variations in environment created centers of endemism in many areas. The Ethiopian highlands, the Central African rainforest, the highland regions in East Africa, and the Mediterranean climate of the Cape of Good Hope have become centers with many unique species of plants, animals, and birds. Conversely, the vast expanses of plains and open woodlands created conditions for the movement of vast collections of animals across long reaches of the continent. In part because humans evolved in tandem with other African animals, African wildlife survived the expansion of human populations better than those in other parts of the world, at least until recently.

The Sahara, the largest desert in the world, and its smaller counterpart in southern Africa, the Kalahari, host some of the most hostile environments in the world for human habitation. (Corel)

One of the most complex aspects of understanding the environmental history of Africa is the effect of climate change. From a long-term perspective, climate has altered greatly across the globe and in ways that sometimes obscure the connections between climates in various parts of the world. Climate, in this sense, has two main features: temperature and precipitation. Both of these vary in annual cycles (the seasons), and both can vary somewhat independently of one another in fluctuations that seem almost random. Although the last 11,000 or so years (the Holocene era) have seen a fairly stable global climate, slight shifts in global temperature and ocean wind and sea currents have brought dramatic changes in African landscapes. The Sahara has contracted and expanded; forests have changed in size and composition; and lakes have overflowed their banks and dwindled down to marshy areas. Even shorter time-scale fluctuations have been important. Since 1850, global warming has occurred in general, and regions in Africa have experienced decades-long declines and increases in rainfall.

Currently (and for the last 14,000 years), sub-Saharan Africa's basic climate pattern is governed by the movement of the Intertropical Convergence Zone

(ITCZ) across the equator each year. The ITCZ marks the meeting of north and south air currents, each bringing varying amounts of heat and moisture. In January, the ITCZ shifts south of the equator, bringing dry continental air to West Africa and moist Indian Ocean air to East Africa. By July, the ITCZ shifts northward, bringing wet Atlantic air to West Africa and dry air to East Africa. The result in much of Africa is a bimodal seasonal system divided between a wet and a dry season. Closer to the coast in both East and West Africa, there is one long "summer" rainy season; further inland, a break often occurs in about the middle of the rainy season. In West Africa, the further north one moves into the Sudan and Sahel, the rainy season's intensity and duration can vary markedly from year to year. As a result, a banding effect in vegetation takes place from rainforest to forest-savannah mosaic to open grassy savannah to arid savannah to desert. In Central Africa, the convergence of ocean currents brings constant and regular rainfall and helps create the Central African rainforest (Fein and Stephens 1987).

In eastern Africa, the highlands along the Rift Valley from Ethiopia to South Africa create rainfall catchments and shadows all the way to the Great Lakes. The highlands, especially in East Africa, are often fertile and well-watered oases surrounded by semiarid plains. Microclimates based on altitude and the direction of the prevailing winds often exist within easy walking distance of each other. Along the great central plateau that stretches from East Africa to southern Africa, the bimodal distribution of rainfall continues with open grasslands and grassy woodlands alternating. In the far south, ocean and air currents combine to create another desert along the southwestern corner of the continent in the Kalahari. In the southernmost tip, the climate abruptly shifts to a Mediterranean climate marked by wet winters and dry summers.

In many parts of lowland Africa apart from the rainforests, the rains vary greatly from year to year. A variety of factors causes these variations. Perhaps the most important factor for eastern and southern Africa is the El Niño–Southern Oscillation, whereas for western Africa it is the North Atlantic Oscillation. Normally, wind and ocean currents in the southern Pacific flow from east to west, bringing warm water and air to Southeast Asia and forcing cold, deep ocean water up along the coast of South America. These connections, in turn, drive the movement of an Indian Ocean low-pressure system westward, bringing the monsoon rains to India and eastern Africa. The El Niño effect occurs when seas warm off the coast of South America. This warming, in reality a stoppage of "upwelling" of colder deep ocean water, starts a chain reaction, called teleconnections, across the Pacific and into the Indian Ocean. As a result, the Indian Ocean low-pressure system exerts much less influence, and drought often occurs in eastern Africa. Egypt feels the effect as the rains in the Ethiopian highlands determine the level of the Nile flood downstream several months later. La Niña effects, an extra strong return to the normal flow of water and air, often bring exceptionally heavy rains to eastern Africa. Various types of evidence, including tree-ring analysis, study of ice cores, and documentary research, have recovered evidence of these conditions for the last 5,000 years (Grove and Chapell 2000, 1–4).

These types of annually measured variations are extremely important for understanding current environmental conditions, but longer-term variations have perhaps been important for determining the potentialities for human action within environments. These periods of long-term climate change are caused by minor changes in the Earth's tilt and distance from the Sun and by variations in the intensity of solar radiation. The changing general temperature of the Earth then causes changes in everything from the amount of evaporation and hence rain to the great global currents that drive the weather. These shifts operate at time spans of tens of thousands of years. The last 14,000 or so years of relatively stable climate mark an interglacial era—between glacial ages—called the Holocene (A. T. Grove 1993, 33–42).

The glacial cycles affected Africa in several ways. During cold phases, including eras of glaciation, the Sahara has dried to a desert. During warmer phases in the Earth's climate, more moisture has reached the Sahara during the summer rains of West Africa. During the wetter times, Lake Chad and several other basins deep in the desert fill with water. Grasses flourish, and the woodlands of the Sudan move north. During drying episodes, the lakes dwindle and vegetation retreats to the highlands of the Ahaggar Plateau and the Tibesti Massif where the altitude draws slightly more moisture out of the arid air. Two of the most important drying episodes occurred around 72,000 years ago and around 6,000 years ago. Both resulted in the Saharan region, in the words of the archaeologist Brian Fagan (1999), becoming a giant pump, pushing people and animals north to the Mediterranean, south to the Sahel, and east to the Nile Valley. The first instance coincides with the most recent estimates of the spread of modern humans out of Africa and into the rest of the Old World. The second, the episode that resulted in the drying of the Sahara as we know it today, coincides with the spread of agriculture and the development of urbanization in both the Nile Valley and sub-Saharan Africa.

Since the beginning of the Holocene era about 14,000 years ago, the Earth's climate, including Africa's, has been relatively stable compared to that during the eras of glaciation and warming that preceded it. This relative stability has contained within it some fairly strong shifts. A warmer, wetter period between 8,500 and 4,500 years ago brought grasses to the Sahara. The final drying of the Sahara did not occur until about 4,500 years ago. The Earth went through a period of much cooler temperatures than today between about 1100 C.E. and 1850

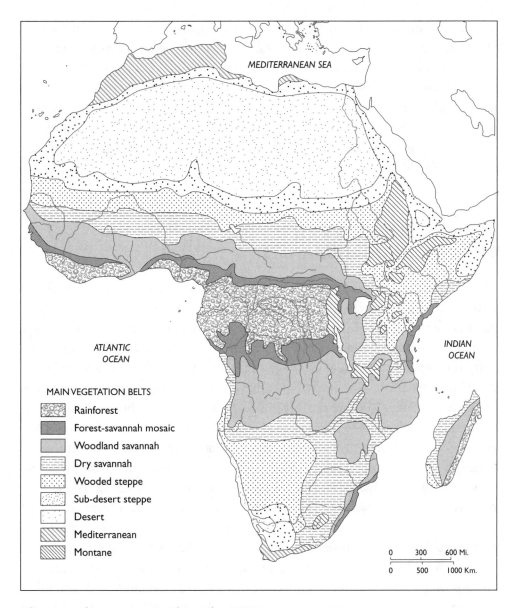

MAIN VEGETATION BELTS

- Rainforest
- Forest-savannah mosaic
- Woodland savannah
- Dry savannah
- Wooded steppe
- Sub-desert steppe
- Desert
- Mediterranean
- Montane

Climate and vegetation in Africa after 2000 B.C.E.

C.E. The Earth has warmed since then, despite a slightly cooler phase in the 1950s, driven in part by the greenhouse effect from the increased burning of fossil fuels associated with industrial growth (A. T. Grove 1997, 35–37).

Although climate change can be thought of as driving change in landscapes and in the organisms that occupy them, over the last 100,000 years at least human agency has also had a hand in altering landscapes. For much of that time, as in the preceding 2 million years when the species ancestral to modern humans had developed in Africa and spread across the Old World, the few bands of protohumans or humans living by foraging had little measurable effect on landscapes. However, humans survived as a species by learning to alter landscapes in order to produce more subsistence from them. Humans, more than any other animal species, were not bound by ecological niche but instead learned to alter their behavior and their environments. Such alterations began with the development of toolmaking technologies and the mastery of fire, long before the domestication of plants and animals. The first landscape to feel the effects of the new lineage's ability to alter the landscapes it lived in was Africa.

AFRICAN ENVIRONMENTS AND THE ORIGIN OF HUMANITY

The origin of humanity in Africa is no longer in much debate. The insight began with Charles Darwin (1859) who noted that most of the species that seemed most closely related to humans lived in Africa, and hence, one could safely presume that humanity originated in Africa. Since the 1930s, the discovery of pre-hominid remains in Africa of great age and large number and diversity, and since the 1950s the dating of these remains, have led to a consensus about Africa as the original "Eden." Debate has continued, however, about the evolutionary order within which to place the various species identified by archaeologists as pre-hominid and about the timing of the departure of some humans from Africa and hence the relationship among modern human populations. Recently, work on the human genome has tended to lend further support to those who argue for a relatively recent expansion of modern humans from Africa (starting about 100,000 years ago) that replaced earlier hominid populations that had spread across the Old World. Earlier theories had noted the spread of a species called *Homo erectus* in Europe and Asia from Africa after about 1.8 million years ago and postulated that populations of *H. erectus* had engaged in parallel (and interconnected) evolution to become *Homo sapiens* (Smith 1997; Stringer and McKie 1996). As best can be reconstructed, the origins of humanity lie in the Rift Valley of Africa. Recent genetic studies of chimpanzees suggest a common ancestor between those animals and humans who lived about 8 million years ago (Smith 1997b; Vrba 1999). Fossils of primates that seem ancestral to hominids have been found dating to about 6 million years ago in Chad and Kenya, but the small number of fossils makes understanding their place in evolution difficult. About 4.5 million years ago, the species called *Ardipithecus ramidus* lived in eastern Africa. It was related to a number of other primates that lived in Africa and had spread to Europe and Asia. As the Rift Valley gradually opened, lakes and mountains developed creating the great variability in landscapes found on the eastern side of the continent. The environment in eastern Africa generally dried, and many of the animals present then went through evolutionary changes that emphasized grazing on open grasslands rather than surviving in woodlands (Stringer and McKie 1996).

From *ramidus* emerged the first *Australopithecus*, a species called *anamensis* that lived between about 4.2 and 3.9 million years ago. Australopithecines, the name given to several species that lived in eastern and southern Africa, survived for about 2.5 million years. *Anamensis*, the earliest found species, walked upright, perhaps to provide cooling for its body in the open grasslands, perhaps to carry food back to a home base. Several species followed, and archaeologists still debate quite hotly the exact relationship between the various species and the emergence of hominids (Smith 1997a; Wolpoff and Caspari 1997). Australopithecines survived by gathering, some hunting, but perhaps especially scavenging. By about 3.9 million years ago, another species had emerged, *A. afarensis*. In general, many think that a slightly larger species emerged from *afarensis*, *A. africanus*. From *africanus* emerged two general lines, *A. robustus*, larger australopithecines adapted to living in grasslands and mostly vegetarian, and a smaller line of gracile creatures (often continuing to be called *a. africanus*) after about 2 million years ago. The gracile australopithecines remained omnivores and seem to have hunted more than other species. Remains of australopithecines have been found from Ethiopia to southern Africa.

From the gracile side of the tree came the first identifiably human creatures on Earth. Archaeologists starting with Richard and Mary Leakey call this species *Homo habilis*, handy man. *Haiblis* remains show a larger brain size relative to body size and were found in eastern Africa with an abundance of stone tools. Some scientists identify a few remains from about the same time as a second species called *H. rudolfensis*. Very quickly, from *H. habilis* another new species emerged, *H. erectus*. *Erectus* was larger than any preceding hominid and had a larger brain size than *habilis*. For a few hundred-thousand years *erectus*, *habilis*, and several species of australopithecines lived in eastern and southern Africa. After about 2 million years ago, only *erectus* survived (Stringer and McKie 1996; Smith 1997b).

A critical question remains as to why hominids emerged and took the evolutionary path they took (O'Brien and Peters, 1999). Some biologists believe that a dramatic drying event 2.5 million years ago (the so-called 2.5 Million Year Event) on a global scale created huge population stress for almost all animals, but especially hominids. Faced with declining sources of food, one group evolved toward greater cooperation in foraging practices and the use of tools in finding and preparing foods. Extinctions of many species of animals occurred in the millennia around this drying.

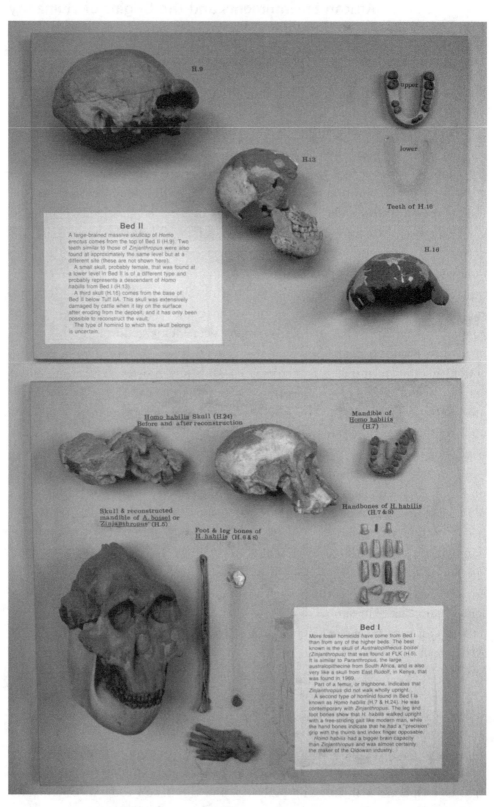

A collection of hominid fossils, mostly skulls, including Homo habilis, Homo erectus, *and* Australopithecus bosei, *a.k.a.* Zinjanthopus, *at the Leakey Museum in Olduvai Gorge. Tanzania, ca. 1985–1995. (Buddy Mays/Corbis)*

Erectus survived because they had developed the ability to use tools—technology—to generate more food. Fossil remains found with *erectus* indicate a mixed diet. Although the large number of animal bones found in spots identified as living floors occupied by *H. erectus* do indicate that *erectus* populations hunted, they also show that they survived by scavenging from the kills of larger predators as well as by foraging. The combination meant generally that *erectus* populations were not tied to specific environments but could adapt to new ones and to changing conditions. Some *erectus* populations proved this by moving down the Nile Valley and on into Asia and eventually Europe. By about 1.8 million years ago, *erectus* populations seem to have at least lightly inhabited most of the tropical and subtropical regions of the Old World. Some specialists suggest the existence of a species intermediate to *H. habilis* and *H. erectus, H. ergaster.* Some also suggest that the hominid population in Africa became a different species from the *H. erectus* populations in Asia and should be called *H. antecessor* or *H. maritanicus* (Stringer and McKie 1996; Blumenschine 1997).

Social cooperation began with the development of tool use. Archaeologists have defined two separate tool kits at sites associated with early hominids. The earliest they call Olduwan (after Olduvai Gorge in Tanzania), and most agree that *Homo habilis* populations made these tools about 2.5 million years ago. Most commonly, it consists of "choppers" from which the toolmakers struck blades. The blades became cutting edges, and the choppers apparently became heavy hand tools. Archaeologists have generally found these tools in limited quantities in eastern and southern Africa. The hominids of Africa continued to make them up to about 1.4 million years ago (Barham 1997).

Archaeologists label the second tool kit Acheulian after the town in France where it was first found. These tools are associated with the larger *H. ergaster* in Africa and *H. erectus* elsewhere. These slightly more complex stone tools have been found not only in Africa but across the Old World tropical belt all the way to Indonesia. They began to appear about 1.4 million years ago and do not disappear until 250,000 years ago. In general, archaeologists label the Olduwan and Acheulian industries the Early Stone Age, and the post-Acheulian the Middle Stone Age.

Based on intensive work particularly at Olduvai Gorge, archaeologists generally agree that the *H. ergaster/H. erectus* populations that made them followed lifeways that broadly resembled those of later forager populations. In particular, they hunted more than scavenged, although plant food remained a major part of the diet. With the evolution of bipedalism, infants were born smaller and helpless. As a result, social coordination had to develop in order to provide extended care for children. Hence, some suggest that cooperation among females increased. Lifespans increased beyond the reproductive years as older members of the band could care for children while younger members could continue foraging. The prolonged maturation of young also seems to have encouraged the long-term pairing between a single male and female (S. Kusimba 2003).

Hominids who practiced this lifeway spread in a variety of different environments. They could adapt their practices to different conditions and learn how to exploit different types of food sources. Indeed, they spread across much of the Old World and endured over twenty alternations between glacial and warmer climates. These hominids knew how to use fire both to cook and to keep other predators away from the base camps they set up. They remained mobile, following food sources in bands.

Control, or at least use, of fire also became one of the first and most enduring ways that humans began to alter the landscapes within which they lived. Eventually, humans learned to set fires to drive animals toward pits and killing zones. Such fires had the effect of reducing low woody growth while encouraging grasses and certain types of fire-resistant trees. Across the savannahs of Africa, bimodal rain distribution meant that conditions for large fires existed during the dry season. When humans began to set fires, the proportion of open grasslands increased over what had existed under a "natural" fire regime. Archaeologists are unsure when human communities began to use fire to manipulate landscapes, but some evidence exists that it could be as long ago as 1 million years. Certainly, by the time *H. sapiens* became the dominant species, human communities commonly used fire where possible to transform landscapes.

The creators of the Acheulian lifeway had their limits. The basic tool kit remained fundamentally the same across the entire range of the species until the very end of its existence as newer hominid species began to displace them. Only the raw material changed with the landscape. Similarly, they seem to have trod lightly on the land, never becoming too large in number or overly affecting the remainder of the natural world in which they lived (Stringer and McKie 1996).

A set of dramatic changes began about 300,000 years ago. These changes mark the biological development of modern humans and the beginning of humanity's ability to remake landscape in an ongoing effort to make it more productive. Constituting the first evidence for these changes was the appearance of new types of stone tools. Archaeologists have labeled this the beginning of the Middle Stone Age in Africa. These new tool kits were more varied and included "microliths"—very small stone blades meant to be hafted into a handle as in a spear or harpoon or arrow. Similarly, a greater diversity of tool types appeared, and regional variation became much more pronounced (Barham 1997).

At the biological level, at least two different new species of hominids emerged. In Europe, *H. neanderthalensis* emerged and populated most of Europe down to the shores of the Mediterranean. In Africa, archaic humans arose, sometimes called *H. heidelbergensis* or *H. rhodesiensis*. Both of these "events" followed the appearance of the new, more diversified tool kits and the spread of hominid populations into new territories. *H. erectus* continued to occupy much of Asia (Smith 1997a).

Most scientists agree that modern *H. sapiens* emerged in Africa between about 200,000 and 130,000 years ago and eventually displaced the previous hominid populations in the Old World. Some specialists, however, believe that the *H. erectus* populations of the Old World remained in sufficient contact with each other to facilitate gene flow, and hence modern humans emerged across the Old World (see Stringer and McKie 1996 for a discussion). The evidence for this view is not totally conclusive. Most scientists favor the idea of a unique origin in Africa for modern humans. The earliest modern human remains have been found only in Africa and in nearby southwest Asia. In Africa, western Asia, and Europe, modern humans lived at the same time as other hominid populations, indicating that they could not be descended from these other hominids. DNA testing of material from Neanderthal remains shows a wide gap of genetic difference, indicating a common ancestor over 500,000 years ago (Sykes 2001). Mitochondrial DNA (a form of DNA that is inherited only from the mother and that mutates at a very rapid rate) testing indicates that modern humans share a common ancestor who lived about 200,000 years ago. Genetic diversity among African populations is much greater than that outside Africa, indicating that humans have lived in Africa longer. However, some scientists, pointing to fossil evidence particularly from East Asia, suggest that *H. sapiens* in some regions may have emerged from breeding between an incoming population and existing hominid populations.

Some specialists link the emergence of a new species to climate change. During the glacial episode that ended about 130,000 years ago, a small group of hominids, made up by possibly as few as 20,000 individuals, were isolated and became the ancestors to all modern humans. Given a sharply drier climate over most of eastern Africa, selection may have operated to push this small population toward larger brains and a greater ability to communicate and think in abstract terms. The population pinch in Africa is supported by a decline in the number of sites that can be securely dated to the time around 150,000 years ago. The earliest clearly modern human remains have been found in three locations. The oldest come from South Africa, dated to about 120,000 years ago. Later finds come from Ethiopia dated to 100,000 years ago and Israel to about 90,000 years ago. The first humans to move out of Africa did not stay permanently, it seems. An extreme cold snap caused by the eruption of Mount Toba in Sumatra about 74,000 years ago caused dramatic global cooling and seems to have resulted in a demographic crash for *H. sapiens.* Populations in Asia became isolated from the original population in Africa and disappeared (Stringer 2002).

Until about 60,000 years ago, then, Africa remained home to modern humans. Only after that time did humans begin to establish themselves outside of Africa—first up the Nile to North Africa and then across the Sinai into the broader Near East. This group quite possibly numbered only in the hundreds. A second group may also have left Africa directly across the Red Sea into Arabia. The descendants of these emigrants gradually expanded their range and in some way or another displaced the hominid populations found in their path. In Africa, after about 100,000 years ago, no more remains for other hominids have been found. In Asia, gradually the *Homo erectus* populations died out, perhaps in part because of the dramatic climate change of the time. In Europe, Neanderthals, adapted to very cold weather, survived deep into the last glacial age of the Late Pleistocene, disappearing only about 30,000 years ago. Archaeologists have no evidence as to exactly how *Homo sapiens* replaced the other hominid populations. They have found no conclusive direct evidence for conflict. However, in some way or another, *Homo sapiens* engaged in a contest for ecological space with other hominids and eventually replaced them.

Two striking features mark the brief outline of human evolution given in this chapter. Compared to earlier views of the emergence of humanity, current research emphasizes that modern humans emerged relatively recently and that some of them left Africa recently. Earlier views had posited a much longer division of human populations based on a process of "parallel" evolution from *H. erectus* populations. Such a model would be unique among species, and genetic analysis contradicts it (see Stringer and McKie 1996; Smith 1997b; Wolpoff and Caspari 1997). An adjunct conclusion is that the isomorphic differences between human populations subsumed under the label "race" are relatively recent and in fact transient developments. They reflect short-term adaptations to general environmental conditions. Most relevant for our case is the greater diversity within African populations than between African populations and populations from outside the continent. Race, then, is a historical construct—real because history makes it real—but it is not a concept that explains history.

CONCLUSION

In one sense, it is easy to view Africa as the home of humanity and then to pass over the development of human populations on the continent after the prodigals left the homeland. Yet, the development of human societies and their technologies that enabled them to survive in Africa continued. The harshness and variability of Africa's environments honed the survival and cooperative skills of humans that allowed them to thrive in other places and at other times and in turn to attain more abundance. Given that the power structure and the methods of producing knowledge have centered on the West for the past two centuries, more information has been available about the development of human societies from outside Africa than within it. Hence, scholarship has often proceeded from the assumption that Africans lagged behind the rest of the world in the development of agriculture, metalworking, and complex, urban societies. In the past 100,000 years, Africans have developed their own means of providing the subsidence necessary for survival and for creating complex societies. As we shall see in later chapters, Africans found it necessary to develop their own forms of food production. And scholars are just beginning to understand those processes.

2

AFRICAN ENVIRONMENTS AND THE DEVELOPMENT OF FOOD PRODUCTION SYSTEMS

For the vast majority of human existence, people survived by eating plants and animals gathered from the wild. Foraging (also called hunting and gathering) brought subsistence to the hominid populations and allowed them to expand out of their African homeland. Foraging meant that human populations depended on the types of plant and animal foods available in a specific environment. The driving force behind human evolution had been the development of the ability to adapt to different environments and to learn to exploit different sources of food. This development allowed human populations to spread across different environments rather than be confined to particular environmental niches. They also developed the ability to alter landscapes in order to make them more productive of the food sources they required, especially through the use of fire. The greater ability to adapt to different environments, reflected in the development of different and more complex tool kits, allowed *H. sapiens* to spread out of Africa and replace the previously existing hominid populations throughout the globe.

By about 10,000 years ago, foraging began to give way to food-producing lifeways based on the cultivation of plants and the raising of domesticated animals. The transition from foraging to food producing was a gradual and multicentered one. Rather than a cataclysmic revolution occurring in a relatively few areas of the globe and spreading in waves of diffusion, the process was one that occurred in many different areas of the globe, with different communities developing different agricultural techniques based on different environmental conditions (Stahl 1984; Sinclair, Shaw, and Andah 1993). Hunting, fishing, and gathering remained important parts of the subsistence strategy of many African peoples into the modern era. African landscapes have proven just as capable of hosting agricultural societies and encouraging innovation as those in other parts of the world.

The development of food production techniques led to the growth of larger populations and the extension of permanent settlements. Agriculture and animal husbandry also meant that human communities began to reshape the landscapes in which they lived. Communities cleared bush, often using fire as their primary tool, and even began the arduous process of turning forests into fields and grazing areas. In doing so, they changed the habitat for many other organisms. The long, gradual process of change did not, however, mean that human communities controlled their environments. Rather, they learned to manipulate elements of their environments in ways that enhanced their chances for survival. However, such manipulations did not always succeed. Climatic variability, climate change, and modifications in the landscape itself induced by human agency all occasionally overwhelmed these strategies and caused dramatic changes in population.

A final technical element in the ability of human communities to wring survival out of Africa's environments came in the form of metalworking technologies, most importantly iron working. These technologies gave human communities not only a new set of tools with which to clear lands and hunt animals, but also new weapons with which to contest control over resources.

Foraging in Africa has a long history. This lifeway survived a cycle of climate change during the last 100,000 years, extending to the end of the last glacial phase about 13,000 years ago and the beginning of the Holocene, or the current era. Human communities in Africa developed a diversity of such lifeways as is evidenced by the artifacts they left behind. After the beginning of the Holocene, African societies developed food production in the form of animal husbandry and agriculture. This chapter will relate the evidence for the development of food production and the formation of identifiable communities on the basis of archaeological findings and historical linguistic data. It will conclude with a discussion of the development of metalworking technology and its impact on the ability of human communities in Africa to affect their environments.

CLIMATE CHANGE AND THE TRANSITION TO THE HOLOCENE

Scientists label the last 2.6 or so million years the Quaternary era. This era in the Earth's history is marked by periods of global cooling leading to the development of glaciers and the expansion of ice packs around the poles interspersed with periods of warming called interglacials. Using data collected from deep marine cores that measure the relative amounts of various isotopes of oxygen, scientists have identified up to thirty glacial-interglacial cycles over the last 2.6 million years. The era lasting to the end of the last glacial era is called the Pleistocene, and the interglacial that began about 13,000 years ago the Holocene era.

Changes in the atmosphere resulting from volcanic emissions up to 2.5 million years ago generated conditions that made possible the cycles of cooling and warming seen in the Quaternary. Once the atmosphere had cooled to a specific

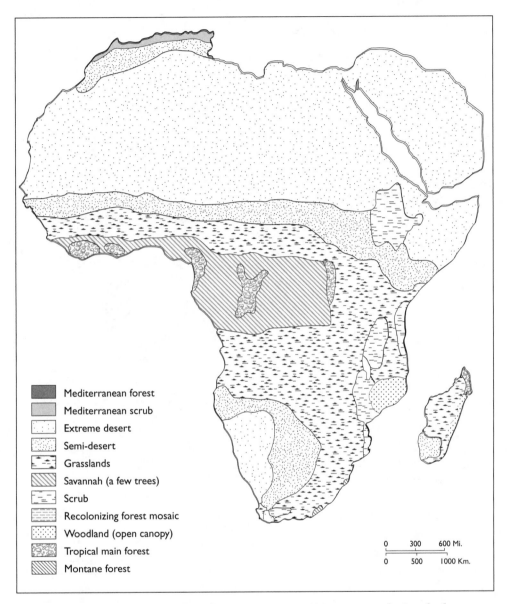

Reconstructed vegetation in Africa from 16,000 to 20,000 years ago during the last Glacial Maximum.

point, systematic variations of the Earth's position relative to the Sun led to changes in climate that were provoked, or "forced" in the terminology of the field, by shifts in deep ocean currents. These cycles, called Milankovitch cycles after the Serbian mathematician Milutin Milankovitch (1879–1958) who first identified them, have operated in an epoch generally much cooler than the one that preceded it. In this view, the Holocene is a warm interglacial. The differing scales of variability mean that the relationship between the various factors affecting climate are complex. Indeed, the present-day global warming, though cer-

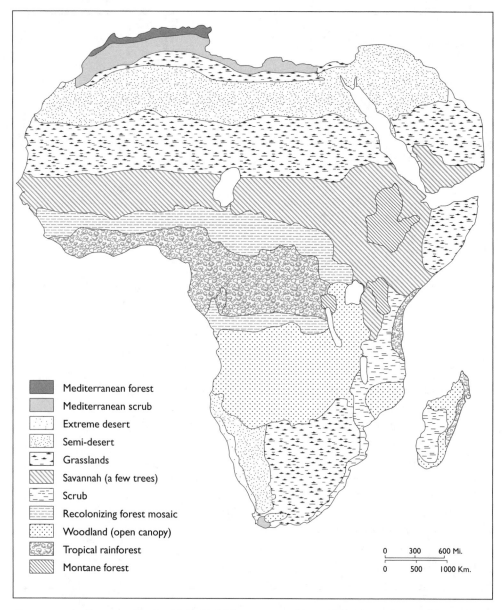

Legend:
- Mediterranean forest
- Mediterranean scrub
- Extreme desert
- Semi-desert
- Grasslands
- Savannah (a few trees)
- Scrub
- Recolonizing forest mosaic
- Woodland (open canopy)
- Tropical rainforest
- Montane forest

| 0 | 300 | 600 Mi. |
| 0 | 500 | 1000 Km. |

Reconstructed vegetation in Africa 9,000 years ago during the African Humid Period.

tainly attributable in large part to human agency in altering atmospheric chemistry, could in fact lead to the early onset of a new glacial era if it reduces the flow of warm tropical water into the northern and southern oceans (Fagan 1999).

For African climates, these cycles of cooling and warming had interesting effects. Glacial conditions both cooled temperatures and reduced moisture in the atmosphere. During the last glacial peak between about 30,000 and 14,000 years ago, glaciers appeared in southern Africa. The Sahara expanded at least 500 kilometers (about 300 miles) further south. Vegetation changes included the downslope spread of Afro-montane vegetation (dependent on cooler temperatures) and the virtual disappearance of tropical rainforest. Only three remnant areas of forest survived. The Senegal, Niger, Shari, and Nile all ponded, and Lakes Victoria and Malawi went dry at about 13,000 years ago. However, this cooler era also meant that tropical water flows changed, and wetter conditions existed in the Kalahari Desert in southwestern Africa (A. T. Grove 1997, 35–39).

The transition to the Holocene climate, as we have noted, began about 14,000 years ago, and it is against this warmer backdrop that human history developed. Even within the generally warmer Holocene, climatic variation has remained common. A wetter and warmer period marked the beginning of the Holocene for Africa from about 14,000 to about 10,500 years ago. Then, for about 1,000 years, a cooler, drier era ensued. At about 9,500 years ago, the Holocene "Optimum" began, setting in motion the warmest and wettest phase of climate during the Holocene. The Great Lakes expanded, and the southern Sahara became a savannah as monsoon rains reached far to the north of their current extent. In the western Sahara, Niger tributaries stretched deep into the desert, and Lake "Mega-Chad" fed it through the Beneu. Although after 7,000 years ago, a general cooling and drying began, the final drying of the Sahara only started about 4,500 years ago. While short-term variation has continued, current conditions were well established by about 2,000 years ago.

The record of climate change and associated change in vegetation provides a backdrop for how human communities survived within Africa. Scholars have often fallen prey to the temptation to relate both biological-evolutionary change and changes in human societies to environmental change. Similarly, for more recent eras, it is easy to link environmental change, especially degradation, to human activity. Although such linkages do indeed exist, they are often not as direct as they may seem. For example, environmental conditions promoted the development of food-producing technologies—agriculture and animal husbandry—but in the long run these technologies expanded because humans chose to adopt them, in many cases in spite of environmental conditions. Human agency has also greatly altered landscapes and environments, whether from clearing forests or emitting greenhouse gases from burning fossil fuels. But the relationship between environmental change, climate change, and the human ability to survive is more complex than would be conveyed by a simple narrative emphasizing degradation.

FORAGING

For the vast majority of *Homo sapiens'* existence on Earth, human communities have survived by gathering plant foods and hunting animals. Such lifeways are

A stone tool found in a sand dune in Namibia. (Gallo Images/Corbis)

called foraging. As omnivores, humans must have a diet rich in nutrients from different sources. *Homo sapiens*, like its hominid predecessors, survived by developing the ability to acquire a variety of different foodstuffs. Acquiring the technology to hunt, fish, and gather these foods required the development of both complex, multipart tools and of social cooperation.

Homo sapiens, in Africa and in the rest of the world, displaced the other hominid species because, being more flexible and creative than other species, it proved a better hunter and gatherer. Modern humans created new technologies and new methods of social organization that allowed them to occupy nearly every variety of environments. In the long run, they succeeded in manipulating the environment in order to increase their food productivity. This process began with the development of tools and continued with the development of strategies for food procurement that sought both to maximize food supplies and to encourage their sustainability. The use of fire to drive animals into hunting zones or traps also gave humans the ability to transform vegetation. We often think of foragers as "living off the land," but as recent research shows, the attempt to manipulate environments began well before the invention of agriculture (van der Veen 1999, 1–10).

In Africa, the transition to the Middle Stone Age about 1.5 million years ago witnessed the development of distinctive regional cultures based on more specialized exploitation of specific resources. During the Middle Stone Age, spears and bows became hunting tools and harpoons, most of which carried fire-hardened wood points, and were used for fishing. It was during this period that the use of fire, at least for cooking and warmth, appeared. The Late Stone Age, which began about 200,000 years ago in Africa, saw the development of smaller and more specialized stone tools. The earliest evidence for specialized fishing tools comes from the Congo River and dates to about 100,000 years ago. Starting about 60,000 ago, roughly coincident with the beginning of the last glacial age, technology underwent further elaboration in Africa: humans now developed the ability to make projectile points, which increased the killing power of spears and arrows. By about 50,000 years ago, the human population in Africa started to expand dramatically and crossed out of Africa (Ehret 2002, 20–21; Driskoll and Motz 1997; Barham 1997).

The lessons learned by humans in Africa were not lost on their prey. When humans expanded into other parts of the world, many animal species became extinct. The movement of humans into the Americas provides the most notable example. In Africa, because humans and game animals co-evolved, many animals evolved techniques that allowed them to remain more numerous than in other parts of the world. In addition, some diseases such as trypanosomiasis, river blindness, and schistosomiasis made it difficult for humans to live in certain African environments and these environments in turn protected animal populations.

Early and Middle Stone Age hunters targeted either big game or slow-moving, easily caught game such as snails. These early hunters also seem to have relied on the scavenging kills of other predators. Plant food played an important role in early human diets (S. Kusimba 2003). Late Stone Age peoples sought faster moving game and fish, and processed plant food more intensively. To exploit smaller food sources, later Stone Age foragers needed harpoons, traps, grinding stones, weighted digging sticks to dig up nuts, tubers, and even small animals, and new ways of cooking food to make it digestible. The need to exploit these sources grew with the increase in human population. Population growth caused the shift to hunting small animals and gathering seeds and nuts.

During the Late Stone Age, diversification of populations continued as the population increased, and areas such as the Congo River Basin, having no humans before, now began to be populated. The first evidence of creative expression also date to this period. The drawings that humans painted on rock surfaces give us glimpses not only of elements of subsistence activities—images of prey animals and gathering activities—but also of symbolic constructions. Archaeolo-

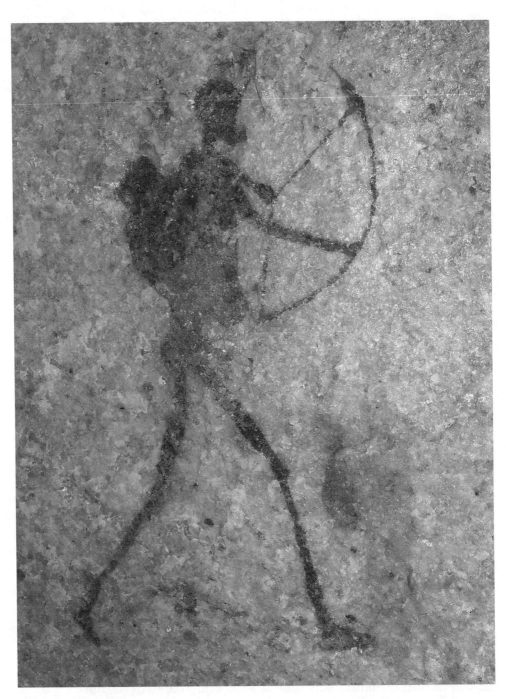

Bushman rock art from southern Africa. Its age is unknown but in southern Africa the tradition of rock painting goes back at least 20,000 years and continued until the mid-nineteenth century. (iStockPhoto.com)

gists have also found items used for decoration, including beads made from shells, pendants made from stone, and clay items. Wood and bone items also probably existed but have not survived (Dowson 1997a; Dowson 1997b).

By the time of the Holocene, humans occupied all of the globe. Some societies had already begun to reside in permanent settlements around watercourses where proteins in the form of fish and vegetable foods were always available. Many used fire to control landscapes. Most human communities had also begun the process of transforming environments—of creating landscapes that supported more people. During the Holocene, people would develop new means of food production that included the domestication of animals and the invention of agriculture. However, hunting, fishing, and gathering remained important in subsistence strategies.

THE TRANSITION TO FOOD PRODUCTION

The development of food production systems—agriculture and animal husbandry—labeled the Agricultural or Neolithic (New Stone Age) Revolution, allowed humans to dramatically transform their existence. In the case of agriculture, it required that people settle down and stay in one place. The revolution also allowed the production of a surplus. Ultimately, this development led craft producers to create new goods, rulers to specialize in governance, religious leaders to maintain relationships with the forces governing the natural world, and warriors to specialize in violence.

The process of developing food production systems took thousands of years, for both animals and plants had to be genetically modified in order to become adequate sources of food. The modifications took place through human selection of seeds or shoots to replant and animals to breed. The process resulted in "domesticated" crops and animals—that is, neither crops nor animals could survive or thrive in their existing form without human intervention (although some could go feral, often evolving new traits to enhance survival in the wild). Domestication meant, for example, that ears of grain stayed on the stalk rather than snapping off and spreading the seeds; that livestock followed human commands. Humans increased the productivity of plants by clearing fields of other plants, now labeled weeds, and providing water through irrigation. They protected livestock from predators and brought them fodder and water. In doing so, they remade landscapes, creating villages, clearing forests, building irrigation works, and hunting animals that preyed on or competed with their own. Human communities continued to practice transhumant foraging throughout the thousands of years involved in this process. Some scholars speculate that people often "planted" food crops and then left them to grow on their own while they traveled in search of wild foods (R. McIntosh 1997, 409–410). Depending on circumstances, fully agricultural people who did not have domestic animals remained dependent on hunting and/or fishing to supply their protein requirements. The expansion of agriculture often resulted in changed landscapes that favored certain species of animals, particularly those adapted to survival in open grasslands over those that required heavier forest cover. Grazers, such as wildebeest and zebra (and the species that prey on them), expanded at the expense of forest dwellers like gorillas.

In several different parts of the world during the early to mid-Holocene, human communities domesticated animals and plants from ancestral species they had long exploited (Harlan 1997b). Archaeologists have long thought that communities living along waterways were among the first to practice agriculture. By about 20,000 years ago, along some waterways in Africa, particularly the Nile, communities had developed a form of permanent settlement that relied on fish, shellfish, and other food items from the river as well as harvesting plants that grew throughout the year along the watercourse. In the case of the Nile, the pre-agricultural people collected sledges and ate the root. These types of settlements spread across Africa as far west as the Niger River and as far south as Kenya. However, research over the last several decades has indicated that riverside peoples seem to have felt little need to intensify production (Haaland 1995; Wetterstrom 1997; MacDonald 1997). Much agriculture and animal husbandry apparently developed in more arid regions where scarce resources forced people to experiment with intensification (McIntosh 1997).

The agricultural systems that developed spread into areas such as river valleys whose favorable conditions made them even more productive. Some domesticates, such as most livestock, proved capable of adapting to many different environments through selective breeding. Plant crops, on the other hand, expanded more in bands based on environment. Temperate crops, such as wheat, could not thrive in tropical environments, and tropical crops, such as millet or bananas, could not survive in colder climates.

The origins of food production—a term meaning both the cultivation of plant foods and the exploitation of domesticated animals—in Africa remains much disputed (McIntosh 1997). Africa's first food-producing region was the Sahara Desert region, where one variety of cattle and a pastoral lifeway emerged. The Nile Valley saw the spread of food production once-sedentary societies developed to take advantage of the regular food supplies provided by the river. The crops that eventually became the basis of agricultural societies in the Nile Valley and North Africa spread from the Near East. In Africa south of the Sahara, the crops of temperate Mediterranean Africa could not survive the tropical pattern of wet summers and dry winters. Hence, a number of crops were domesticated, and several were adopted from across the Indian Ocean. Although livestock spread all the way to southern Africa, cattle diseases prevented their expansion in large parts of the continent (Gifford-Gonzalez 2000).

Two occurrences probably compelled the transition to food production. The first was a general and gradual increase in population that put pressure on food sources as humans filled up the landscape. Although the expansion of agricultural communities gradually transformed landscapes, human population densities remained very low. Early farmers practiced long fallow agriculture in which a field would be allowed to regenerate bush as a means of restoring fertility after as few as one year under crops. Just as humans developed new hunting technologies to harvest different types of animal food, using traps and snares for small game and increasing the exploitation of aquatic sources, so they developed new technologies to exploit plant foods more effectively. They began to gather small grains from grasses and grind them and cook them. These processes required new tools—grindstones in various guises and pottery. Similarly, digging tubers required a digging tool; in Africa, a stick with a stone weight at the bottom was commonly used to pry the tuber out. Such innovations occurred in the Late Stone Age in Africa. Grindstones appear in the east African Microlithic culture that dates to about 20,000 years ago and in the cultures of South Africa. They are also found throughout the Sudanic regions of Africa south of the greatly expanded Sahara Desert of the late Pleistocene dating to about 40,000 years ago (Ehret 1998).

During the peak of the last glacial age, the human population in Africa shrank and retreated back to the old cradle of humanity in eastern and southern Africa, except along the Nile. There the sedentary foraging community continued, with people living along the river and collecting the roots of sledges that grew in the river as well as fish and shellfish (Stringer and McKie 1996). With the return of the rains after about 18,000 years ago, the practice of intensive gathering spread across both North Africa and along the gradually shrinking southern frontier of the Sahara between the Red Sea and Lake Chad. By about 13,000 years ago, the Great Lakes had refilled, and in the case of Lake Chad had expanded to far beyond its current much-diminished area. The "Green Sahara," covering what is now the southern half of the desert, filled with grasses and game, and people followed gradually from the refuges to which they had been forced to retreat.

A distinctive sedentary lifeway also developed along the upper Nile that extensively exploited the fish and grasses of the riverside environment. Starting at about 11,000 years ago, the African aquatic civilization spread along the Sudan from river to lake, reaching the Niger River a few hundred years after the tradition developed. Examples of aquatic settlements were located along the Nile up to the Ethiopian Highlands, along the rivers that ran from what is now the Sahara to Lake Chad and the Niger River, and around Lake Chad. This lifeway supported many people but remained dependent on the increased rainfall in the Sudan and Sahel (MacDonald 1997).

Historians using comparative linguistics have proposed links between the sketchy archaeological record and the spread of specific agricultural traditions. Languages spoken today can be thought of as belonging to families. Languages split and diverged as groups speaking the same language moved apart from one another. Drift of this sort can be identified by fairly regular and uniform changes in phonetics and grammar. Languages also absorb new words from other languages. Using both the comparison of rates of common words and spatial analysis of the distribution of both retained words from protolanguages and borrowed or innovated words, linguistic historians have made several broad claims about the relationship of archaeological cultures to ancient protolanguages. Christopher Ehret (1998) has argued that speakers of Afro-Asiatic languages occupied all of northern Africa by 15,000 years ago. In the east, from the Nile Valley to the Red Sea, they became intensive foragers and gathers of wild grass grains and tubers. One group moved across the Sinai, taking with them intensive gathering techniques. The languages spoken by this group eventually became the classical and modern Semitic languages. The group farthest east in Africa, the Kushitic group, eventually spread into the Ethiopian Highlands and further south into East Africa.

Most scholars working in historical linguistics accept the placement of the origin of Afro-Asian languages in Africa and its spread into southwest Asia (Blench 1997; Nettle 1998). They note that all but one of the Afro-Asiatic subfamilies consist of languages spoken almost entirely within Africa and that internal comparative evidence strongly indicates that the Semitic branch spread from Africa. Others, using additional evidence, suggest that Afro-Asian languages originated in southwest Asia and that the subgroups then differentiated as they spread within northern Africa. The geneticist L. Luca Cavalli-Sforza (1994) has argued that Afro-Asiatic languages diffused into Africa in response to the spread of food-producing technology first developed in southwest Asia. Other scholars have noted that the genetic markers indicating Asian origin for some North African populations probably come from the movement of people from Asia into the region over the last 2,000 years, not 10,000 years ago (Nettle 1998). Ehret's argument completely turns this contention on its head. He suggests that the techniques of intensive gathering that formed the base of the development of agriculture in both Africa and southwest Asia spread from Africa (Ehret 1998).

Ehret and Wendorf (1998) have also claimed that Africa was home to one of the earliest domestications of livestock in the world, and that unlike the case in

Cattle at the end of a trek across the Kalahari Desert are led to a water hole in small groups to prevent stampeding, Ghanzi, Botswana. (Peter Johnson/Corbis)

other parts of the world, domestication of livestock preceded the development of agriculture. The people moving into the Sahara gradually domesticated a species of wild Auroch, *Bos primigenius,* perhaps as early as 11,000 years ago in western Egypt. Ehret suggests that this group spoke a language of the Nilo-Saharan family; however, cattle pastoralism seems to have gradually spread among the people of the green, southern Sahara, whatever language they spoke. The Auroch, a humpless breed of domesticated cattle, became the basis for a pastoral, food-producing lifeway that spread across the southern Sahara. Associated with sites showing evidence of domesticated cattle are grindstones, pottery, and limited indications of the exploitation of sorghum and millet as foodstuffs dating to 9,000 years ago. The earliest sites identified as pastoralist suggest that the first pastoralists continuously moved between desert and Nile and rarely ate cattle, but they used the animals' milk and blood. The African tradition of using cattle for wealth may have started in this area.

Wendorf's recently confirmed evidence of domestication in western Egypt has some striking implications for the history of African environments and humanity. First, this evidence comes before other domestications of livestock. DNA evidence indicates the existence of three separate centers of domestication of cattle: India, the Middle East, and Africa. In Africa, unlike the Middle East and India, pastoralism may have come before the development of agriculture. DNA analysis also indicates substantial interbreeding over the years: very early on, Africans began to create new breeds, incorporating cattle brought up the Nile, across the Sahara, or across the Red Sea or Indian Ocean (Holl 1998a).

Second, pottery found in the Sahara is the earliest ever recovered. It was sometimes found at pastoralist sites and at other times at sites associated with foraging. This indicates that pottery (whose existence generally implies the importance of both cooking and the storage of plant foods) developed and spread separately from pastoralism (Ehret 1998).

Finally, the discovery of proto-agricultural settlements in the Nile Valley and their spread into the Sahara, as well as the early domestication of cattle in northern Africa, has led to an interesting debate amongs archaeologists and historical linguists over early food production in Africa and the Near East. In particular, in the area of Egypt's Nabta Playa in the Western Desert of Egypt settlements that date to almost 10,000 years ago contain evidence of cattle and extensive collections of grindstones showing that both sorghum and millet were harvested. After a brief period when the playa, a depression in the area that collected water during the rains, dried out completely between 9,800 and 9,600 years ago, people returned to the area as the rains returned. They built sizable stone homes and dug extensive storage pits and wells. The whole area of the Egyptian Western Desert and further west and south continued to host herding settlements of this sort for the next 3,000 years (Wendorf and Schild 1998).

At the same time, to the east, the Nile Valley underwent its own connected but different transformations. From a proto-agricultural base, Egyptians adopted the crops that had been domesticated, notably varieties of wheat and barley, in the Near East and imported the herding of sheep and goats by about 8,000 years ago. They quickly developed the techniques of flood-recession cultivation whereby crops are planted as the waters recede and grow using the moisture left in the soil by the flood. Whereas goats and sheep spread rapidly to their Saharan neighbors (as well as a new breed of cattle that the pastoralist bred with their own), the crops did not spread (Hassan 1997a). The Mediterranean crops evolved within a dry summer/wet winter environment, but the monsoon system south of the Sahara has wet summers and dry winters—on both sides of the equator. Hence, despite extensive contact, the crops did not diffuse, and so Africans living on the southern side of the Sahara had to develop their own crops.

Between about 12,000 and 8,000 years ago, both pastoralism and sedentary communities along bodies of water spread all the way across Africa south of the

Sahara. During this wetter phase, lakes such as Chad expanded, and tributaries drained into the Niger from the north in what is now the heart of the Sahara Desert. Their former channels became pathways into the desert. Both the cattle herding and aquatic settlements in these communities exploited grains extensively. The most important food grains came from grasses that eventually became the domesticated crops of Africa. Sorghum, *sorghum bicolor,* shows most prominently, and plant biologists have long known that its ancestor grew wild in the area of the southern Sahara and Sahel. Pearl millet, *Pennisetum glaucum,* was also domesticated in the Sahara. In addition, the middle Nile Valley seems to have been home to the domestication of cotton around 7,000 years ago, as evidenced not only by signs of plants ancestral to cotton but also by clay spindle whorls used to make thread from the cotton fibers (McIntosh 1997).

The evidence that these Africans had made the transition to fully agricultural societies is not undisputed, however. Some archaeologists suggest that the Africans of the Sahara and Sahel practiced an extensive form of gathering that proved quite successful in supporting an expanding population. They note that grain processing tools are common but evidence of grains is rare. Hence, it is not possible to determine whether the grains so processed came from domesticated or wild varieties. They argue that these communities probably engaged in a kind of semicultivation that perhaps cleared fields and planted but did not result in the total genetic transformation of the plants themselves (van der Veen 1999).

In particular, sorghum has a tough stalk that is not easily cut without metal blades; most archaeologists believe it was harvested by shaking the ear into a basket rather than reaping. The lack of large numbers of polished stone axe heads, they maintain, indicates that grains were not harvested but rather were stripped from plants. In most sites, too, there is evidence of other definitely gathered foods. Hence, they suggest that these food-producing communities could best bear the label of intensive gatherers who practiced limited cultivation. They argue that grain agriculture did not appear in the Sudanic regions south of the Sahara until 5,000 years ago and only became widespread about 4,000 years ago. This expansion, they suggest, occurred because the Sahara began to dry up about 6,000 years ago, pushing people south (and north and east to the Nile). By 4,000 years ago, this agropastoral way of life had reached as far as southwestern Mauritania (Muzzolini 1993; McIntosh 1997).

Ehret and the plant biologist Jack Harlon (1993), maintain that archaeological evidence combined with linguistic data prove that people first cleared fields by burning, then planted, and harvested the land. They suggest that these practices rather than the method of harvest or the degree of genetic change induced by breeding better define agriculture and, further, that more research will reconcile the differences over the chronology and conditions of the development of agriculture just as the differences as regards the discovery of ancient animal domestication in Africa have been reconciled (Harlan 1993, 1997b; Ehret 1998). A later domestication in Africa occurred in the region between the Inner Niger Delta in what is now Mali and the Senegal River in Senegal. There, at some point after about 5,000 years ago, African farmers began to cultivate a rice plant. They "ennobled" *Oryza barthii*, a relative of the wild ancestor of Asian rice, into *Oryza glaberrimi*, African or red rice. The *Oryza* genus is found in most tropical regions of the world as wild grasses. Only in Asia where *Oryza sativa* and in Africa have examples of Oryza been domesticated. The farmers of West Africa developed a number of different varieties of the crop—some "floating," some grown in paddy-like fields in the floodplains of the Niger and Senegal rivers, some dry land, matching the varieties developed of Asian rice. Rice cultivation eventually spread west and south through the rainforest regions and east to Lake Chad (McIntosh 1998).

The Sahara/Sudan region was not the only area where Africans developed food producing and eventually agricultural systems. In the Sahara, enough archaeological research has been done that when this research is used in conjunction with climate change data, linguistics, and biological evidence, we can argue about timing and processes. This is not the case in the forest/savannah frontier to the south and in the Ethiopian Highlands, where too little archaeological evidence is available.

The Ethiopian Highlands, like the other highlands of eastern Africa, represent a major region of plant endemism. This region receives fairly regular rain from the monsoon system and has many microenvironments based on altitude. The highlands themselves are surrounded by arid plains. African communities in this area domesticated four crops. Two, coffee and finger millet, *Eluesine coracana*, eventually spread beyond the region, and the two others, tef and ensente, remain little grown outside Ethiopia. Tef is a small grain used to make Ethiopian *njera* bread, and ensente is a bananalike plant whose stem marrow is edible. Linguistic evidence indicates that domestication perhaps started as early as 6,000 years ago. By 4,000 years ago, a process of intensification had begun in the highlands that would lead to the development of a remarkable civilization (Brandt 1984; D'Andrea, Lyons, Haile, and Butler 1999).

In the forest/savannah frontier region, the grain crops of the drier savannahs did not grow as well. In these areas, people gradually domesticated the oil palm and several varieties of African yam. Again, the archaeological evidence is spotty, but linguistic data indicates that such crops were heavily exploited by about 8,000 years ago. The earliest archaeological evidence comes from sites of the Kintampo culture located in what is now Ghana and Côte d'Ivoire dating to about 5,000 years ago, but evidence of oil palm exploitation comes from earlier dates through analysis of pollen residues from cores taken from lakes in the forest-savannah border area. In general, exploitation of the true forest remained mostly in the hands of foragers until the coming of metal. Agriculturalists lived in the frontiers of more open tree cover and had only limited success in cutting or burning back the forest. Climate change remained more important in determining human land-use and settlement patterns, with the forest expanding during wet phases, such as the one lasting from about 12,000 to about 6,000 years ago, and retreating during dry phases (Ehret 1998; Anquandah 1993).

Throughout the early phase of food crop domestication, people continued to supply much of the protein they needed through hunting and fishing. Wild animal populations remained high because agricultural societies occupied relatively little of the landscape. Livestock, both cattle and sheep and goats introduced from southwest Asia, could not expand into areas with concentrations of tsetse. These areas, covering much of the tropical savannahs as well as forest regions in Africa, remained home to substantial game populations. Domestic fowl also began to supply some of the protein needs of agricultural populations. West Africans domesticated guinea fowl at some point before 5,000 years ago, and the raising of these spread throughout the savannah. Chickens, originally from Southeast Asia, did not make their appearance in Africa until the beginning of the common era. People also worked to develop livestock breeds that could survive in their climates. Goats in particular became widespread, and by about 4,000 years ago a specialized breed of cattle had been developed that could live at least in the forest-savannah borderlands (Clutton-Brock 1997).

DESICCATION AND THE EXPANSION OF AGRICULTURAL SOCIETIES

A critical climatic change began about 8,000 years ago. The Earth became somewhat cooler, and as a result the deep ocean currents shifted, causing the Intertropical Convergence Zone to shift south toward the equator. A gradual and irregular reduction in the reach and intensity of the monsoon rains in the Sahara resulted in a reversion of the area to desert. Lake Chad began to decline in size, dwindling to the point where today it is hardly a lake at all. The northern channels of the Niger and Senegal river systems dried up. By about 4,500 years ago, the Sahara had reached roughly the conditions and size it holds today. Since then, the variations in climate and in the extent of the desert have been minor.

The drying of the Sahara precipitated a series of events that produced dramatic changes across Africa and the Near East. Starting in the northern and eastern areas, the pastoral and agricultural settlements retreated south and east. The Nabta Playa settlement that provided the most detailed evidence for the herding communities of western Egypt was abandoned by 6,300 years ago. Some archaeologists have suggested that herders from the west moved into the Nile Valley and that the conflict generated between the settled agricultural communities of the Valley and the herders helped generate the political centralization that led to the development of Egyptian civilization. They note the importance of cattle iconography in predynastic and Old Kingdom Egypt, despite relative lack of importance of cattle in predynastic livelihoods.

In the Sudan, a general expansion of the territory within which cattle remains and evidence of intensive exploitation of grains occurs. In Ehret's terms, the Sudanic agropastoral culture expanded greatly. By 5,000 years ago, cattle, sheep, and goats reached the forest-savannah borderlands. During this time, communities engaged in further experimentation with mixes of crops and continued to breed both crops and animals to fit into different environmental conditions. They appear to have begun to use fire both to clear lands for agriculture and to burn back bush, allowing for the growth of new grass for livestock. They also faced continual variations in the climate and began to develop flexible strategies and social organization to deal with these variations (Ehret 1998).

As agricultural and stock-keeping communities expanded about 5,000 years ago, they also began to face new diseases created, in a sense, by the remaking of the landscape. These diseases affected both people and their livestock. Although agricultural communities faced greater threats from diseases than mobile foraging communities, in Africa reordering the landscape proved particularly daunting. Many common diseases arose by parasites crossing from one species to another. Generally, deadly diseases are new introductions of hosts; disease (or microbes) can live in one host fairly peacefully but be deadly in another. Living in close contact with animals created more opportunities for such transmissions to occur. Smallpox, for example, moved from cattle to humans early in the history of domestication.

As cattle keepers expanded their range southward in Africa, they began to face new threats to their livestock. Five deadly diseases in livestock had already evolved and proved able to shift to cattle in Africa: malignant catarrhal fever, Rift Valley fever, East Coast fever, foot and mouth disease, and trypanosomiasis. The last-named proved perhaps the most important. The *Trypansoma* microbes that cause the disease evolved in a cycle between wild animals and a biting, blood-dependent family of flies known as tsetse (sp. *Glossina*). The five species of *Trypansoma* that cause death in humans typically do not cause disease in the fly, and in at least one wild animal, this means that a wild-animal reservoir always exists for the parasite. Several of the larger ungulates, especially buffalo in Africa, harbor the microbe as a relatively benign parasite. The microbe, however, causes death in livestock, especially cattle and horses, presenting a classic example of a disease that evolved with its hosts, which then became virulent when presented with a new host. Tsetse flies live in bush and can live off blood from most types of larger animals, including humans. Tsetse flies live in tropical environments lying between 20 degrees north and south. Humans may have generally avoided tsetse areas such as the rainforest until this era (Ford 1971; Giblin 1990).

The expansion of stock-keeping communities seems to have slowed because of the impact of these diseases. In one sense, stock keepers followed the gradual drying of the environment after about 6,000 years ago that removed habitat for tsetse and the ticks that carry East Coast fever. Yet, they also eventually expanded into areas that still proved capable of harboring tsetse. They learned both to read the landscape, finding areas where conditions made them free of tsetse, and to alter the landscape. Clearing land for settlements and fields reduced the cover tsetse needed, and larger-scale burning cleared bush. Finally, tsetse-resistant breeds of cattle were created in some areas (Gifford-Gonzales 2000).

Only after about 1,000 years did stock keepers apparently develop the full range of strategies and tools that allowed them to expand across the continent. Stock-keeping communities first occupied the area around Lake Turkana in northern Kenya by about 4,000 years ago. It was only about 3,000 years ago that evidence of stock became available from southern Kenya and northern Tanzania. Initially, sheep and goats appeared, and then cattle followed. Sheep and especially goats both survive trypanosomiasis better and are less favored as food sources by the fly. Initially, immigrant communities in eastern Africa speaking Sahelian and Cushitic languages came with their stock. By about 3,000 years ago, some Khoisan speakers had adopted stock keeping. By about 2,000 years ago, livestock reached southern Africa in advance of agricultural communities (Gifford-Gonzales 2000).

The environments capable of hosting trypanosomiasis continue to mark settlement and land-use patterns in Africa. Cattle keepers occupied drier lands, used fire to keep bush down, and kept their cattle away from areas that harbored tsetse. Settlements in regions that hosted tsetse kept bush down in surrounding areas and relied on small stock for meat. Areas that harbored tsetse flies also contained larger populations of wild animals. Shifts in territory open to cattle could occur as climate varied and as human actions that cleared bush waxed or waned.

Human diseases also greatly affected the spread of agricultural communities in Africa. The mosquito-borne malaria parasites, transmitted by several species of the *Anopheles* genus of mosquitoes, appear to have evolved with humans and their mosquito vector before humans left Africa. Both parasite and vector traveled with humans out of Africa. As was the case with tsetse and trypanosomiasis, humans survived in part by avoiding areas where mosquitoes thrived. However, many scholars believe that the development of agriculture greatly increased exposure to malaria and that it became a major cause of death among humans in Africa (and elsewhere in the tropics) with the spread of agricultural communities (Bradley 1991). Human settlements moved toward wetter areas that supported both plant and mosquito survival. In addition, one particular species of mosquito, *Anopheles gambiae,* thrives in the broken soil and new plants of agricultural fields.

Foraging communities would have avoided the environments that produced human sleeping sickness, and human communities developed biological defenses against malaria. They needed these defenses because of the large range of the various species of *Anopheles* mosquitoes. Exposure to these mosquitoes was constant in the wettest regions of Africa, whereas exposure in most of the rest of the continent tended to be seasonal. An indication of the effectiveness of biological defenses is that many African populations carry the Duffy negative factor that inhibits vivax malaria. Vivax malaria has a wide distribution in Asia, but that form of the disease seems to be very limited in Africa where *Plasmodium falciparum* is the most common type of malaria (Bradley 1991).

The development of agriculture gradually created more settled populations and led to a greater propensity for the spread of viral and bacterial diseases. Settled populations allowed some diseases to become endemic. Several important diseases either moved back and forth or made the evolutionary jump from domestic animals. Similarly, settled populations could both control some diseases more easily and created the environment for the concentration of other disease vectors. Clearing land for agriculture in areas where tsetse could survive reduced the danger to humans (although not necessarily as much for domestic animals, which still required grazing lands), while it created more opportunities for the *Anopheles gambiae* mosquitoes that carried malaria. Some scholars have speculated that the movement of Bantu-speaking farmers into some parts of eastern and southern Africa may have been aided by the spread of sedentary diseases through foraging populations. Several scholars have made the case that African societies developed the means to control exposure to tsetse for both humans and animals (Ford 1971; Giblin 1990). Such methods of control and exposure work mostly in savannah-type environments where human clearing is not overwhelmed by rainfall. In addition, water-borne diseases such as river blindness and schistosomiasis (caused by a flatworm transmitted from snails to humans) inhibited settlement along tropical waterways.

Changes in population density caused by political disruption or climatic variation such as drought could result in the loss of the ability to maintain such control locally. Cyclical events could lead to the repeated ebb and flow of population between drier areas generally free of tsetse and wetter areas better suited to agriculture. These regular disruptions helped create long-term webs of connections across environmental frontiers.

THE MONSOON EXCHANGE

People in Africa never lived in isolation. As John and William McNeill (2003) have observed, they developed contact systems akin to networks of spider webs that kept expanding as humanity expanded. Over the 100,000 years of human expansion, ideas, technology, goods, and diseases have moved from one node to another, even though individuals might never leave their own node and only contact people from neighboring nodes. The intensity of transmission has varied with time and circumstance. Occasionally, nodes lost contact with neighbors, as happened when humans entered the Americas and parts of Oceania in the Pacific.

We can further complicate this spider web analogy by thinking perhaps of our webs as occupying different environments as they spread. Those living in similar environments would wind up looking similar to each other, while those in different environments would be different despite continuing connections across the broader network. In this sense, then, we can understand why Africa south of the Sahara might have more in common, and share more technology and agricultural practices, with other parts of the tropical Old World such as southern India and Southeast Asia than it would with the Mediterranean world with which it had much more direct and intimate contact.

After about 5,000 years ago, as global contact among agricultural societies or evolving food-producing societies increased, crops and technologies spread. People experimented with crops that they acquired from neighbors. In European and Western views of diffusion, sub-Saharan Africa has often been portrayed as a recipient of diffusion often associated with a Near Eastern origin. Modern scholarship has sharply revised these views. Recent research suggests that some of the most fundamental steps in creating food-producing societies were taken in Africa. Later, some of the elements of food production that became basic to African lifeways did originate outside Africa but not just in the Near East. And African domesticated plants spread very rapidly outside of Africa to become important staples in other parts of the tropical Old World. In short, diffusion of agricultural practices can be said to have occurred in rough, environmentally determined bands across the globe, where innovations can spread from niche to niche as contact permits.

In sub-Saharan Africa, the most important acquisitions from the Near East were sheep and goats, which arrived in the Nile Valley about 8,000 years ago.

Camels in the Sahara Desert. (Emma Peascod/iStockphoto.com)

They spread into the "green" Sahara and from there all over Africa. However, Africans in the Nile Valley and the Sahara had already domesticated cattle. These herders of the Sahara seem to have readily realized the utility of mixed livestock herds and the benefits of selective breeding. As stock rearing spread south of the Sahara over the next few millennia, sheep and goats often spread in advance of cattle because of their greater resistance to disease, and several breeds of cattle that emerged adapted to different environmental conditions. These cattle drew not just on the original African domesticates but on Near Eastern and especially Indian cattle.

Horses also crossed the Sahara from the Near East. In classical times, people sometimes used horses to cross the Sahara. They became part of the herding and eventually military practices of Sahelian and Sudanic Africa. People never adapted them for use in areas with tsetse, and their spread to southern Africa had to await the Maritime Revolution of the 1400s.

At a much later date, another domesticated animal would have a large impact on Africa as it spread from the Middle East. By about the beginning of the

Banana plants at a plantation in Transvaal, South Africa. (Philip Perry; Frank Lane Picture Agency/Corbis)

common era, the camel became the main means of transport across the Sahara. The Berber-speaking peoples of the desert quickly adopted camels. The introduction of the camel also helped begin contact with Arabia, whose influence reached its culmination with the rise of the first Islamic Empire.

As important as these innovations were for Africans, the vast majority of their food crops before the modern era were domesticated in Africa itself. Sorghum and millet quickly spread along the shores of the Indian Ocean. Sorghum has been found in Arabia dating to about 4,500 years ago and millet in India to about 4,000 years ago. Given that biologists have long agreed that the ancestor of these crops is found only in Africa, these early dates for finds outside of Africa have tended to support scholars such as Ehret who argue for an early indigenous agricultural revolution in tropical Africa. In addition, coffee, a variety of cotton, and sesame spread from Africa into the rest of the world.

The most important crops with external origins in Africa before the Maritime Revolution came from tropical Asia. Bananas and plantains arrived in Africa early in the first millennium of the modern era. They probably came to the East African coast with voyagers from Southeast Asia, some of whose descendants helped give the Malagasy of Madagascar their language. Southeast Asians seem to have come to the coast of what is now Tanzania and Kenya during the beginning of the common era. They eventually spread what must have been small trading settlements to the uninhabited island of Madagascar. On the coast, the settlements eventually merged into the developing Bantu-speaking coastal communities that became the Swahili. On Madagascar, their language survived and became the dominant language of communities made up of immigrants from Southeast Asia and from the African mainland. Along with them came chickens and sugar cane. Chickens spread quickly from East Africa into Central Africa and supplemented or displaced guinea fowl as the main domesticated fowl. Later in the first millennium of the current era, Asian rice and sugar cane became important crops in eastern Africa (Ehret 1998).

Bananas became an extremely important food crop in several parts of Africa after their introduction in early C.E. These crops spread inland along two routes after their introduction; these routes are marked by the development of different varieties of banana. "Soft" bananas spread to the highland regions of East Africa, starting perhaps with the Upare Mountains in eastern Tanzania. They then spread to Mount Kilimanjaro, Mount Meru, Mount Kenya, and eventually to the highland regions of the Great Lakes. They would become a main staple in these regions after about 1000 C.E. "Hard" bananas, usually called plantains, spread along a more southern route around the Rovuma River to Lakes Malawi and Tanganyika. By about 300 C.E., they reached the Central African rainforest, where they may have helped spark a dramatic intensification of agriculture in the forests and an increase in population densities. With proper care, both types of bananas when under permanent cultivation produce a high yield. As a result, in regions that could support them, population densities increased dramatically (Rossel 1996).

CONCLUSION

The "monsoon exchange" reflects the importance of environmental conditions in facilitating certain types of innovations and limiting other possibilities. Earlier scholarship on the emergence of agriculture in Africa and the development of Neolithic societies more or less saw the Mediterranean as a center from which radiated innovation. This paradigm has gradually begun to collapse. Modern views describe local environmental conditions as promoting or inhibiting innovation and support the idea of multicentric exchanges of people, technology, and ideas across networks. Sub-Saharan Africa needed no contact in the development of food production; relationships with the Mediterranean world were important, but in many ways contact along the Indian Ocean was more important. And innovations launched in Africa played a large role in developing agricultural societies outside Africa.

3

AFRICAN ENVIRONMENTS AND THE DEVELOPMENT OF COMPLEX SOCIETIES

Africans developed food production systems out of their own environments, domesticating their own plants and animals, and developing their own techniques for increasing food production. Where appropriate, they adopted crops and production techniques from outside the continent. After about 5,000 years ago food production spread throughout sub-Saharan Africa. Gradually, Africans developed more intensified methods of production and began to build and live in urban areas. Such actions were not solely the result of manipulating the natural environment toward food production; rather, they occurred through specialization and trade and through the development of new technologies. They also resulted in an even more intensive transformation of African landscapes.

Food production spread through the Central African rainforest and down the eastern quadrant of Africa after about 5,000 years ago. In Central, eastern, and southern Africa, the spread of food production was accompanied by the development and spread of iron working after about 3,000 years ago and by the spread of Bantu languages. The development of metal working helped spread more intensified production and urbanization. Africans in Nubia, Ethiopia, the Inner Niger Delta, the East African coast, and Zimbabwe created these societies within local environments. They all faced the difficulties of sustaining large populations without degrading the ability of the environment to remain productive. Examination of the development of complex societies will also focus on the importance of long-distance exchange in these societies.

THE SPREAD OF FOOD PRODUCTION

The spread of food-producing societies in Africa south of the Sahel is much better understood than the origins of food-producing techniques in the Sahara-Sahel region. Two main types of evidence account for this greater understanding. First, after about 2,500 years ago, iron working spread rapidly through sub-Saharan

Africa. In general, only farming and herding communities made iron implements (although foragers could acquire them through trade). Evidence for iron working can be easily dated. Similarly, in Central and southern Africa, the spread of food-producing communities coincided with the spread of the Bantu languages. Historians using linguistic methods have rather clearly dated the spread of these languages. These dates coincide with archaeological evidence for the spread of food production and iron working in those regions. Livestock and pottery seem to have expanded in advance of the spread of iron and agriculture.

Throughout sub-Saharan Africa, the period after about 4,000 years ago marks a dramatic increase in the density of population as food production spread. Such an increase is clearly seen in the archaeological record; it has been noted that for every pre- or proto-agricultural settlement site found, there are a hundred agricultural ones (S. Kusimba 2003). Archaeologists suggest that such increased density results from several factors. Agricultural settlements tend to be more permanent, whereas forager sites tend to be only temporary. This permanence refers to both more substantial structures (houses instead of tents) and larger accumulations of refuse (the archaeological equivalent of a gold mine). But archaeologists also suggest that the greater number of sites does represent increased population density. Evidence from study of sediments in swamps and lakes also indicates declining tree cover in some areas and an increased presence of grasslike plants in others.

Anthropologists have long debated whether the spread of food production resulted in higher standards of living. Most agree that given enough space, foraging represents a fairly optimal way of living, but that pressure on resources caused by both population growth and climate change led people to innovate to ensure adequate supplies of food (McIntosh 1997; S. Kusimba 2003). In Africa as elsewhere, agricultural societies expanded as population grew. This growth put new pressures on land fertility, on water supplies, and on supplies of game near settlements.

African societies had to develop mechanisms for allocating access to resources, controlling conflict over resources, and ensuring sustainability. In the face of population growth, African societies faced two choices: either increase production or have some of the people in an area move to another. The success of the second alternative depended on the availability of new land to colonize. Colonization did not mean the land had no people before the immigrants arrived, but only that a niche existed that the newcomers could exploit. The mobility option remained the preferred one throughout much of Africa before the modern era. In fact, scholars, most notably the anthropologist Igor Kopytoff (1987), have suggested that the "frontier" is an integral component of understanding African social organization. At their lowest level, many African societies before the modern era were organized into "houses" or kin groups. In the normal course of events, such houses, led by "big men" who "earned" their position, and kin groups, led by hereditary, senior leaders, would naturally grow and divide as children reached maturity and sought to establish their own homes.

The process of fission and movement thus became ingrained in many aspects of African social organization. It provided a means of avoiding conflict, and in the context of a relatively low population density, it meant that control of territory often became less important than control of people. It also provided a means for maintaining soil fertility. Swidden agriculture, whereby people would clear land and plant crops on it for a few years before allowing it to regrow bush as fallow, became the norm in regions with abundant land. Such practices, whether in forest or savannah, also meant that open space remained for wild animals. As a result, hunting was an important subsistence strategy for many African communities.

Such expansion had its limits, of course. First, in areas of very productive land such as some river valleys and in well-watered highland areas, land could remain productive for long periods of time. Then communities had to develop the means of allocating and maintaining control over these resources. The relatively favorable conditions encouraged population growth. This growth in turn helped encourage intensification of production. In other parts of the world, the development of hierarchical societies and states is associated with river valleys or control over other productive resources. In Africa, a variety of means developed for promoting innovation and controlling resources that did not necessarily include hierarchical states. River valleys, such as the Niger, and well-watered highland areas, such as the Eastern Arc Mountains of East Africa and the highlands around the Great Lakes hosted systems of intensive production and population stability. However, many river valleys in Africa remained inaccessible for cultivators because of water-borne diseases such as schistosomiasis (caused by a flatworm carried by snails) and river blindness.

All along the frontier between forest and savannah from the Atlantic to Central Africa, agricultural expansion began about 4,000 years ago. As the climate dried, the forest itself gradually retreated, leaving a more open woodland environment. In this environment, Africans experimented with yam cultivation, oil palms, and grains such as sorghum and rice. They probably did not extensively clear the land, but more people farmed in the openings that appeared in the forest. In fact, the forest-savannah mosaic created by human occupation probably encouraged tree growth by protecting stands of oil palm and other trees planted around settlements from fire. These groups kept at least goats and possibly sheep, although tsetse fly prevented the holding of cattle. In the far west of the region, a variety of rice that Europeans would come to call Guinea rice be-

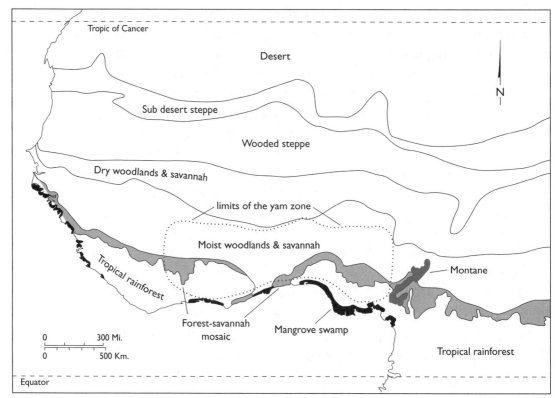

West African environmental zones showing area of major yam cultivation.

came the staple crop. From about the modern Côte d'Ivoire over to what is now Cameroon, yams remained the staple.

In the east of this zone, around what is now the border between Nigeria and Cameroon, the participants in this culture began an expansion to the east. The people engaged in the original expansion spoke a language we have come to call Bantu (or more accurately, "Proto-Bantu"). Within 3,000 years, languages "descended" from Proto-Bantu had spread across Africa to the Indian Ocean, through the Congo Basin rainforest and down the east coast of Africa all the way to South Africa. These languages supplanted most of the other languages spoken by earlier people in these areas. They remain remarkably close to one another, as close as French is to Spanish or English is to German.

Historians have debated why and how these languages expanded so rapidly. The development of a new agricultural system based on yams and oil palms fueled the initial expansion across the forest-savannah mosaic toward the Great Lakes and down into the rainforest itself. The speakers of these languages gradually absorbed many of the foraging populations they encountered, in the long run overwhelming them with numbers, more than anything else. By about 3,000 years ago, speakers of one of the daughter Bantu languages had reached the Great

Lakes region and came into close contact with herders and grain farmers speaking both Nilotic and Cushitic languages (Ehret 1998; Schoenbrun 1998).

Around the middle of the last millennium before the current era, a further technological innovation helped facilitate agricultural expansion and intensification. In northern Nigeria at a place called Nok dating to about 600 B.C.E. and in the area around Lake Victoria dating to sometime between 700 and 500 B.C.E., evidence for iron working has been recovered. Gradually, iron provided the basis for increased efficiency in agriculture (and in warfare). Iron axes cleared land; iron hoes dug deeper; and iron blades harvested grain—all with less effort (Schmidt 1996). Some archaeologists argue that Africans could not fully domesticate sorghum with its thick stalk until they had iron blades for cutting them down (Muzzolini 1993).

Many archaeologists believe that knowledge of iron working diffused into Africa from the Mediterranean (Clark and Brandt 1984). Iron working was invented in Anatolia around 1500 B.C.E. and spread throughout the Mediterranean as well as into Asia through India to China. Some posit that trade contacts between Carthage, in North Africa, and West Africa eventually led to the spread of iron working across the Sahara. The earliest evidence for iron working south of the Sahara comes from the region that gave rise to the Nok culture in northern Nigeria. From thence iron spread across Africa. Historians argue that little to no evidence exists for a process of experimentation in metal working that would lead to the development of iron technology in Africa (Schmidt 1997b).

Peter Schmidt, however, had led those who have argued that Africans independently invented iron working. There has been a great deal of controversy about dates from some ironworking sites; a few radiocarbon dates suggest ages for charcoal well before the beginning of the last millennium before the current era. Most archaeologists, however, dismiss those dates as erroneous, caused by corruption of the testing material or the use of "old wood" in the smelting process. Nonetheless, two sites in the Sahara have yielded evidence of copper working dating to perhaps 4,000 years ago. Archaeologists maintain that working other metals like copper gives metalsmiths the experience necessary to experiment with the more complex process of smelting iron. Hence, the lack of evidence for copper or bronze working in Africa before the appearance of iron working has always been taken as a sign that the technology diffused into Africa. This evidence suggests that around 3,000 years ago somewhere in the region between northern Nigeria and Lake Victoria north of the forest, iron working was invented by Africans. This independent invention is a better explanation for the appearance of iron working around Lake Victoria by 700 B.C.E. and at Nok than diffusion. As it stands now, the debate cannot be settled on the basis of available evidence; only further discovery and analysis of more sites will provide an answer (Scmidt 1997b; de Maret 1996).

No matter how it reached Africa, iron working after 500 B.C.E. spread rapidly through most agricultural communities in sub-Saharan Africa. As noted above, it potentially improved productivity in agriculture. Yet, evidence seems to indicate that the impact of iron working was felt only gradually. Although the raw material for iron occurs abundantly in sub-Saharan Africa and the technology diffused rapidly, most Early Iron Age sites also contain a large number of stone tools. Agricultural communities had expanded rapidly before the introduction of iron and continued to do so afterward.

Iron production had a significant impact on resource use, requiring large amounts of wood. Some have suggested that the rapid expansion of iron production contributed to the decline of forest cover in the Great Lakes region as measured by palynological analysis of cores from Lake Victoria around the beginning of the current era. Palynological analysis measures layers of pollen deposited each year in the beds of lakes. The layers provided a way of measuring the types of vegetation in surrounding areas and can be dated using radiocarbon measurements. These cores from the Great Lakes region show evidence of a decline in the pollen of tree species and a rise in both agricultural and grass species. However, the era between about 300 B.C.E. and 300 C.E. was a particularly dry one, a condition that could also play a part in these findings (Taylor and Marchant 1996; Schmidt 1997a).

After about the beginning of the current era, iron use advanced with food producers in Central, eastern, and southern Africa, and its use "filled in" throughout the rest of Africa. In northeastern Africa, modern Sudan, Eritrea, and Ethiopia, metal working probably arrived via Egypt and the Near East. The city of Meroe in the Sudan became a major center of iron working by about 400 B.C.E (Shinnie 1996). In eastern and Central Africa, most scholars believe that the rapid expansion of iron-working sites from what is now Kenya and Tanzania all the way down to South Africa by about 600 C.E. reflects the expansion of Bantu-speaking communities (Hall 1997).

In western Africa, the gradual spread of iron does not seem to have had a revolutionary impact on the continuing expansion of agricultural societies. Kintampo sites, named after the site of the discovery of this ceramic and stone tool set of artifacts, spread over present-day Ghana and Côte d'Ivoire and date back to about 1000 B.C.E. These communities cultivated African or aerial yams and used oil palms extensively. Iron working appears in sites dating to before 600 B.C.E. in Niger and then further south in Nigeria (Anquandah 1993; Shaw 1984). Although iron facilitated the exploitation of forest regions, its first critical use in the region may have been in breaking the heavy clay soils in the area around the gradually shrinking Lake Chad (Breunig, Neumann, and Van Neer 1996). Similarly, the gradual shrinking of the area under permanent water in the Inner Niger

Delta led both to the development of rice production and its expansion through-out most of the western part of the West African coastal regions (McIntosh 1998). The lack of archaeological work means that we must rely on linguistics and logic to piece together the history of the spread of agriculture in the West African forest. A few sites testify to iron working in the eastern parts before the beginning of the current era.

The forest itself contains a constantly changing mosaic of microenviron-ments based on local topography, soils, and especially past use. As such, settle-ment patterns in the forest vary from nucleated settlements to more dispersed ones. Given the regular supply of water, once agriculture became established in the rainforest, it led to relatively dense populations despite the greater incidence of diseases such as malaria and yellow fever and the lack of cattle because of tsetse. Communities also made extensive use of fire to expand the area under cultivation. At what point density passed that of the savannahs and forest-savan-nah mosaic is hard to say, but certainly sometime after the sixteenth century the most densely populated parts of West Africa were to be found in southern Nige-ria, Dahomey, and the southern Gold Coast.

After about 500 c.e., settlements intensified in both savannah and forest in West Africa in part because of the return to slightly wetter conditions after about 300 c.e. Wetter conditions expanded the area capable of supporting agriculture and livestock. As population increased in West Africa, a remarkable system of caste and ethnic specialization that stretches all across the region developed. It built on the layering described by archaeologists working in the Inner Niger Delta. Associated with speakers of languages in the Mande family, it seems to have begun with the development of specialized villages linked together. It even-tually developed into a widespread system of caste and ethnic specialization that included pastoralists, farmers, traders, religious specialists, and craftspeople (McIntosh 1998).

In the Congo Basin, a slow filtering process continued through the rainfor-est. Iron working became a part of this process. In the forest, farming communi-ties continued to exist beside foraging ones. Farmers did not clear the forest but farmed in spaces opened within it. They gradually developed an almost symbi-otic relationship with existing foraging communities. Eventually, the foragers completely adopted the languages of the farmers with whom they lived. It is too simple to see the existing Baaka and Batwa foragers as remnants of the pre-agri-cultural populations. Some of their practices have come down from those popu-lations, and both they and farming communities generally consider them as de-scendants of the earlier occupants of the forest. More accurately, their communities emerged out of specialization within broader forest environments and in structural opposition to farming communities. The continuity arises out of a lifeway, not a straightforward biological descent, and the shared language between foragers and farmers represents over 2,000 years of interaction (Vansina 1990, 1996).

At some point before 1000 C.E., plantains reached the Congo River Basin. African farmers took up plantain cultivation easily, and the relatively productive crop became a staple or important supplement throughout much of the forest area. It spread more slowly north into the West African rainforest and resulted in a gradual transformation of the forest itself. Perhaps, however, it did not cause a great decline in vegetative cover. Plantains grew along riverbanks and in clearings created or expanded by human settlement (Schoenbrun 1993; Rossel 1996; de Langhe, Swenne, and Vuylsteke 1996).

Around the beginning of the common era, Bantu-speaking agriculturalists reached the southern limits of the forest. Agricultural communities then began a dramatic expansion through what is now Angola, southern Congo, and Zambia. They brought iron but not the first domesticated animals. Sheep and goats had filtered down from eastern Africa in advance of agriculturalists reaching Botswana. Their Khoisan-speaking foragers began to develop a pastoral lifeway that gradually spread over much of southern Africa. This lifeway did not totally drive out foraging, and a general symbiotic relationship between foragers and herdsmen developed (Ehret 1998).

In eastern Africa, two of the great food-producing systems of early Africa converged by about 800 B.C.E. After several centuries of interaction, people there produced a new synthesis that in turn spread rapidly in Africa. Before the arrival of Bantu-speaking agriculturalists, Cushitic-speaking farmers and herders had gradually moved south from Ethiopia starting about 5,000 years ago. The spread of cattle from northern Kenya to northern Tanzania took over 1,000 years in part because of a very difficult process of acclimatization to new disease environments. They had to build up a density of settlement large enough to suppress the bush that harbored tsetse. These farmers and herders arrived just as the first Bantu-speaking farmers entered the basin created by the Great Lakes in what is now northeastern Congo, southern Uganda, Burundi, Rwanda, and northwestern Tanzania. The highly diverse environment of the broader Rift Valley created the opportunity for experimentation with new combinations of agricultural and animal husbandry practices. Bantu speakers at first seem to have stuck to well-watered highlands of the region, leaving the lowlands to the agropastoral communities. By about 500 B.C.E., Bantu speakers seem to have adopted elements of grain production and cattle keeping and began a massive expansion into the grasslands around the lakes (Ehret 1998; Schoenbrun 1998).

From the Lakes, Bantu speakers seem to have moved rapidly across the landscape of eastern Africa. One branch occupied the highlands of Tanzania and

Kenya's Rift Valley near Nakuru. (Paul Almasy/Corbis)

Kenya. Other groups moved into the drier lowlands of the Rift Valley, including the Serengeti Plain, more slowly because these areas were held by agropastoral communities and foragers. Cushitic-speaking communities continued to occupy small regions in Kenya and especially central Tanzania. By about the beginning of the current era, another wave of expansion began from what is now southern Tanzania.

In the Great Lakes region, the expansion of agricultural and iron-working communities appears to have had its costs. Peter Schmidt, an archaeologist who has long worked in the area, has argued that by about 300 to 500 C.E. agricultural expansion and use of fuel wood had created an environmental crisis in much of what is now northwestern Tanzania and Rwanda. Relying on mixed-grain farming and cattle keeping, the area saw its total production decline. Population fell until the development of a banana production system created a new way of exploiting the area after about 1000 C.E. (Schmidt 1997a).

In south-central and southern Africa, the spread of ironworking and farming communities began in the first few centuries of the current era. On the eastern side of the continent, agricultural communities expanded on either side of Lakes Tanganyika and Malawi, developing two different styles of pottery. Both "streams" moved fairly rapidly, with the western or Kwale ware branch reaching South Africa by the 300s C.E. and the eastern or Nkope branch reaching Zimbabwe by the 500s. Until they reached the very driest regions of Botswana, Namibia, and South Africa, they tended to absorb the forager-herder communities they encountered. Livestock, especially cattle, remained important to the first agriculturalists, but as they encountered different environments, many with the potential to harbor tsetse, they adopted different mixes of husbandry and agriculture. In some areas in southern Congo, Zambia, Malawi, and Mozambique, cattle had to become relatively less important. In large parts of southern Angola, Namibia, Botswana, Zimbabwe, and South Africa, cattle were critical to the development of early agricultural settlement (Hall 1987; Maggs 1996).

On the western side of the continent, agriculturalists using iron and pottery making associated with what is called the Kalundu tradition spread south through Angola. They reached Botswana and Zimbabwe by about the 500s C.E. Gradually, throughout the region a rough geographic division of communities began to emerge. In the higher and drier regions of northern South Africa, Botswana, Zimbabwe, and Namibia, the societies that developed centered around control of cattle. They often had hierarchically ranked settlements. In the wetter lowlands bordering the Indian Ocean, communities were smaller in population but more numerous. These societies also used cattle as a central store of value but perhaps held fewer (Garlake 1973; Ehret 1998).

In the vast dry reaches of northwestern South Africa, western Botswana, and Namibia, the landscape dried to the point that agriculture could not be practiced. In these areas, Khoisan-speaking pastoralists and foragers maintained their hold on territory. As in the drier reaches of the Rift Valley and in the Congo rainforest, the presence of foragers into historic times has long tempted outsiders to see them as remnants of truly ancient populations (Klieman 2003). This tendency toward assigning a special place to such foragers, whether romanticized as "original people" or dehumanized as closer to other animals, is especially pronounced where foraging groups have maintained their own languages, as is the case of the Khoisan speakers of southern Africa and the Hadza and Sandawe of Tanzania who also speak a Khoisan dialect. Such a view is false on several levels. First, historic San (Bushmen) and Khoi herders are no closer to our common ancestors than anyone else alive. More importantly, from the moment iron-using agriculturalists appeared, foragers began to change their lifeways in response to new conditions. They lost territory to herders and agriculturalists, and they tried to supply goods in exchange for products, such as iron, from the agriculturalists. Many came under the domination of settled peoples, whereas others retreated deep into almost-desert areas to avoid domination. Especially after the conquest of the Cape region by Europeans in the seventeenth century, they were forcibly integrated into a colonial economy. Like the foragers of the forest, some elements of social and technical practice do come down from the distant past, especially since the Khoisan speakers of southern Africa maintained their own languages, but we must be very wary of equating their condition or actions with those of people even 2,000 years ago (Elphick 1977).

As we have seen in this chapter, Africans developed agricultural societies based primarily on crops domesticated in Africa itself and on technologies developed in Africa. Such packages gave communities the ability to adapt to the widely differing soil and climate conditions they found. They also provided the basics for developing complex societies through intensified manipulation of the environment and increased production of foodstuffs.

PATTERNS OF INTENSIFICATION

Between about the beginning of the current era and about 1500 C.E., African societies built on the accumulated fund of knowledge represented in the African Iron Age to intensify production and support fairly sizable urban communities and sometimes large state infrastructures. Despite their technological sophistication, these societies remained vulnerable to both changing environmental conditions and to overexploitation of resources. The oldest centers of urbanization in sub-Saharan Africa lie first in Nubia and a little later in Ethiopia in northeastern Africa. The Inner Niger Delta in West Africa also saw the development of urban societies before the current era. Intensification and urbanization in eastern Africa included both the rise of the Swahili cities and the development of extensive agricultural production based on bananas in the highlands of East Africa. The "rise and fall" of the Zimbabwe civilization in southern Africa will provide the last example of early urbanization from Africa.

As we will see in this section of the chapter, African societies increased their ability to alter the landscapes in which they lived. They all dealt with the problems associated with intensification caused by increased populations and the concomitant pressure on resources. Trade did become increasingly important to African societies after the beginning of the common era, but much of this trade took place within Africa. Specialized production of metals and cloth encouraged trade. In several regions, salt became the most important commodity in trade. External long-distance trade gradually became important, drawing on the specialized resources of Africa. However, Africans also exported craft-produced manufactured goods in certain places and times.

Most of the regions and states discussed below went through periods of expansion and decline, not just in political terms—dynasties and regimes—but in terms of population and degree of urbanization. They faced crises of production and survival. In each case, scholars have sometimes pointed to environmental causes for the collapse of particular societies. They have suggested either that intensification became unsustainable or that it often left societies vulnerable to crisis caused by climate change (Schmidt 1997a; Fagan 1999). Such explanations, however, sometimes seem a result more of a lack of evidence than clear indications of environmental crisis.

The oldest example of intensified production and accompanying urbanization comes, of course, from Egypt. Egypt became an early center of agricultural intensification based on recessional cultivation of the Nile Valley as the flood receded each year. The confined amount of land that Egyptians could cultivate as the Nile ran through the desert created the conditions for development of a hierarchical society and the incentive to develop and adopt means of extending productivity. These methods included irrigation, water wheels to lift water from the Nile even at low flood, and ox-drawn plows to turn the soil deeply for cultivation. The system proved exceptionally productive and stable. Urbanization seems to have developed first to the south of Egypt proper in the Dongola Reach of the Nile in modern Sudan. Urbanization and intensification in both Nubia and Egypt received a boost from the end of the Holocene Optimum and the drying of the Sahara (Hasan 1984, 1997a; Haaland 1995).

Some scholars have suggested that Egyptian consolidation occurred at the expense of the Nubian state of Ta-Seti in about 3300 B.C.E. (Ehret 1998). With consolidation a rough ecological boundary emerged between Nubia and Egypt at the First Cataract at Aswan. Nubia remained in close association with Egypt throughout the era of its greatness, but it remained distinct, if not always independent. Part of this distinction was geographic; Nubia also retained trade relations with communities around it and remained closely linked to the agropastoral peoples who ranged far to the south. Part of the distinction was also ecological. Although the earliest Nubian agriculturalists likely grew barley and emmer wheat that originated in southwest Asia, the growing season favored the African summer crops. They quickly became the dominant crops in the region after the beginning of the current era. In general, Nubian political structures, ide-

The largest site of Kush civilization burial pyramids lies north of Khartoum, along the Nile River in ancient Meroe, Sudan. (Jonathan Blair/Corbis)

ological bases, and even ethnic and linguistic identity changed over time, but like Egypt to the north, it appears to have gone through no great period of population crisis.

Urbanization within Nubia remained linked to the Nile. The earliest signs of urban settlement predate the emergence of a unified Egypt. After Egypt's unification, the town of Kerma became a center for the Nubian kingdom, which Egyptian sources generally refer to as Kush. During the New Empire period of Egyptian history from the sixteenth to the thirteenth centuries B.C.E., Egypt conquered and ruled Nubia. After the slow decline of Egypt's power, Nubians reasserted their independence in a state centered on the town of Napata, and in about 750 B.C.E. a dynasty from there conquered and ruled all of Egypt for about 100 years. After the Assyrian conquest of Egypt, Napata gradually reoriented itself to the south along the Nile, and the town of Meroe, south of the confluence of the Blue and White Niles, became the center of the culture. It has become famous as an iron-working area, and the shift south may reflect both the growth of an iron industry and the adoption of African crops coupled with irrigation methods that would have increased the productivity of the southern reaches of the Nubian Nile. While the Napata state maintained trade relations down the Nile with Ptolemaic and Roman Egypt, it also began to conduct trade directly through the Red Sea. As Christianity spread in the Mediterranean world, it also spread into Nubia. By about 400 C.E., two new states had emerged in the area of Nobadia in the north and Alodia based on the town of Soba in the south. Even with the rise of Islamic states along the Nile after 700 C.E., fueled in part by revolts by peasants and pastoralists against an oppressive Christian nobility, the fundamental structure of the Nubian agricultural system remained in place, producing a fairly sizable surplus in a very narrow niche along the Nile using irrigation and mixed cropping (Shinnie 1996; Connah 2001).

The highlands of modern Eritrea and Ethiopia provide the second main site for intensification and early urbanization in sub-Saharan Africa. These highlands in the northeastern corner of Africa provide an ideal place for the development of intensive agriculture and urbanization. The highlands rise, surrounded by dry lowlands. They tilt toward the west so that much of the water that runs off goes down the Blue Nile, eventually helping ease the annual floods of that river. The soils mostly resulted from volcanic activity at some point before 10 million years ago, and as a result they generally are fertile. The highlands are cut by deep gorges that create very sharp combinations of microenvironments based on elevation within a short distance. Rainfall generally reaches about 1,000 millimeters (about 40 inches) in the highlands, but diminishes sharply and becomes much more variable in the lowlands. As a result, there exist three distinct environment zones in the highlands: land above 2,400 meters (about 7,800 feet) which has a temperate climate, land between 1,800 and 2,400 meters (about 5,900 feet) that is warmer, and land below 1,800 meters that is hot (Nyamweru 1997).

As a result of the Ethiopian and Eritrean Highlands' distinct position, they became a center of plant endemism, which created the potential for developing a distinctive version of agriculture. Archaeological evidence for early agriculture is extremely limited, but linguistic reconstruction and biological evidence indicated that speakers of Cushitic languages domesticated several plants beginning perhaps as early as 7,000 years ago. Tef, a small grain little grown outside Ethiopia, seems the most important locally. Ensente, a relative of the banana, was domesticated in the southern part of the highlands. Finger millet, which did spread across Africa, comes from the lowlands. Coffee also comes from Ethiopia (Brandt 1984; Phillipson 1993).

Evidence for agricultural intensification becomes clearer after 1000 B.C.E. The northern portions of the highlands in Eritrea and northern Ethiopia are very similar to the adjacent regions of south Arabia just across the Red Sea. During this period, farmers in the area began to use ox-drawn plows and to grow wheat

Overview of agricultural highland which provides major employment in Ethiopia. A subsistence teff crop related to millet is produced in the region. (Earl and Nazima Kowall/Corbis)

and barley in addition to the Ethiopian crops. By about 500 B.C.E., South Arabian immigrants had established settlements in the northern highlands where they appear to have introduced terracing along the escarpments they found. Gradually, a new synthesis emerged that combined elements of South Arabian influence, including a newly derived Semitic language, Ge'ez. This language gave rise to the modern Ethiopian and Eritrean languages of Tigrinya, Tigré, and Amhara, while itself remaining a ritual language of the Ethiopian Coptic Church. The new culture combined oxen-drawn plows with extensive terracing and irrigation in the highlands. It grew a wide variety of crops ranging from wheat and barley in the high altitudes to teff and finger millet in the lower ones. It drew on livestock herded by pastoralists in the lowlands and reached its early peak in the era between 100 C.E. and about 700 C.E., with a state centered on the large town of Askum in far northern Ethiopia (Connah 2001).

The agricultural system spread with the Aksumite state south into the highlands. It took with it the plow-and-terrace agricultural system, language, and, after about 330 C.E., Christianity. The power of the state waxed and waned, particularly after 700 C.E., but the agricultural system remained. By 500 C.E., farmers had brought the vast majority of cultivable land using this system under agriculture. This process had reduced the forests of the highlands to remnants and had created shortages of wood for fuel, for fire, and for construction. Food shortages often struck the broader region, caused by variations in rainfall, too much cold in the highlands, or especially locust invasions. State authorities, when strong, often sought both to control surpluses to fight shortages and to promote protection of timber. The general productivity of the system within its context created what one of its students has called a "Malthusian demographic scissors" effect (McCann 1995). Population grew as the system expanded into new territories. But once growth had forced people into more marginal areas further down the slopes that were more vulnerable to crop failure or caused a reduction in fallow because of land pressure, then regions became vulnerable to demographic shocks caused by crop failures or disease. This cycle, seen in other premodern societies, operated right into the modern era.

Across Africa, in the West African Sudan, another center of intensification emerged along the Inner Niger Delta. The savannah region of West Africa might seem to defy predictions of areas for the development of intensified agriculture. It is a broad swath of land running from the boundary of the rainforest in the south to the Sahel and the Sahara in the north. It receives rain from the monsoon caused by the movement of the Intertropical Convergence Zone. Both the amount of rainfall and the length of the rainy season decline as one moves northward. This phenomenon has led to the banding effect of West African environments, a banding that is ancient even if the boundaries have shifted as climate has changed. From the coast, the environment changes from mangrove swamp to rainforest, to moist forest-savannah mosaic to dry forest-savannah mosaic, and to arid steppe to desert. This broad banding meant that from the earliest days of agricultural settlement opportunities existed for exchange and that in the savannah seemingly little in the way of specialized land resources existed (McIntosh and McIntosh 1984; McIntosh 1998).

As a result of the banding of environments especially at the northern edges of the system, even short-term climate variations could have dramatic effects on the landscape. Rainfall variation in the Sahel and northern savannahs was very high, and relatively small variations in climate over the course of decades and centuries, even after the establishment of the modern climatic regime, could cause dramatic movements in ecological zones in West Africa. Just as the "great Saharan pump" (Fagan 1999) had perhaps set off the process of plant and animal domestication and the development of urban societies in parts of Africa, so a smaller, more regular Saharan pump regularly allowed herders to move their animals farther north and farmers to open fresh fields and then sent them fleeing back south, leaving dried up wells, scorched crops, and dying animals. By the beginning of the current era, farmers and herders developed complementary relations across the ecological bands. Conflict often erupted in such periods of stress, as camel-driving nomads moved into the territory of cattle keepers and as cattle keepers encroached on the lands of more sedentary farmers. It was on the basis of his experience in North Africa that Ibn Khaldun, the great Arab historian of the fifteenth century, developed his theory of history as driven by conflict between nomads and sedentary peoples. His examples came from the northern side of the Sahara, but his theory reflected the reality of climate effects dating to a time farther back than even he realized (McIntosh 1998).

The intensification process in the West African savannahs was a gradual one, driven by the variations in climate. The development of urban settlements along the inland Niger Delta provides a clear example of this process. As noted before, the period between about 300 B.C.E. and 300 C.E. was an extremely dry one in the Sahara. The boundaries shifted southward, compacting farmers on top of one another and moving cattle nomads much farther south. Some historians believe this period saw the collapse of trans-Saharan contact that had previously utilized horse-drawn chariots for transportation. In what the archaeologist Roderick McIntosh (1997) has called a "pulse," people were driven to seek to exploit new resources with ever greater intensity. The same dry period saw large parts of the inland Niger Delta become less inundated. Previously, fisher folk had occupied the banks of the river channels and lakes; now, iron-using farmers joined them, bringing with them cattle, sorghum, and millet. They experimented with planting behind the receding flood of the Niger. They also developed the cultivation of African rice and subscribed to a broad network of subsistence strategies that stretched across the savannah, marked by specialization and exchange.

THE INNER NIGER DELTA

This specialization represented a generalized strategy not just to take advantage of the environmental differences, but to deal with climatic variability. It included an ethnic specialization so that Berber-speaking, camel-keeping nomads, who appeared after about 300 C.E. differed from Fulani-speaking, cattle-keeping nomads, who differed from sedentary populations speaking Wolof or Mande languages or Songhai or Hausa. Among especially the Mande peoples, the idea of caste specialization developed, sometimes merging into ethnic identity.

Along the Niger caste specialization developed into an urbanized tradition. At Jenné Jeno and other sites along the Niger and its channels, town populations grew to the tens of thousands by 500 C.E. In the Senegal Valley, the towns did not appear to be as large but were perhaps more evenly spread across the valley. On the eastern side of the Niger bend, Timbuktu and Gao also date their formation as towns to well before 1000 C.E. Farther west, the foundation of the Hausa towns in the savannah of what is now northern Nigeria dates to about the beginning of the second millennium.

It is against this context that the famous internal and trans-Saharan trade systems developed in West Africa. Gold and slaves have garnered the greatest historic interest in this system, but its origins probably lay in strategies designed to cope with environmental diversity and climatic change. The pastoralists' exchange of milk, hides, and animals for grain and cotton cloth from the farmers made up perhaps the most fundamental part of this system. Casted craftspeople produced cloth, leather goods, and metals for the system. Berber nomads developed the trade of salt from mines in the Sahara to salt-deficient regions in the savannahs. The Berber community of the desert took slaves from the Sudan to the mines, and this exchange probably formed the base of the trade (Brooks 1993).

Rains seemed to have moved farther north with more regularity after about 300 C.E. Agricultural settlements "pulsed" out from centers of intensification like the Niger Delta and moved generally north. At about the same time, camel-keeping nomads began to increasingly dominate the central Sahara, occupying many of the highland oases. They intensified trade with the Sudan and with the Mediterranean, purchasing slaves, gold, and leather with salt.

Scholars have long suggested that this trade, especially after it intensified following the Muslim conquest of North Africa, sparked the development of urban societies and empires in West Africa. The empire of Ghana, dating to about the tenth century C.E., with its capital at Kumbi Saleh in modern Mauritania, exemplifies this process. However, it now appears that urbanization in West Africa predated the development of the trans-Saharan trade. The flowering of Ghana, Mali, and Songhai that controlled the trans-Saharan trade from the south, the development of new Jenné Jeno and Timbuktu as centers of Islamic learning, and the complexity of the Hausa city-states were built on this flexible and integrated system of coping with environmental diversity.

The system did experience some crises, however. In general, the climate remained relatively wet, and the boundary of the desert was farther north up to about 1400, although extended periods of dryness continued to occur. After that time, a general period of drier climate set in (albeit marked by occasional periods of wetter weather). The boundaries of the bands shifted south once again, and population fell. Slave raiding from the desert intensified as horses became increasingly important in the warfare of the savannahs. Some scholars believe that the outbreak of the plague in the fourteenth century contributed to this decline.

Certainly after about 1600, both environmental and political conditions generally worsened. Even then, however, the basic equation that began developing almost 3,000 years ago continued to provide sustenance for most of the communities of the broad area, even as the communities continued to experiment with new crops and techniques (McIntosh 1998).

Intensification followed somewhat later in the West African forest than in the savannahs. The most prominent examples of early urbanization in the forests come from southern Nigeria. There the cities of the Yoruba country and Benin, both to the west of the Niger River, date to about 1000 C.E. In both cases, long, thick walls progressively enclosed more and more territory. The territory included dwellings as well as some fields. In the case of the Yoruba cities of what is now southwestern Nigeria, they stretched into the forest-savannah borderlands. The largest of the prenineteenth-century towns, especially the "imperial" capital of the Oyo federation, were located along the forest border. However, the center of Yoruba and other southwestern Nigerian cultures' cosmologies lay at Ile-Ife deeper in the forest. There, an urban complex dates to the eighth century C.E. Further to the east, Benin also developed an extensive system of enclosed towns and fields. In both cases, a logical explanation for the walled town system probably lies in the relative value of the land so enclosed. Archaeological studies indicate that almost all the land within the old walls had been cleared and cultivated, probably in a bush fallow system (Connah 2001).

To the east and west of the Yoruba country and Benin, agricultural expansion took a more extensive form. Farther east, in the Niger Delta of southeastern Nigeria and stretching into southern Cameron, more extensive settlement patterns prevailed. Extensive trade in the area dates to well before the beginning of the current era as people used the many rivers to move goods around. Towns developed slowly and concentrated on riverbanks tied to trade. Farther west, in what is now Ghana and the Côte d'Ivoire, towns date to about 1000 C.E. Agricultural intensification took the form of the extension of settlement in the forest regions and were linked to the extension of trade routes into the region, primarily in order to move gold north to the savannah markets (Andah 1993).

In the forest, however, by sometime after 1000 C.E., population density in the forest regions began to exceed that of the savannahs. The biggest reason for this sustained period of population growth lies in the regularity of rainfall in the forests. Similarly, forest soils, though varying in fertility, could be very fertile given enough fallow. These advantages apparently outweighed the disadvantages, which had kept human populations in the forest regions low before the introduction of agriculture. The forest probably had a deadlier disease environment. Malaria and yellow fever were endemic in forest regions, and human populations in these areas developed the sickle cell trait as a means of combating

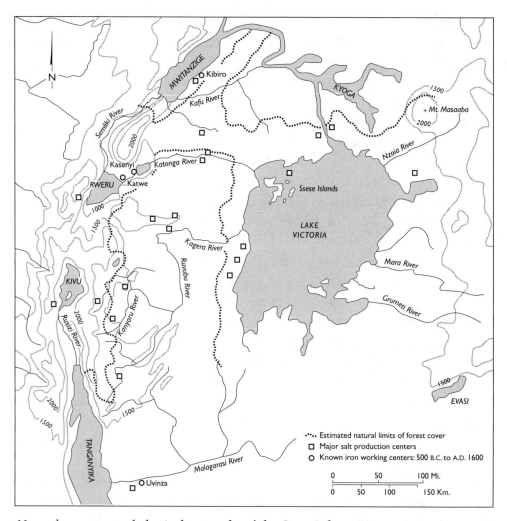

Natural resources and physical geography of the Great Lakes region.

the diseases. Tsetse flourished and with it trypanosomaisis. Although people controlled human sleeping sickness by clearing the land around fields and settlements, the fly prevented the keeping of much livestock. Hunting remained important for forest communities, with a wide variety of smaller game hunted or trapped. Goats, chickens, and fish were all important sources of protein. In the savannahs, the role of livestock and the regular failure of the rains kept the agricultural population spread out and towns concentrated along trade routes.

Agriculture arrived later in eastern and southern Africa, but intensification and urbanization still began in several areas early in the first millennium C.E. By the beginning of the second millennium, African farmers and herders had transformed landscapes using new crops and technologies, and by 500 C.E, some regions had begun to face crises in sustainability.

The Great Lakes region of eastern Africa includes the northeastern corner of the Democratic Republic of the Congo, Rwanda, Burundi, western Uganda, southwestern Kenya, and northwestern Tanzania. The area lies between the eastern and western arms of the Great Rift Valley and generally is about 1,000 meters above sea level. On the west, the Ruwenzori Mountains of the Congo mark both its boundary and the divide between the Nile and Congo watersheds. In the east, the escarpment of the eastern arm of the Rift Valley marks its extent with drier grasslands stretching north, east, and south. Several of the African Great Lakes lie within it, including Victoria, Kyoga, Mwitanzige (Albert), Rweru (Edward), Kivu, and Tanganyika. As discussed in the previous section, in the last millennium C.E. the region became a major point of contact between the expansion of the West African planting tradition and Cushitic agropastoralists. The region has also produced some of the earliest evidence for iron working in sub-Saharan Africa. It was in this region in about 300 B.C.E. that a rapid expansion of farming communities began (Schoenbrun 1998).

In the Great Lakes heartland, a gradual filling up of the land resulted in pressure for a dramatic transformation. In the better-watered highlands around the central plains of the area, people continued to grow yams and oil palms and to keep some stock. Moving down the slopes toward the plains, they planted grain crops of various sorghums and millets. They kept cattle in the almost tsetse-free central plains. In all of these areas over the course of the first centuries C.E., they cleared land for agriculture and cut timber for fuel, especially for metal working.

By the end of the recent Holocene, "optimum" in rainfall in about 950 C.E., a full crisis seems to have struck the Great Lakes region. The communities in the region responded by reorganizing production first to agriculturally exploit lands left for pastoralists in the past, and then, somewhat later, to develop extensive banana cultivation. In doing so, they remade social relations and developed the basis for a hierarchical society with ramifications felt even today (Schoenbrun 1998).

Scholars still disagree over whether the dramatic changes that developed in the Great Lakes resulted from environmental stress caused by human action in clearing woodlands or from climate change. Evidence exists for increased land clearing and iron production in the centuries around 500 C.E., during a period that had a relatively moist climate. Drying associated with the end of the recent Optimum did not set in until around 950 for the Great Lakes region. Rainfall in Ethiopia and the Great Lakes region can be estimated from the "Rodah nilometer," a gauge set up at Rodah in Egypt where records of the Nile flood were kept. The height of the annual flood reflected the intensity of the Ethiopian rains, whereas the level of the river at its lowest reflected the rains in the Great Lakes region. Yet Peter Schmidt in particular has argued that a subsistence crisis devel-

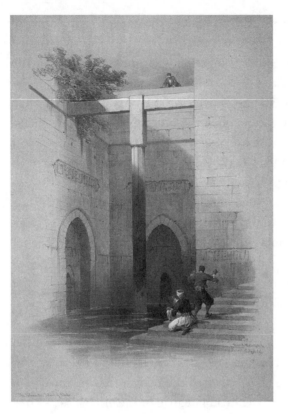

The nilometer, Island of Rhoda. (Library of Congress)

oped in the region after about 600, with some areas, such as the Bukoba region in Tanzania west of Lake Victoria, losing much of its population (Schmidt 1997; Nicholson 1981). The first part of a solution to this crisis appears to have developed at the beginning of the dry century around 950. The dry period coincided with an intensification of cattle keeping and grain farming in the central lowlands of the Great Lakes region. From the beginning, this extension of agricultural settlement into the lowlands remained linked to the better watered highlands that encircled it (Schoenbrun 1993).

Farmers in the Great Lakes regions seem to have cultivated bananas as a minor crop following their introduction shortly after the beginning of the current era. Farther west in the Congo Basin, one version of the banana, the plantain, quickly became a staple. Farther east, in the highland regions of the Eastern Arc, bananas gradually became a more important crop. Farmers used the bananas as a staple, often boiling it with beans and other vegetables. With adequate rainfall distributed fairly evenly over the year and with warm temperatures, in reasonably fertile soil, bananas can yield enough food to feed a family on a relatively small amount of land. The trees yield up to thirty years, and output can increase using both green and animal manure. Other shade-tolerant crops can be planted around the trees. In the Eastern Arc Mountains, farmers developed irrigation systems to channel water to the groves and ensure steady moisture. They regulated such systems through neighborhood associations. Households and kin groups also owned land at various elevations in order to diversify crops. The mountains of Upare, Usambara, Kilimanjaro, Mount Kenya, and Mount Meru all became home to very dense populations based on the permanent cultivation of bananas mixed with grain farming and stock keeping on lower slopes (Rossel 1993; de Langhe, Swenne, and Vuylsteke 1996).

In the lands around the lakes, banana cultivation began to dominate the highland areas starting around 1200. Farmers cleared the last of the forests on the ridges and converted grain fields and fallow into banana groves. Very quickly, the highlands became almost entirely covered in groves, and population grew dramatically. In the plains which the highlands surrounded, population also grew as mixed farming communities increased grain cultivation and cattle numbers to supply the people of the highlands. By the 1500s, a new social order emerged in which the cattle owners based originally in the plains dominated the banana farmers with their fixed, permanent fields. A large, centralized state arose in Rwanda dominated by Tutsi nobles; smaller states existed in Burundi, eastern Congo, and northwestern Tanzania. In the northern part of the region, in what is now southern Uganda, territorial chiefs came to dominate political life based on their power over the farming communities. They created another complex of states including Bunyoro, Ankole, and Buganda (Schoenbrun 1998).

In all of these cases, agricultural intensification resulted in sustained, increased production. The landscapes of the Great Lakes region and of the Eastern Arc Mountains had been totally reshaped early in the second millennium. The expansion of production created the possibility of, and perhaps the necessity for, large-scale political organization to regulate access to resources, especially the land used for banana groves and cattle. The argument that intensification led to political complexity seems perhaps commonplace, for it reflects a typical pattern of state development. However, older scholarship has often associated both agricultural intensification and political centralization with external influence. As this case shows, such Africans did not need such influences, and even when external contact played a role in the development of political elaboration, it operated in a context generated by processes of development based on local resources.

The development of the Swahili civilization along the East African coast provides an example of the development of urbanization connected to outside contact. The port towns of this area owe their existence to the ancient seaborne trade between East Africa and the rest of the Indian Ocean world. The monsoons that bring the rains to East Africa also make possible relatively easy transport. They blow south along the coast for six months and then turn and blow north for six months. By the early years of the current era, relatively regular commerce existed between East Africa and the Arabian Peninsula. From there trade moved on both to the Mediterranean world via the Red Sea and to the Indian subcontinent. By about 500, sites from Mogadishu in Somalia south to at least the current Tanzania-Mozambique border shared a material culture, with subsistence based on fishing in addition to agriculture. Trade goods from the Indian Ocean and Mediterranean trade networks circulated in this system. The culture eventually extended to the Comoro Islands and Madagascar (C. Kusimba 1999).

In the early centuries of this commerce, the Swahili ports likely exported iron, grain, mangrove poles, ivory, and slaves. Through the ports came manufactured goods, including ceramics and textiles. In addition, bananas, sugar cane, and rice made their way into the interior of East Africa. The Swahili ports developed up to about 900 as a network linked together by a common material culture and language.

After about 900, Islam began to play an increasingly important role in the development of Swahili society. While eventually the elites of the major Swahili ports such as Lamu, Mombassa, Zanzibar, and Kilwa claimed descent from immigrants from Arabia or Persia, the language and the culture remained rooted in a local progression. Even the development of building in stone drew from local motifs of housebuidling, as well as incorporating elements from the broader Muslim and Indian Ocean world (C. Kusimba 1999).

After the rise of Islam in the Middle East gave the Muslims control of the access point for East Africa trade, the goods traded began to change. Slave exports increased after about 800, and gold began to flow out of especially the southern Swahili ports. On the coast Kilwa rose to prominence by the thirteenth century and held it until the sixteenth. The gold came from Zimbabwe. There, a stratified society emerged on the Zimbabwe Plateau famous for its capital, called Great Zimbabwe, and other settlements containing stone buildings.

African cattle keepers in southern (and eastern) Africa had long used stone to build enclosures for cattle, often near water sources. In southern Africa, this general technology combined with a readily available source of easily worked granite and the development of social stratification to allow the construction of large buildings and enclosures in stone used by a political elite. Although sites dating from before 1000 are widespread, the largest early urban site is at Mapungumbwe on the Limpopo River. By about 1200, Great Zimbabwe eclipsed all the other towns in the area and became the center of a large kingdom. The rise of the Zimbabwe complex, which includes well over 100 sites, has usually been associated with the gold trade to the Indian Ocean. Like all of the cases mentioned here however, the Zimbabwe settlements grew based on elaboration of the local agricultural system (Garlake 1973).

The mining and trade of gold date to about 500 C.E. Over the course of the period before about 1100, mining activity moved southward on the plateau. The most intensive mining began in about 1100 and lasted until about 1700. The timing of the Great Zimbabwe complex coincides with the intensification of mining in 1100. By about 1500, however, Great Zimbabwe itself went into serious decline and lost most of its population. The state centered on Great Zimbabwe collapsed, and several successor states known historically divided the plateau. Although the gold trade played an important role in the Great Zimbabwe state, its social structure was rooted in the ecology of the plateau, and its collapse may be rooted there also.

In the case of Zimbabwe, the central plateau is a relatively tsetse-free area surrounded by areas dominated by tsetse-infested bush. Cattle appear to have become not only a means of subsistence but as elsewhere a means of controlling people. The builders of the Zimbabwe settlements placed many of them on the borders of the highlands where they could exploit dry season grazing in the lowlands when tsetse were cleared for part of the year. During the first millennium, a hierarchy of sites centered at Mapungubwe, in what is now northern South Africa, developed in the area. Starting in about 1100, the town at Great Zimbabwe began to grow and became the center of a network of sites that stretched into the northern portions of the Zimbabwe Plateau as well as into northern South Africa and Botswana. The elite of this state used the stone walls and buildings as sites of power. For example, at the core of Great Zimbabwe itself, huge amounts of cattle bones from relatively young cattle indicate that the elite dined regularly on meat. Commoner houses show much less reliance on meat generally and a greater proportion of their meat coming from small stock.

Great Zimbabwe's collapse may have been caused by climatic variation coupled with increasing pressure on resources. The site itself had a population of up to 18,000 at its height. Very quickly fuel wood and game disappeared from the area. Eventually, both agricultural land and grazing would decline as fallow was reduced and the constant movement of cattle to the capital continued. With the population decline during the fifteenth century, power flowed from the site (Garlake 1973; Beach 1980).

Not all researchers agree on an environmental causation for the decline of Great Zimbabwe. Pikirayi (2003) argues that the development of trade connections farther north from the coast toward the Malawi kingdom and the first Luba kingdom in the Upemba Depression drew trade away from Zimbabwe and weakened the power of its rulers. Two major states rose on the plateau to compete for control of the gold trade. In the north, the state of Mwene Mutapa controlled access to the gold trade when Portuguese traders reached the area. In the south, the state of Torwa with its capital at Khami continued the political traditions of Great Zimbabwe.

To the north, at a slightly later date, a process of intensification began along the Lualaba River sometime before 1000. The Upemba Depression is a low-lying area in the savannahs of the southern Congo. The Lualaba, a major tributary of the Congo, runs through it, filling it with small lakes and streams. The region became an early center of settlement by Bantu-speaking farmers, who quickly began to exploit the fish of the wetlands. At some point before about 1000, they began to construct dykes to contain the streams of the area and protect house sites. They left other areas open to allow the river to flood for agriculture. An extensive trade network had developed, moving salt, fish, and especially copper for the great African copper belt about 300 kilometers (about 180 miles) to the south. By the eleventh century, a state developed ancestral to the Luba Empire of historical records. Once again, as we saw in several of our other examples, the Luba people of the Upemba Depression developed techniques to exploit a specialized landscape and the potential for trade that included the development of political hierarchy (Connah 2003).

CONCLUSION

The sixteenth century marked the beginning of a radical transformation in the relations of African societies to the rest of the world. An older scholarship portrayed Africa as relatively undeveloped and its natural resources as unexploited; however, as this chapter has argued, more recent scholarship has shown how societies in Africa developed a wide variety of approaches to producing subsistence and surplus. They developed methods of intensive cultivation where necessary, and they utilized extensive methods where possible. In some cases, they seem to have overexploited their resource base and generated crisis. The dramatic changes that accompanied the coming of Europeans during the Age of Exploration operated within societies already in a process of development.

4

AFRICAN ENVIRONMENTS AND THE ERA OF THE COLUMBIAN EXCHANGE

The fifteenth century witnessed the beginning of the process of sharing biota across the globe in a new way. Alfred Crosby (1986) named this process the Columbian exchange after the confused Italian mariner who eventually talked the newly unified Spanish crown into backing his effort to reach Asia by sailing west from Europe. After 1500, the previously isolated Western and Eastern Hemispheres began to share their plants, animals, microbes, and people. The result was disastrous for the indigenous population of what became known as the Americas, which declined dramatically in the hundred years after 1500. Africa and African peoples played critical roles in this exchange. The effect on African landscapes, though perhaps not as dramatic in the early centuries as the effect on the Americas, was also large.

In assessing the impact of increased and direct communication between Africa, the Americas, and Europe, two scholars have proposed overarching explanations for the relationships of African landscapes to the rest of the world before and after the 1400s. Crosby has argued that Africa participated in what he calls the Old World Iron Age. In short, African societies shared many elements of a technological "kit" common to the Eastern Hemisphere. As we have noted in earlier chapters, Africans participated in some of the earliest developments of agriculture, animal husbandry, and metallurgy. Crosby also noted that the Eastern Hemisphere generally constituted one large intercommunicating zone, with the exception of some very isolated areas. Technology, goods, ideas, crops, domestic animals, and microbes moved across this zone, often in station-to-station relay fashion. Some things were environmentally specific. Wheat, a crop that relies on wet winters in general, never became a staple in tropical regions. Diseases such as malaria, which are dependent on insect vectors, could not move outside areas where their hosts could not survive. But other things made their way from the tip of southern Africa to the Polar Circle and from the Atlantic to the Pacific.

Jared Diamond (1997) has added a caveat to Crosby's argument. He suggests that the most intense transfer of both technology and biotic matter occurred in latitudinal bands across the Eastern Hemisphere. This factor explains the spread of technology from Europe across Asia. Africa, as he sees it, was isolated from this main expansion and, like the Americas, burdened with a geography that oriented it north-south, cutting across environmental zones. As such, intercommunication and transfer gradually lagged behind that of the northern Eastern Hemisphere. Diamond suggests that, like the Americas, Africa was fated to be dominated in the short term by societies from the Northern Hemisphere.

Although Diamond's argument about the relative ease of transfer of technology and biota in latitudinal bands has some merit, he gets the argument backwards. Africa had participated in another massive exchange of technology and biota in the centuries before 1492 that shaped it (and shaped the rest of the world) as profoundly as the Columbian exchange would. African societies were always in greater or lesser contact with other regions of the Eastern Hemisphere. In particular, exchange across the Indian Ocean in some ways prefigured the Columbian exchange in the movement of plants, animals, and people into new landscapes. This "monsoon exchange" brought African crops, animals, people, and goods throughout Asia and introduced new ones to Africa, thereby transforming African landscapes.

Yet the centuries after 1500 did bring dramatic changes to Africa. One of the biggest such changes was the movement of millions of Africans to the Americas through the trans-Atlantic slave trade. Until shortly after 1800, a majority of the immigrants to the Americas from the Eastern Hemisphere were Africans who came almost entirely against their will. A partial explanation for the preponderance of Africans lies in disease environments. Philip Curtin (1968) has posited that the spread of Eastern Hemisphere diseases, both infectious and insect borne, created an environment in the tropical parts of the Americas that decimated indigenous populations while at the same time proving deadly to Europeans from temperate latitudes. This combination created a survival advantage for Africans in the tropical Americas, which led Europeans to choose to repopulate those regions with captives taken from Africa. These captives took with them more than their "unrequited toil"; they brought with them knowledge of how to live in different environments that shaped the ways that the Creole societies of the Americas developed.

During the era of the Columbian exchange, Africa received new types of plants and animals that gave African societies new options in surviving within their environments. The introduction of maize, cassava, and tobacco reshaped agricultural practices in different regions of Africa, while increased exchange brought greater supplies of previously well-known goods such as metals and horses. Greater contact, especially along the western seaboard of Africa, circulated diseases more rapidly and in a few exceptional cases led to American-style demographic crises where a majority of the population dies shortly after coming into contact with Europeans and the diseases they bring.

This chapter looks at how expanded and reoriented contact with the outside world impacted Africa. Whereas previous chapters have focused more tightly on African ways of living within their environments, this chapter explores the environmental history of contact between Africa and the rest of the world. It begins with a discussion of the exchange of plants, animals, and peoples across the Indian Ocean, and then it turns to the period of maritime expansion of European trade and conquest, including the beginnings of European settlement in Africa at the Cape of Good Hope. It also discusses the environmental causes of the slave trade and the transition to commodity trade that followed.

THE MONSOON EXCHANGE

Looking at the exchange of biological agents across the Indian Ocean demonstrates the dynamic of the biological and environmental consequences of contact. Exchange across the Indian Ocean unfolded over a longer time frame than the Columbian exchange, but many of the mechanisms were similar. Examining this exchange first perhaps unfreights some of the present-oriented biases of earlier views of exchange, including the often implicit notion that Europeans wound up being biologically or environmentally dominant (McNeill and McNeill 2003).

Before the 1400s, Africa remained in contact with the rest of the world through two principal routes. One, of ancient importance, crossed the Sahara; the other, which developed by about the beginning of the common era, used sea routes across the Indian Ocean. The Saharan frontier was of critical importance from the time of the spread of the first hominids out of Africa. Its expansion and contraction served as a catalyst for the spread of modern humans out of Africa and for the development of food production on the African continent, as shown in the previous chapters. Certainly, however, after the beginning of the Holocene it also served as a filter, limiting the types of contact and exchange that could occur. Over its broad expanse it marked the boundary between an important ecological division between Mediterranean climates with wet winters and tropical climates with dry winters. Whereas animals could move both ways, especially along the Nile corridor, and people could exchange goods, agricultural plants generally could not make the transition (with the partial exception of the Ethiopian Highlands). Wheat never spread south of the Sahara (except in Ethiopia), and sorghum and millet never made a big impact in the Mediterranean world. Cattle, sheep, goats, and camels crossed the Sahara, with cattle perhaps being first domesticated in the Sahara and spreading northward as well as southward.

From the very distant past, however, the seaboard of the Indian Ocean

Book illustration of warriors and women in Madagascar by Gallo Gallina. This illustration was published in a book by Giulio Ferrario in the early 1800s. A principle reason why African and European merchants were attracted to Madagascar were the high-quality cotton, raffia, and silk textiles produced by indigenous artisans. (Stapleton Collection/Corbis)

served as a route of contact and exchange. African domesticated crops crossed very early through Arabia and on into Asia. Both sorghum and millet have been dated to at least 2000 B.C.E. in the Indian subcontinent. Undoubtedly, they gradually diffused across climates for which they were well suited.

The evidence for adoption of other crops and domesticated animals in Africa from Asia is less clear. Several crops developed in Asia became known and even important in Africa before the 1400s; these included bananas, rice, sugar cane, and chickens. Bananas and chickens seemed to have spread into Africa around the beginning of the common era. Only later did Asian rice and sugar cane became important, possibly after the development of Indian Ocean trading routes connected to the Muslim world. The early transfers of bananas and perhaps chickens seem to have come with speakers of a Malagasy language. These people, carrying trade around the Indian Ocean, had contacts along the east African coast and established settlements on the island of Madagascar, which before their settlements seem to have been uninhabited. Their language would become the dominant one on the island, whose population eventually included other immigrants from eastern Africa (Sutton 1990).

The Malagasy influence indicates that contact across the Indian Ocean gradually increased after the beginning of the common era. It received a large boost with Islam's rise to dominance after about 700 C.E. In addition to crops, this gradual and uneven expansion of trade also brought new demands on African resources. Africans became producers of gold in Zimbabwe and ivory throughout the region. Mangrove swamps along the coast became the source of much of the timber traded into the northwestern corner of the Indian Ocean. In addition, East Africa became the source of people brought to the heartland of the Islamic world to work as slaves on agricultural estates. This slave trade was nowhere near the size of the later Atlantic slave trade, but all the same it represents the same sort of core/periphery relationship (C. Kusmiba 1999).

In the long run, the introduction of African crops into the rest of the Old World and the introduction of Asian crops into Africa may have had a greater impact on the African landscape than the Columbian exchange, but obviously the cultural and political impact was much less. Outside ideas, especially Islam after 700 C.E, did spread into Africa, but African societies remained autonomous. In the Sudan all the way across Africa, Islam gradually became dominant; yet the societies affected remained independent and primarily African. In East Africa, before the nineteenth century, Islam remained a religion of coastal enclaves and traders. Even trade in East Africa to the interior remained in the hands of the peoples of the interior. On the other hand, banana and plantain cultivation, as well as root crops such as taro, transformed agricultural practices all across Africa, allowing for more intensive settlement in highland regions and in the rainforests. Over the course of the centuries, it proved a mutually beneficial exchange.

AFRICA AND THE COLUMBIAN EXCHANGE

The geography of Africa and the reality of its peoples and cultures played a critical role in the process that created the Columbian exchange. Africa south of the Sahara, like Europe, was generally a near periphery of the Islamic world throughout most of the period between 700 and 1500. North Africa became integrated into the Islamic world, and its urban areas became "Arabized." Although East Africa provided iron and food and West Africa provided leather goods that merchants took to the Islamic heartlands and beyond, Africa was most well known to the Muslim world as a source of highly prized commodities. These included timber from East Africa, ivory, captives, and especially gold from West Africa and from Zimbabwe. The creation and elaboration of trade routes had in the long run brought new crops that transformed landscapes but had not dramatically transformed production to meet external demands. Production for external demand remained in the hands of a few specialized producers, and aside from salt from the Sahara, the primary imports remained luxury goods.

The development of Sahara's salt trade in return for gold, leather, textiles, and some captives illustrates why this was the case. Across the Sahara and the Sudan, merchants used animal power for transport. Camels carried the main trade goods across the Sahara, while donkeys and horses were used to some extent in the Sudan. Traders used river transport in some areas. The Niger carried a great deal of commerce connected to the Saharan trade, and to a lesser extent the Senegal did as well. Elsewhere, long stretches of the Congo and many lesser rivers in Central and eastern Africa carried more localized commerce. However, away from the rivers, trade goods had to be carried by humans. Tsetse flies carrying trypansomiasis limited the use of the major domestic animals in large parts of the forest and wetter savannahs. Hence trade remained localized, with goods imported from longer distances fetching extremely high prices (Brooks 1993).

The story of the Age of Exploration is a very familiar one. In general, by the fifteenth century, merchants in southern Europe knew what types of trade goods were available in Africa and in the great trading cities of Asia. The Portuguese took the lead in trying first to find routes to tap the gold trade in West Africa, next to reach the Indian Ocean and the gold trade of East Africa, and then to follow the well-established routes across the Indian Ocean to reach India, the East Indies, and China. The Portuguese expeditions began to reach the coast of West Africa south of the Sahara in the 1450s. By the 1470s, they had established commercial relations with African societies as far south as the modern nation of Ghana. In 1481, the Portuguese reached the mouth of the Congo River and established contact with the kingdom of Kongo. In 1487, Bartholomeu Dias led an expedition that turned the Cape of Good Hope and reached the Indian Ocean (Crosby 1986).

The nautical key to this enterprise was the discovery of the trade winds. A prevailing wind and current blew south along the coast of Morocco. Dating back to the Phoenicians of antiquity, several attempts had been made to travel by sea along the Atlantic coast, but no records of a successful return exist. Local communities along the Atlantic coast of Africa used smallish boats to fish and transport goods over relatively short distances, but did not succeed in developing the techniques of open sea navigation. Portuguese mariners first discovered the system of trade winds around the equator; they learned that by sailing northwest from the Atlantic islands of the Canaries they could reach the "westerlies" that would carry them back to Europe.

Before Columbus's voyages, the first direct contact between Europeans and Africans in Africa had little impact on the landscapes of Africa. Although some early Portuguese voyages tried to imitate the program of French, Spanish, and Portuguese attacks on the native peoples of the Canaries, they brought with them little that gave them the advantages Europeans held of the Guanches of the Islands. The African societies they encountered in the Saharan and Sudanese sections of Atlantic Africa had stable food production systems that included domestic livestock. They produced and used metal goods, including weapons, and they had horses. They lived in societies that were at least partially Islamicized or connected to the Islamic world (which was one of the reasons for the Portuguese attacks). Most importantly, perhaps, they shared many of the same diseases in some form or another. The Portuguese quickly decided to pursue a policy of trade and began to draw at least some of the gold out to the mouth of the Senegal River (Crosby 1986; Diamond 1997).

As they sailed further south and east, Europeans not only had no great advantage over African communities in the fifteenth century; on the contrary, they operated at a disadvantage. They had little to no immunity to the tropical disease endemic in the moister savannahs and rainforests, and their horses could not survive the disease environment either. From the beginning, in the Gold Coast and on the Congo, they pursued a policy of trade.

In West Africa, the Portuguese found much that they wanted but had little to offer in return. They could obtain gold both at the mouth of the Senegal and along the Gold Coast. In both cases, they dealt with Mande-speaking merchants connected to the Empire of Mali. To the east of the Gold Coast, they could purchase Malabar pepper and cotton textiles. The Portuguese paid a local ruler on the Gold Coast for the right to establish a permanent trading post that became known as Elmina. Further south, they found little in the kingdom of the Kongo that they wanted but discovered that they could purchase captives with textiles (often of African origin) and eventually cowry shells that served as the currency of the kingdom and surrounding area. Along the Gold Coast, in addition to textiles they found a market for those captives, as well as on the Atlantic islands such as Madera where sugar production had begun to increase. The value of the trade in captives remained less than the value of other exports from Africa carried by Europeans well into the sixteenth century, and the numbers of captives captured or bought by Europeans (reaching perhaps 5,000 per year in 1500) probably had little demographic impact on the general population of Africa from the Senegal River to the Cape of Good Hope. In short, for the first fifty or so years of direct European contact with sub-Saharan Africa, the effects were mostly economic (Lovejoy 2000; Manning 1990).

The events of 1492 changed all that. Columbus's voyages to the New World for Spain, followed quickly by Portuguese claims to Brazil, began the process of transforming Africa's relationship to the rest of the world. The short-term result was the introduction of new crops into Africa. The medium-term result was to turn Africa into a hunting ground for people to help repopulate the New World. The long-term result was to turn the subcontinent into a producer of raw materials for world markets. This long-term process remade African landscapes. Populations moved into new areas, and people developed new ways of exploiting available resources. The slave trade and attendant violence uprooted millions. The development of commodity production reconfigured productive systems and made African societies dependent on world markets in a way that earlier forms of trade had not.

Initially, and very rapidly, African agricultural systems gained new crops from the opening of contact with the New World. Maize had the earliest prominence. It spread so rapidly in Africa that many scholars have attempted to prove that Africans and perhaps Arabs had knowledge of the crop before 1492. However, two factors in the evidence tend to support the conventional explanation. First, the earliest documents describing maize in Africa date to after 1500. Second, the earliest accounts of maize find it in the coastal West Africa locations where it would have spread through Portuguese contact. The spread of maize into eastern and southern Africa occurred only in the late sixteenth or early seventeenth century. Neither of these pieces of evidence is decisive, and neither precludes a just barely "pre-Columbian" introduction by unrecorded voyages in the fifteenth century; however, they are strongly indicative (McCann 2005; Miracle 1966).

Maize proved to be quite well adapted to many African environments. It grows well in warmer and wetter environments. The learning curve for African farmers familiar with sorghum and millet cultivation was short. Maize grows much like the African grains, and its grains can be processed into food in similar ways. It can be shucked, hulled, and ground into flour and then boiled with water. It can also be roasted or boiled on the cob. It yields well in moister environments such as the rainforest or moist savannahs, better than most sorghums and millets. It also matures faster. However, most sorghums and millets withstand drought better than maize can. In the long run, maize gave African farmers another crop that they could fit into varying environmental conditions.

The Portuguese, more than the Spanish, brought both maize and cassava to ports and trading posts along the African coast. The initial area of introduction seems to have been the Cape Verde Islands, where maize was grown in sizable amounts by 1535 to 1550 according to an anonymous Portuguese account (Miracle 1966). Numerous European accounts attested to the growing of maize along the coast of West Africa. People along the Gold Coast especially took it up. This region hosted several permanent Portuguese trading posts by the early sixteenth century. The Portuguese also brought maize to the region that became known as Angola in southwestern Africa (McCann 2005).

On the basis of records showing that different varieties of maize were adopted in the drier Sudan, some scholars suggest that maize also spread from Spain across to North Africa and then across the Sahara to the West African Sudan. Such a route is possible, but there is no direct evidence for it, and the distribution of types of maize in modern times may reflect the properties of the different varieties more than the history of their spread. "Flint" maizes with smaller grains that withstand dry weather better have become more common than softer "starch" maizes in both Iberia and the Sudan (McCann 2005).

From West Africa maize spread rapidly throughout the forest regions of West Africa. The development of Portuguese trade for gold in the Gold Coast, in combination with the introduction of maize, apparently triggered a sharp rise in population in that region. Portuguese and later Dutch traders report bringing captives from the Niger Delta and the Congo/Angola regions to trade for gold at Gold Coast ports. This rise in population, prefiguring a rise in population densities in most of the forested regions of West Africa by the seventeenth century, resulted in expanded clearance of the rainforest.

Outside of West Africa, maize spread more slowly in the sixteenth and seventeenth centuries. Maize was closely associated with the Portuguese trading diaspora in the sixteenth century in Africa. The Portuguese used maize as a food for their trading vessels, especially those carrying slaves, and they took maize with them around the Cape of Good Hope into East Africa. Maize is first mentioned in East Africa in accounts of Portuguese settlers on Zanzibar and Pemba off the coast of East Africa growing maize to supply Portuguese outposts on the coast such as Mombasa. Although it may have been known in the interior of East and southern Africa by that time, it does not seem to have been a staple crop. Sorghums and millets remained far and away the most important crops in the savannahs, while bananas remained important in the highlands. The intensification of trade in the nineteenth century may have promoted the expansion of maize production in many areas of eastern and southern Africa. In the savannahs, however, periodic drought limited its adoption. Sorghums and millets withstand drought better even if they yield less in wetter years.

Manioc (cassava) also became an important crop in parts of Africa as a result of Columbian contact. It first gained a hold in Africa in the Congo/Angola region. Again, its early appearance coincides with the establishment of a Por-

Nineteenth-century print of workers grating manioc. (Gianni Dagli Orti/Corbis)

tuguese presence and the development of a trade in slaves during the sixteenth century. Many varieties of manioc require special processing to remove toxins before consumption. It grows well in a variety of climates from tropical rainforests to semiarid regions, and it also withstands drought well. Manioc has a relatively low nutritional value, however. It spread in Africa gradually, becoming an important crop in the Central African rainforest. In many cases, it displaced or supplemented grain crops. In the twentieth century, it spread in areas where population pressure reduced fallow time for fields (Jones 1959).

Other New World crops gradually spread in Africa. Among them were chilies, which spread in tandem with maize and became commonplace in African farming systems. They likely spread to the eastern Mediterranean from eastern Africa and then through the Ottoman Empire into eastern Europe. Tomatoes also spread in Africa, though much more slowly. Peanuts (also known as groundnuts) came from South America; like maize and the African grains, peanuts grow similarly to bambara nuts and hence African farmers easily adopted them. They very quickly became a crop grown in the West African savannahs and eventually spread into eastern and southern Africa. Tobacco also came with contact, and it, too, gradually spread in Africa (Brooks 1993).

Of course, the Columbian exchange was not limited to new crops. African-origin microbes played a major role in the conquest and depopulation of the New World, and African people played a major role in the repopulation of the New World. The Africans brought to the New World became laborers in the dehumanizing institution of racial slavery, bringing with them knowledge and technology that their owners exploited for profit.

African-origin diseases, especially malaria and yellow fever, played a major role in the depopulation of tropical America. Malaria was present in southern Europe also, but it struck there with less intensity. Hence, even in Italy and the Iberian Peninsula, human populations never developed the sickle cell trait to the extent found in African populations. Since it established itself fairly rapidly, the *plasmodium falciparum* that causes the most severe form of the disease in humans apparently both found new host mosquito species in the New World and came with African *Anopheles* species that traveled with Africans and Europeans to the New World. It seems to have established itself in some of the earliest settlements of Europeans in the Caribbean during the sixteenth century. Thus established, it played a major role in depopulating the tropical Americas of the Caribbean, Central America, and the Amazon Basin. In many cases, as in Africa, density of settlement and extensive clearing for agriculture helped improve conditions for transmitting malaria. This transmission helped make the tropical New World deadly for Europeans, although not nearly as much as for Native Americans (Bradley 1991).

Yellow fever established itself somewhat later. Yellow fever is transmitted by the *Aedes aegypti* mosquito, a species that reproduces only in still water. Unfortunately for humans, the casks and cisterns used to store water proved ideal breeding grounds for the insect. The disease became particularly associated with seamen. It did not become serious in the Americas until the mid-seventeenth century, when an outbreak was reported in Havana. Once established, it became epidemic.

African diseases had a profound effect on the course of conquest of the New World. While Old World infectious diseases generally decimated New World populations, the spread of Old World tropical diseases especially wreaked havoc on tropical populations in the New World. Through this combination, the tropical regions of the New World were almost totally depopulated. The Caribbean Islands lost almost all of their population, and Central America and the Amazon Basin lost the vast majority of their inhabitants. Spanish and Portuguese military action during and after conquest (and later in North America, British and French conquest) played an important role in reducing these populations, but in more temperate zones such as central Mexico or the Andes Highlands in South America, a core of the population remained. In large parts of the tropical world, however, none remained (Crosby 1986).

The malaria and yellow fever infestations in the tropical regions of the New World made those regions deadly for Europeans. Those who had been exposed to these diseases in childhood, and survived, acquired a degree of immunity to them; generally, Europeans did not have this immunity. Death rates for Europeans in the tropical New World soared above those in either Europe or the more temperate parts of the New World. As the Portuguese and then the Spanish, Dutch, English, and French began to establish sugar plantations in the New World, they found that mortality rates for European servants made their use on sugar plantations uneconomic. During the sixteenth century, they increasingly turned to African captives whom they held as slaves. The Africans' slight epidemiological advantage in the tropical New World, as Philip Curtin (1968) has argued, laid the basis for the development of racial slavery in the New World.

The Africans' survival advantage can best be understood as the "tipping factor" or decisive element in establishing racial slavery, but it was not the only factor. The Portuguese had engaged in raiding and trading for captives on the African coast during the fifteenth and sixteenth centuries before either the Old World tropical diseases or sugar plantations became known in the New World. The Portuguese had used these slaves on sugar plantations in the Mediterranean and Atlantic islands, where these captives joined slaves from other parts of the Mediterranean world; the Portuguese had even sold them from one region of Africa to another. The Spanish and Portuguese had imported African slaves into the New World for use as servants and laborers in urban areas and in mines. Ironically, in the silver mines which the Spanish established in Mexico and Peru, Africans often served as the skilled labor while Native Americans provided unskilled labor under forced conditions. However, the gradual establishment of sugar plantations in Portuguese Brazil and then in the Caribbean, where both the indigenous population had disappeared and the European death rates due to disease were increasing, shifted the balance strongly in favor of the use of African slaves. As a result, the demand for African slaves shot up, and the system of solely enslaving Africans became clearly established by the end of the sixteenth century. From its base in the tropics, the system was taken up by the Dutch, English, and French as well. Each initially experimented with using Europeans as laborers, but each in turn shifted to African slaves in the tropics and adapted the system for use outside the tropics. Racial slavery expanded outside the tropics not because of a disease advantage but because the system already existed and proved more beneficial to the emerging colonial elites in places like Virginia and Louisiana (Lovejoy 2000).

Africans brought a great deal of what we may think of as environmental knowledge with them to the New World. In the sixteenth century, people from the West African Sudan helped establish the extensive cattle-raising system in parts of Mexico and Argentina, bringing knowledge of range management with them. They also introduced rice cultivation to the New World in places like the Carolinas. In the Carribean, they brought knowledge of banana cultivation. And peanuts apparently made a roundabout voyage from Brazil to West Africa and then back to North America (Carney 2001).

Africans made massive contributions to the modern world. Patrick Manning (1990), in his study of the slave trade's demographic impact on Africa, has described the historic sacrifice that Africans, at enormous costs to them as a people, made for the New World. African labor made the New World profitable for Europeans and gave emerging American societies not only people but much of what is distinctive about their cultures.

THE ENVIRONMENTAL IMPACT OF THE SLAVE TRADE ON AFRICA

Any discussion of the Atlantic slave trade's impact on African environments must begin with a basic understanding of the scope of the movement of millions of people out of Africa into the New World over a 450-year period. First, more Africans were brought to the New World than was necessary to repopulate the areas of the New World left underpopulated by the effects of conquest and disease after 1600, given a normal pattern of reproduction. Until the nineteenth century, most slave populations had negative natural growth rates. Africans, as we have noted, had a survival advantage in the tropics over Europeans, and Creoles (those born in the New World regardless of parents' place of origin) had a large survival advantage over immigrants. But the Africans brought to the New World often lived in conditions that made it very difficult for them to reproduce and maintain positive natural growth rates. Because the slave owners much preferred males for fieldwork, they imported two males for every female. This un-balanced sex ratio in itself inhibited natural reproduction rates for peoples of African origin in the New World. Second, while Africans had better survival rates against tropical diseases than European immigrants, the constant move-ment of people in slave ships spread infectious diseases much more rapidly. Ev-ery arrival of a new slave ship in an American port with its cargo of hungry, ill-treated, and weakened Africans heralded an increased potential for outbreak of cholera or typhoid or some other fatal disease. In addition, in areas where pro-duction of sugar in particular was most intense, owners quite literally calcu-lated that it was cheaper to buy a slave, work that slave to death, and then buy another slave. This equation did not hold everywhere or at all times, but it was operative long enough that imports of Africans, especially in the Caribbean and Brazil, remained high into the nineteenth century. Certain factors could inter-rupt this pattern of high mortality and high imports. For example, once the pro-ductivity of land declined, imports of new slaves declined. Often, the African-origin population would fall, eventually stabilize, and begin to grow again. In healthier regions, such as parts of North America, Africans survived better and began to experience positive natural growth rates in the eighteenth century (Manning 1990).

The result of all these factors meant that European slaveholders maintained a constant demand for new slaves right up to the middle of the nineteenth cen-tury. The total number of Africans forcibly enslaved and brought to the New World will never be known with perfect accuracy. Shipping records, import records, censuses, and statistical estimates have given historians general figures. Such figures take into account both lost and nonexistent records as well as smuggling. However, historians agree that these figures can only be regarded as the minimum number of Africans who landed in the New World (Lovejoy 2000).

TABLE 4.1 Atlantic Slave Exports

Period	Number
1460–1600	367,000
1601–1700	1,868,000
1701–1801	6,133,000
1801–1900	3,330,000
TOTAL	11,698,000

Source: Lovejoy, *Transformations in Slavery*.

Unfortunately, reporting such figures serves to routinize the horrors of the slave trade and at the same time does nothing to help us understand the effects of the trade on Africans, African societies, or African landscapes. As Table 4.1 suggests, during the earliest years of direct contact between the Europeans and Africa the volume of the trade remained relatively low. It rose dramatically after 1600 as sugar produc-tion began to dominate the Caribbean, and it stayed high throughout the eigh-teenth and into the nineteenth centuries. Several events caused the demand for slaves to begin to decline after about 1800. The Haitian Revolution of 1791–1803 and the abolition of slavery by many of the new independent republics emerging out of the Spanish Empire, along with the British and American abolition of the trade in 1807 and 1808, respectively, all played their part. In the 1830s, the British abolished slavery throughout their empire. Brazil and Cuba, however, continued to import large numbers of slaves until the 1830s, and smuggling continued to occur into the United States and a few other colonies until the final abolition of slavery in 1865.

Most Africans taken to the New World were purchased by European merchants from African traders. Some of the earliest Portuguese voyages to the Atlantic coast of West Africa ended in slave raids, but they and other Europeans quickly came to realize that they could not expect to maintain any kind of trade relations if they engaged in random raiding. The Europeans did not hold the advantages over Africans that they held over the native peoples of the Americas. Africans had settled agricultural societies that produced enough food to support relatively dense populations where environmentally possible. And once they came into permanent contact with Europeans, they did not succumb to new disease; indeed, European mortality in Africa made military operations there difficult. In the fifteenth and sixteenth centuries, European firearms did not provide any great advantage over spears or bow and arrows. If Europeans wanted the gold, pepper, cloth, and eventually the people of Africa, they had to trade with Africans. Although the Portuguese founded a string of trading posts in West Africa, created a small colony in Angola, had a number of presidios in what became Mozambique, and managed a precarious hold on some of the Swahili cities in East Africa, in none did they establish thriving colonies. Even their largest concentrations in Angola and Mozambique remained small and isolated. The outposts survived in part because a Luso-African population emerged that formed the bulk of the people accepting allegiance to the Portuguese crown. It was not until the nineteenth century that they were able to conquer sizable African states. As other European powers followed the Portuguese to Africa, they made even less effort to create outposts. The Dutch settlement at South Africa was the exception to this pattern, but as we shall see, conditions there adhered more to the American model than to that of the rest of Africa.

The question remains then as to why Africa could supply so many captives, especially since the majority of captives were not taken in raids by Europeans. Broadly speaking, there are two schools of thought as to why Africa could supply so many captives. One school holds that the European demand for labor in the New World meant that European merchants paid extraordinarily high prices for captives. If African states or merchants would not meet the supply, European merchants quite happily traded with either enemies or factions within society. The infusion of wealth and weapons generated conflict and gave ambitious men the opportunity to build up followings and political structures based on the benefits of the trade. The generally small scale of African states along the West African coast especially, but even in the Congo/Angola region, meant that over the centuries Europeans had plenty of opportunity to engage in these practices. As the West Indian historian Walter Rodney (1982) has argued, the slave trade helped set the terms for the underdevelopment of Africa. By the mid-1600s, African societies in both West Africa and the Congo/Angola region exported little else but slaves. While local production of textiles and metals did not totally disappear, Africa became locked in the New World capitalist system as a provider of labor only. This process of underdevelopment would continue after the demand for raw labor in the New World came to an end and the colonial conquest of Africa began in the nineteenth century (Lovejoy 2000).

A second view on why Africa could supply so many captives comes from the historian John Thornton (1998). He argues that African societies could supply captives at the beginning of European demand for slaves and could increase the supply as demand increased because institutionally human beings represented the most common form of productive and movable wealth in Africa. The relatively low population density throughout most of Africa, he suggests, perversely became the reason Africans could export so many people. According to Thornton, population densities remained low in Africa because of a variety of environmental constraints. In the savannahs, periodic drought meant that societies had to rely on cattle and mobility in addition to agriculture for survival. In the more humid forests, diseases kept population densities down. In addition, many African soils were very fragile. Some are laterite and form a hard crust if exposed too long to water. Others, including many in the rainforest regions, lose fertility rapidly. As a result, many African agricultural systems remained extensive, using long fallow periods of up to twenty years. Land remained relatively available, and implements and investments in land improvement were simple and easily replaceable. Consequently, in most African societies, the wealthy elites invested in people. Dependents came in many forms. Wealthy men in many societies could marry more wives. Younger men and women, often relatives, could become dependents. Captives taken in war could be integrated into households, and larger numbers could be settled in separate settlements. Given the value of people as producers, in many societies some captives were always available for trade. This type of slavery, sometimes called lineage slavery, existed in the large states of the West African Sudan and in the decentralized polities of the rainforest. In those areas of Africa influenced by Islam, such as the Sudan and the East African coast, slavery could be more formalized. In most cases, the end result, even in areas where Islamic law operated, was the integration of marginal newcomers into the community, though often in a permanently junior position (Thornton 1998; Lovejoy 2000; Miller 1989).

In Thornton's view, the demand for captives from the coast that began with the Portuguese drew on established mechanisms for supplying the internal and external demand for people and turned it into a pump that drew people out of the continent. While Thornton's view has some merit, it is not totally satisfactory. For example, as noted earlier, the Gold Coast imported people into the sixteenth century in exchange for gold. The ancient state of Benin, located just to the west of the Niger River in what is now southern Nigeria, eventually withdrew from the slave trade entirely. In both cases, not only was labor more valuable than the goods potentially obtained by trade, but land in the rainforest for agriculture required a massive investment in labor. In short, Thornton's generalized condition for the ability to supply captives does not hold. Yet, after 1600 the Gold Coast became an important source of captives for the slave trade. What had changed was the European demand for slaves. The price paid by Europeans overwhelmed the value of captives locally and set off cycles of violence that perpetuated the supply of captives (Lovejoy 2000).

Different regions of Africa were victimized by the slave trade at different times, and areas often became "slaved out." Within regions, a process that Joseph Miller (1989) has described as the "slaving frontier" often operated. For example, on the Gold Coast, the appearance of Portuguese traders brought Mande-speaking traders to the coast. These African traders, known in the region as Dyula, had begun to trade gold from the Akan gold fields long before, in the fourteenth century. They carried the gold to the Empire of Mali and from there it was traded across the Sahara. In the late fifteenth and early sixteenth centuries, the Portuguese were hard-pressed to find a commodity that they could trade with the Akan peoples of the forest in exchange for gold. The Portuguese had no great supply of textiles or metal goods to trade. Their firearms were, at the time, of only little advantage, although firearms would eventually became a major trade item. The Akan peoples had access to textiles produced locally and to imports of cotton goods from the West African Sudan.

The Portuguese could acquire captives in other parts of Africa, particularly the Congo/Angola region, from about 1500. These found a ready market in the Gold Coast. Ivor Wilks (1977) maintains that population in the forest regions of the Gold Coast increased dramatically during the sixteenth century as imported labor helped mine gold and clear forest fields for agriculture. Up to the end of that century, gold remained the most valuable export from West Africa.

After 1600, however, the price Europeans paid for captives rose dramatically. The Dutch, English, French, and even minor states like Denmark now entered the slave trade; they competed with each other for access to African ports as well as for colonies in the New World. On the Gold Coast, two small states located just beyond the coast began to supply growing numbers of captives. A cycle of

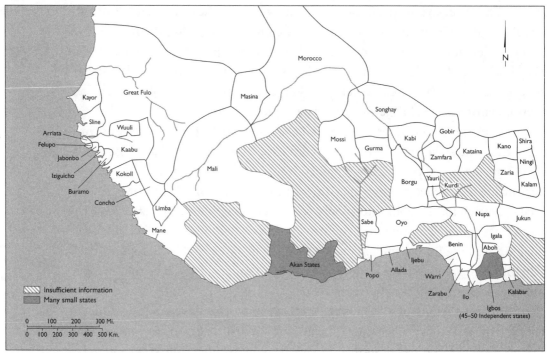

West African states, 1625.

warfare and raiding began, victimizing people living farther north. Partially in self-defense, in the 1680s in the region around the town of Kumasi, a new alliance merged under the leadership of the ruler of Kumasi. Taking the title Asantehene, Osei Tutu created a federation that used alliance and military force to eventually conquer almost all of what is now the modern state of Ghana. The kingdom of Asante traded captives from its wars of conquest for trade goods and guns. It imposed tribute on communities to its north, continuing to integrate many people as laborers, while exporting enough to maintain its position against potential internal and external enemies. Exports of captives peaked in the first decades of the 1700s and then declined to a more modest level. In short, a slaving frontier had moved through the region, creating havoc until a new political order emerged that could at least control the effects of the trade (Kea 1982).

The same process played out in many regions of Africa. In the Sudanic regions of West Africa, early Portuguese trade in the later fifteenth and early sixteenth centuries created disruption and temporary depopulation in the areas around the Senegal and Gambia rivers. In this case, by the early 1600s an Islamic reform movement took hold that campaigned against selling captives to infidel Christians. Later in the 1700s, the political violence set off by the defeat of Songhai by a Moroccan invasion force created a climate of instability that saw a new trade in captives, mainly Mande speakers, coming to the coast. French and

Gold Asante jewelery, in the style worn by senior court officials in the eighteenth century. Under the leadership of Osei Tutu, the Asante people dominated the gold trade of western Africa, in an area comprising much of modern-day Ghana, eastern Ivory Coast, and western Togo. (Werner Forman/Corbis)

British traders became the most important merchants for this trade in the eighteenth century (Lovejoy 2000).

Further south, in the area around the mouth of the Congo River, the Portuguese helped establish the system that would provide the single largest source of captives for the entire era of the Atlantic slave trade. When their first voyages reached the mouth of the Congo, they made contact with the kingdom of the Kongo. Over the course of the sixteenth century, the Portuguese succeeded in converting part of the royal family to Catholicism. A cycle of conflict began with Portuguese merchants from Portugal and from Saõ Tomé trading with different factions. Eventually, the kingdom of the Kongo weakened and neighboring states

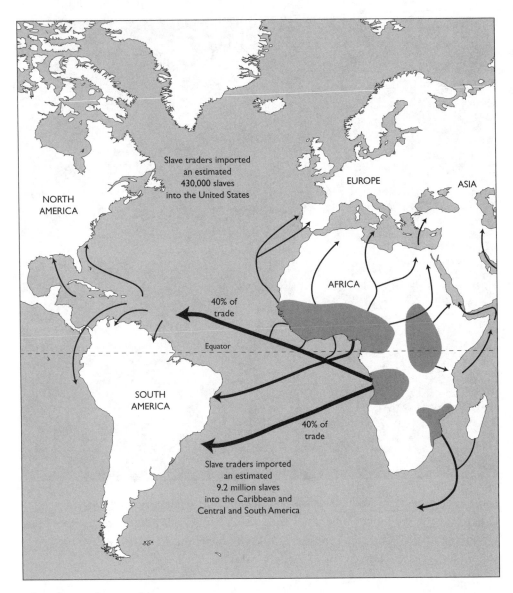

Slave traders imported
an estimated
430,000 slaves
into the United States

NORTH
AMERICA

EUROPE

ASIA

AFRICA

40% of
trade

Equator

SOUTH
AMERICA

40% of
trade

Slave traders imported
an estimated
9.2 million slaves
into the Caribbean and
Central and South America

The African slave trade.

gained strength. The Portuguese established a permanent base at Luanda, to the south. Although the Portuguese never succeeded in extending their direct control far from the port cities of Luanda and Benguela, Luso-African traders, called *pombeiros*, carried the trade to the interior. In the seventeenth century, raiders called Imbangala laid waste to large areas of modern Angola. In the eighteenth century, the Lunda Empire farther west became the major supplier of slaves. Again, the frontier effect seems clear. (Thornton 1998)

Eastern and southern Africa became integrated into the broader Atlantic world less intensely. East Africa's connection with the Indian Ocean trading system made it a prize in the early Portuguese expansion. As in West Africa, the first Portuguese efforts in East Africa involved conquest. They sacked the port city of Kilwa on their first voyage to the Indian Ocean and tried to occupy the leading Swahili cities in order to control trade. Their efforts proved futile in the long run, but they did succeed in establishing posts in Mombasa, Zanzibar, and especially in what would become Mozambique. They tried to control the gold trade from the Zimbabwe region, especially trading with the state of Mwene Mutapa, and they established presidios along the Zambezi Valley in an attempt to control this trade. They began to export captives from the region in the eighteenth century, when East Africa became one of the few areas of which they had a strong position. The French joined them and established plantations on Indian Ocean islands to produce sugar (Manning 1990).

As observed earlier, general evaluations of the effect of the Atlantic slave trade on Africa and African environments are difficult and in some ways misleading because the effects of the slave trade were localized in regions, and the experiences of regions differed greatly. In addition, the slave trade occurred simultaneously with the spread of American food crops into Africa, thereby making it difficult to draw out the sequence of cause and effect. In general, as Patrick Manning (1990) has argued, the population drain caused by the Atlantic slave trade probably caused short-term demographic decline in western Africa and inhibited long-term population growth. This statement seems especially true in the context of the violence that accompanied the slave trade. The production of slaves did not resemble the production of any other commodity. Most victims of the slave trade were taken by force in wars or raids. Even those judicially condemned were subject to violence. Holding slaves between capture and sale to overseas merchants, and especially the trip to ports on the coast, also caused many deaths from malnutrition and greater exposure to diseases in new environments. Scholars have agreed on a figure of around 13 million people carried out of Africa between 1450 and the middle of the nineteenth century. The number of deaths caused by the process of enslavement could easily double that figure (Lovejoy 2000).

In the absence of any real data to reconstruct demographic trends, scholars have also debated the potential effects of this destruction on African populations. Most historians believe that African population overall did not grow as fast as it could have or as fast as populations in the rest of the world grew in the aftermath of the Columbian exchange. While no area of Africa was absolutely depopulated, growth in West Africa was probably lower than it would have been given the introduction of New World crops, and in the Congo/Angola region population may have actually declined until the first decades of the nineteenth century. East Africa saw the effects of an intensive slave trade later in the nineteenth century, and the broad belt of Sudanic Africa from the Atlantic to

Ethiopia may have experienced, overall, moderately rising populations. Southern Africa presents a different case: the arid regions of southwestern Africa saw a New World–like demographic collapse, whereas the wetter regions of southeastern Africa probably witnessed a general growth in population.

The slave trade was not the only influence on the rate of population growth. First, as noted above, the spread of New World crops gave Africans new crops to mix in with their existing ones in specific environments. The introduction of maize and cassava in particular allowed people to intensify agricultural production in the wetter parts of Africa. In the long run, the introduction of these crops made possible a rapid expansion of population in the rainforests of West and Central Africa so that some of these regions became and remain among the most densely populated in Africa.

Second, the demographic structure of exports affected the ability of African populations to recover from a period of intense slaving. Because, as we have stated, Europeans generally demanded more men than women for their plantations, about two-thirds of exports were male. Evidence indicates that men and women fell victim to capture at about the same rate. Many of these women who were captured, rather than being sent to the New World, remained in Africa usually as slaves, but were often integrated into local communities as wives or concubines. In situations where women outnumber men, even though the total fertility rate per female declines, still total births would have been higher than in a population where losses were more evenly balanced between the genders (Manning 1990).

Operating against these factors would have been the rapid spread of disease. The movement of people toward the coast from the interior in fairly sizable numbers would have brought people into new disease environments, causing heightened mortality. The packing together of people in trading towns and ports would have assuredly increased the speed at which infectious diseases circulated. And as we have seen, the ships themselves brought diseases when they called at various ports.

SOUTHERN AFRICA IN THE AGE OF CONTACT

The very southern tip of Africa presents an exception to the general story of the Europeans' failure to occupy African spaces. There, a slight difference in the microenvironments meant that a small European settlement could gain a toehold and then overwhelm the African population in the immediate surrounding vicinity starting in the mid-seventeenth century. About two and a half centuries later, the descendants of the original Dutch settlers would begin the actual conquest of a

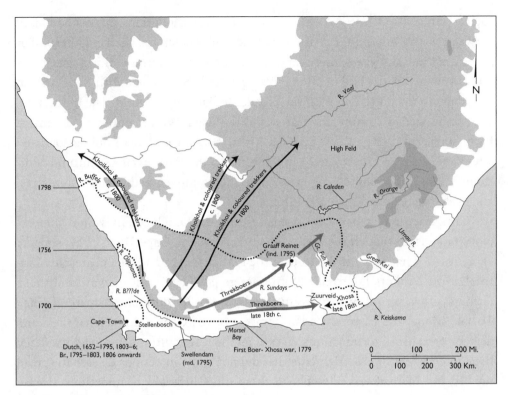

Southern Africa, 1625–1800.

much larger territory with a much larger population of Africans, creating what is today the Republic of South Africa. The original conquest during the first European imperial phase of history provided the basis for a later of expansion of white domination during the second European imperial phase in the nineteenth century.

During the sixteenth and into the seventeenth centuries, the Portuguese had generally tried to sail around the southern tip of Africa as fast as possible. The southwestern coastal regions had little water, and their trading focus lay on the route to the gold of Zimbabwe and the Swahili trading towns. They occasionally stopped at Table Bay near the Cape of Good Hope to acquire water and animals from Khoi herders there. Dutch and British ships began to use the area as a resting place in the late sixteenth century. As the Dutch began to use the Dutch East India Company to challenge Portuguese control of trade with Asia, they decided in 1652 to create a permanent base at Table Bay.

Cape Town lay at a unique ecological position. It dips down into a latitude that creates a Mediterranean-type environment with dry summers and cool wet winters. To the immediate north and east stretches a long expanse of arid and semiarid territory, the "dry veld" in South African terms. Further to the east and northeast, the climate changes to a more typically African monsoon climate with wet summers and dry winters. Running roughly parallel to the shore along the eastern coast of South Africa, the Drakensberg Mountains mark a boundary between wetter and hotter coastal lowlands and cooler and drier highlands. For at least a thousand years, these lands had been home to ironworking agriculturalists who kept cattle and maintained at least indirect contact with their African neighbors to the north and with the broader Indian Ocean world. The frontier between the wetter summer rain area and the drier, winter rain area became a fairly stable frontier at about the Fish River. To the east lived Khoi pastoralists, who kept cattle and sheep, and San foragers. To the West lived agricultural communities speaking Bantu languages. Individuals moved across the boundaries, and exchange occurred regularly (Elphick 1977).

This situation created two anomalies that the Dutch colony would later exploit. First, even though the area immediately around Cape Town could support agriculture, it had none at the time of the Dutch settlement. The nearest agriculturalist settlements lay over 100 kilometers (about 60 miles) away, and those farmers' crops would not thrive in the dramatically different climate of Cape Town. Second, the pastoralists and foragers of the Cape did not have the same degree of exposure to infectious disease as their agriculturalist neighbors. They lived in very scattered settlements that did not allow infectious disease to spread rapidly, but instead served to choke them off. Outbreaks of infectious disease emanating from their agriculturalist neighbors or from ships stopped for a few days or weeks to take on fresh water and buy meat might have caused deaths in one or two small communities, but then the disease would disappear for lack of fresh victims. Hence, the Khoi and San communities, like the Native American ones, were relatively free of disease before the appearance of a relatively large number of disease-ridden settlers from Europe, Asia, and Africa, constantly reinforced by fresh microbes brought by vessels plying between all three continents.

The Dutch intended to plant a settlement that would grow vegetables, wheat, and vines and to purchase cattle and sheep from the Khoi. The environment surrounding Cape Town proved as congenial for these Mediterranean transplants as the Dutch expected; however, the Khoi proved much less amenable to Dutch plans.

The Dutch imported "indentured servants" from Europe; they were mostly Dutch but also included French and German Calvinists who were escaping persecution. The Dutch East India Company expected to work these servants on plantations, and they also planned to purchase cattle and sheep from local pastoralists to supply meat for the colony and ships calling. However, local herders, practicing subsistence pastoralism rather than commercial stock keeping, generally sought to sell many fewer stock than the Dutch wanted to buy, and usually sought to sell poorer quality stock. Thus, when the Dutch felt supplies were critically low, they often resorted to raids. Gradually, as some European servants fin-

Boers halting during a trek across Africa, ca. 1877. (Hulton Archive/Getty Images)

ished their contracts, they took up cattle and sheep rearing in the arid regions around the Cape, action that led to further conflict with the Khoi. In 1713, a smallpox epidemic swept through the Khoi and San peoples of the Cape, decimating their populations (Elphick 1977).

A multicultural society emerged in response. White settlers, well on their way to becoming the Boers, spread across the arid regions around the Cape. In front of them went small mixed communities of Khoi, San, and European outcasts. Like the Boers, these groups, often called Baastards or Griqua, generally spoke the emerging local dialect of Dutch called Afrikaans. The Boers themselves kept "servants" captured from local communities. At the Cape, the Company began to import slaves to work on the wheat farms and vineyards from its trading posts in the East Indies as well as from Madagascar and East Africa.

In many respects, the landscape changed little with the replacement of indigenous Khoi and San with Boers and Griqua. The new settlers continued to practice extensive pastoralism and lived in mobile small bands. However, some important changes emerged by the mid-1700s. The Boers claimed large expanses of land as "loan-farms" from the Company government. While they may not have used the land very differently from the Khoi, the claim of exclusive owner-ship marked a dramatic new way of thinking about the land. The Boers' gradual ability to enforce that claim presaged the imposition of colonial conquest throughout South Africa (Beinart and Coates 1995).

That conquest did not occur until the nineteenth century. By the mid-eigh-teenth century, the Boers generally dominated the area of winter rains in South Africa. Their domains stretched to the Fish River in the east and to the arid re-gions around Kimberley to the north. However, in the better watered lands to the east and northeast, African communities practicing African agriculture lived in more dense settlements. The Xhosa peoples of the east and the Sotho and Tswana peoples of the northeast also appear to have been in enough, if indirect, contact with the peoples to their north to have had some exposure to diseases like small-pox. Despite the Boers' efforts throughout the eighteenth century and into the nineteenth, they were unable to dislodge these peoples. A series of conflicts halted their expansion. It would require the British takeover of the colony in the early nineteenth century and the Industrial Revolution to give the Boers the abil-ity to expand their settlement through conquest (Beinart and Coates 1995).

CONCLUSION

The Dutch colony's small success in South Africa both highlights what did not happen in the rest of Africa before the nineteenth century and points out what would happen to the rest of Africa by the end of that century. South Africa, un-like the rest of tropical Africa, was, in Alfred Crosby's term, almost a neo-Eu-rope. It had a subtropical climate, and in the Cape region its population was iso-lated and scattered. The settlers brought with them "guns, germs, and steel" and had the time to acclimate themselves to their new land. However, they could not expand their area of control into the regions where African communi-ties with their own version of the "Old World Iron Age" had thrived. These communities did not die out and, like their counterparts in the rest of Africa, eventually with the addition of New World crops to their repertoire, they ex-panded demographically.

In the rest of Africa, the particular combination of disease environments with the depopulation of the New World created the conditions for the Atlantic slave trade. With the end of the plantation boom in the New World during the nineteenth century and the stabilization of the African- and European-origin population of the New World, the slave trade gradually ended. In its place, a new division of labor developed that replaced the export of people with the export of commodities for markets connected through the European world system. In some ways, the development of commodity exports from Africa during the nineteenth century did more to reorganize African landscapes than the slave trade had. Production patterns changed dramatically to meet new demands associated with the rise of industrial production. This production also undercut the production of textiles and metals in Africa by Africans. The Industrial Revolution also gave Europeans the military means to conquer more dense populations, and medical advances improved survival rates for Europeans in the tropics by the end of the century. Competition for control of African markets and African resources led to the "Scramble for Africa" and the dramatic reshaping of African landscapes in the late nineteenth century.

5

AFRICAN ENVIRONMENTS AND THE REORGANIZATION OF SPACE UNDER COLONIAL RULE

Despite the decline of the Atlantic slave trade after about 1830, throughout the nineteenth century the value of exports from Africa rose. With increasing industrialization in Europe and other areas, demand for commodities of all sorts continued to increase. African societies became increasingly enmeshed in commercial networks that supplied the widening variety of manufactured goods available. In return, they had to reshape their productive activities to supply groundnuts, palm oil, coconut products, cloves, rubber, ivory, and minerals. Until the late nineteenth century, outside of South Africa, this transformation into commodity-producing regions in the international capitalist division of labor was guided mostly by indigenous institutions. The most important outside influences in this transformation (again with South Africa as the major exception) came, as they had for many centuries, from Asia, specifically the Middle East and the Indian subcontinent. In northeastern, eastern, and Central Africa, external trade flowed through these regions and people from or connected to these regions continued to influence the development of new productive systems.

The development of commodity production in Africa during the nineteenth century took many different forms, all of which involved more intensive use of resources. In parts of West Africa, export trades in agricultural commodities developed in areas where rivers like the Senegal and Niger provided inexpensive transport. In Central and eastern Africa, ivory became a major export. From the east coast of Africa, plantation products such as sugar and cloves became major trade goods. In most cases, as was typical of peripheral regions in the world economy, an elite class monopolized most of the benefits from this expanded production and used violence to coerce labor from the rest of the community. Slavery and other forms of unfree labor expanded throughout the nineteenth century. The existence of slavery in Africa became one justification used by European imperialists in the scramble for Africa. Ironically, these same European powers found it expedient to continue to tolerate the existence of slavery well into the twentieth century and to use forms of forced labor that resembled slavery in all but name in the development of their new colonies.

South Africa followed an accelerated path, but one that ran along the same lines. Before the 1860s, major struggles arose over the control of people and land between African states such as the Zulu kingdom created by Shaka in the early nineteenth century, Boer "republics" created by Afrikaner settlers seeking land and freedom from the British, and the British colonies in the Cape and Natal. After the discovery of diamonds in the 1860s and gold in the 1880s, the stakes were raised. The British finally won mastery over both African states and Boers in 1900 and then proceeded to create a system of imperial and racial domination that became in many ways a model for colonizing efforts throughout Africa. The Union of South Africa allowed whites to claim control over all the productive resources of the country—minerals and agricultural land—and to turn the African population into a perpetually disempowered and impoverished working class.

Colonial rule throughout the rest of Africa was perhaps not as thoroughgoing as in South Africa but nevertheless helped completely reshape the relationships of African communities with their environments. While not as severe as the depopulation crisis that took place in the New World at the time of conquest, the extreme ecological crisis throughout much of Africa during the last decades of the nineteenth century posed many economic and social difficulties. Warfare, most notably "pacification" by the French, Germans, British, Portuguese, and Belgians, devastated large regions of the continent and disrupted agricultural production. New diseases, especially the cattle disease rinderpest, which was long present in Europe but never in Africa, spread into the continent and killed upwards of 80 percent of the cattle it infected. Old diseases circulated also much more rapidly, causing millions of deaths. African communities faced new and sometimes extreme demands on their labor and resources. By the end of World War I, the total population of the continent seems to have declined; certainly the new colonial rulers worried about growing evidence of "depopulation" (McCann 1999b; Davis 2001).

In the midst of this destruction, the new European overlords set out to construct a system of "rationally" exploiting Africa's resources and labor. In some regions, colonial regimes expected Africans to become a dispossessed labor force, working for extremely low wages on white-owned farms and plantations, in mines, or for concessionary companies. This pattern of development spread from South Africa to what is now Zimbabwe, Zambia, Malawi, parts of the Congo, Kenya, and parts of Tanzania. In other areas, colonial regimes relied on "peasant" production of agricultural commodities. In all cases, colonial regimes used often heavy taxation as a means of forcing production from African populations.

Colonial regimes also claimed control over the natural resources of their territories. From the very beginning of colonial rule, officials feared the depletion of forest resources and the destruction of "wild" Africa by expanding agricultural production. Conservation measures began early in the twentieth century, and most of these measures involved depriving Africans of access to resources such as new land, forest products, grazing, and game meat on which they had long relied. For many rural Africans, conservation became another term for exploitation.

THE COMMERCIAL REVOLUTION OF THE NINETEENTH CENTURY

The nineteenth century opened with the Atlantic slave trade still the dominant feature of Africa's interaction with the outside world; by the end of the century, the export of people as captives had ended, and most of Africa had become entwined in the world economy as commodity-producing areas. This transformation occurred in many regions of Africa before colonial conquest in the late nineteenth century. It included not only the ending of the Atlantic slave trade, but probably an increase in slavery in the western regions of Africa, as well as the rise and fall of an eastern African slave trade destined both for external markets in the Indian Ocean region and for new plantations within Africa. This process culminated with the epic disasters of the late 1800s and the conquest and, in many cases, the dispossession of African peoples.

The rise of commodity production in Africa represented the transformation of Africa's place within the emerging international division of labor created by the capitalist world system. It occurred not because the Atlantic slave trade ended, but rather because it represented the continuation of the globalization process the slave trade had started. The areas that first responded to the increased demand for commodities from Europeans in the nineteenth century generally were those that had been involved in the slave trade before. In many cases, the same merchants who sold slaves responded to changing demand by increasing the commodities they produced.

In the Senegal and Gambia valleys, the transition to commodity production occurred in conjunction with the slave trade during the eighteenth century. Slave traders brought captives from the interior to the river valleys. The rise of the Bambara state of Karatu had led to an increase in conflict in the old Mande-speaking heartland around the headwaters of the Niger and Senegal rivers. Merchants purchased captives in the area for transport to the coast. They also began to rent land in the valleys from the local population and put their captives to work growing rice and millet. Some went to feed the captives, but much was sold to the towns along the coast of what is now Senegal and Gambia. There it fed the townspeople as well as producing provisions for the ships that called to take captives to the New World (Brooks 1975).

After 1807, when the British formally made it illegal for their citizens to participate in the slave trade and sought to repress the trade during the first part of the century, the number of slaves exported only gradually declined. Groundnut production began in the Gambia Valley in the 1830s and spread gradually south into Sierra Leone and north to the Senegal Valley. British merchants took the lead in purchasing the groundnuts for oil, but American merchants also began to buy the crop for consumption in the United States. P. T. Barnum's circus peanuts in the antebellum period often came from Gambia until a tariff in 1842 made it unprofitable to import them. Exports through British-controlled ports in West Africa reached 11,000 tons of unshelled nuts by 1851. By the 1840s, French merchants took the lead in beginning to buy groundnuts for use in the production of edible oil and soap. The French willingness to intervene in political conflicts in the area and their deep connections with the Sonnike-speaking traders connected to the Senegal Valley meant that French merchants became the leading buyers of groundnuts in West Africa. Exports from Senegal reached new highs by the 1840s (Brooks 1975).

Production took place in several different ways, all of which involved new, and often commercialized, systems of agricultural production. Many merchants, political leaders, and military commanders in the numerous small states of the region settled captives on land and required that they grow groundnuts in addition to their own subsistence. This form of "village" slavery has sometimes been compared to plantation slavery. In addition to captives, labor migrants began to appear. Particularly associated with the Sonnike people, these mostly younger men would come to the river valleys and either rent land or work as paid labor on groundnut plantations. They would then purchase imported manufactured goods for sale further into the interior. Finally, Koranic schools required their students to pay for their education with labor in agricultural fields. These schools became some of the leading producers of groundnuts in the region, and the clerics who headed the schools have remained important (and wealthy) public figures (Brooks 1975).

In the wetter forest regions, palm oil became a major export even before the slave trade began to decline. The increase in production, destined primarily for Britain, began from southeastern Nigeria by the end of the eighteenth century, and the expansion of palm oil production in other West African regions began late in the century. David Northrup (1978) has argued that the trade began there because it had the highest concentration of oil palms. The trees grew after the clearing of the forest for agriculture and were assiduously protected by yam farmers. Rather than either competing with or driving out the slave trade, the

Making palm oil at Whyda, Guinea, Gold Coast. Africa Magazine *illustration published in* Ballou's Pictorial Drawing Room Companion. *(Corbis)*

two branches of commerce increased in tandem. Northrup suggests that farmers sought to acquire trade goods brought by merchants and chose to produce provisions for the slave trade and palm oil for export. The trade expanded dramatically after the slave trade began to decline because demand increased from the British, and local producers increased supply. The export of palm oil from the Bight of Biafra to Britain increased from about 40 tons a year in the 1780s to 150 tons per year in the 1790s to 3,000 in 1819, 8,000 in 1829, 13,600 in 1839, 20,000–25,000 in the 1850s, and 40,000–42,000 in the 1860s. By the middle of the nineteenth century, the expansion of palm oil production had led to increased clearing of the forest for palm oil planting (Northrup 1978).

Other regions in West Africa began to develop export trades in commodities early in the nineteenth century. In Liberia, the establishment of a colony based on free American blacks led to increased commerce from the region after 1821. Coasting ships based in Liberia traded along the coast for camwood, which is used in red dye, and palm oil. The trade expanded through the middle decades of the nineteenth century as Liberian merchants both dominated the trade along the Windward Coast and engaged in trade with American ports (Syfert 1977).

In Dahomey to the east of the Gold Coast, palm oil exports rose dramatically in the 1840s and in the Gold Coast accelerated in the 1830s. The kingdom of Asante also increased gold exports and exports of kola nuts to northern trade routes during the first half of the nineteenth century. This expansion of commerce in the aftermath of the decline of the slave trade had the effect of transforming productive practices and landscapes perhaps more than the slave trade itself had.

In the area claimed by Portugal in the colony of Angola, the gradual decline of the slave trade took place only in the second half of the nineteenth century. Demand for labor in Brazil remained strong, and Portuguese officials winked at an illicit trade, even transforming it legally into the "migration" of "contracted" servants. However, even there, after 1850, the development of commodity production took place, often taking the form of small- to medium-sized plantations worked by slaves and owned by Portuguese and Brazilian settlers or Luso-Africans. Coffee became a main export from the region of Cazengo, a small but hilly and productive region just to the east of Luanda. By the end of the nineteenth century, the entire district was "owned" by immigrant farmers and worked by servants or slaves (Birmingham 1978).

In eastern Africa, too, the slave trade flourished during the first two-thirds of the nineteenth century but did not prevent the development of either natural resource exploitation or commodity production. Most of the captives taken in eastern Africa were not part of the Atlantic trade, and many remained in Africa, but the effect of the trade in the broad region was similar to that on western Africa in the preceding centuries. A slaving frontier moved across the region from the Indian Ocean to the Great Lakes. In the south, Portuguese and South African demand for labor drove it; in the central regions, an Arab-dominated demand created a new economic order; and in the north, the revitalization of Ethiopia during the second half of the century helped create conditions for the production of captives for the Indian Ocean trade (Lovejoy 2000; Sheriff 1987).

In East Africa proper, a more intensive slave trade developed in the nineteenth century that first met expanded markets in the Indian Ocean and then an expanded demand for agricultural labor along the East African coast. Coupled with this rise in the export of captives was an increase in the production of ivory. As in all of Africa, long-distance trade before the colonial era remained constrained by high transport costs. In eastern Africa, no large river traffic developed because none of the region's rivers contained long, navigable stretches comparable to the Niger or Congo rivers. Hence, only goods that carried an extremely high value—such as ivory or the gold that had come from Zimbabwe—or goods that walked themselves to market could bear the transport costs. Regional and local networks existed throughout the area, and some spanned great distances. Even foodstuffs could be transported cheaply around some of the Great Lakes. The reliance on human transport limited extensive trade to either short distances or areas quite close to the lake shores.

The Swahili cities had long exported ivory, especially to India. During the nineteenth century, demand escalated as commerce generally increased. Throughout East Africa, specialists had long dominated hunting, especially elephant hunting. These specialists used both poison and traps to bring elephants down. Elephants, being difficult to hunt, probably made up only a small proportion of the total hunting take in the region. Before the nineteenth century, at least some of the ivory had been "found" when an elephant died. Merchants began to bring larger, more powerful firearms into East Africa by the beginning of the nineteenth century that transformed hunting for ivory. These guns had a better chance of bringing down elephants, and caravans could use massed firepower on elephants because they had less concern for the meat from the carcass than for the tusks (Sheriff 1987).

Over the course of the 1800s, an ivory frontier moved across East Africa that probably greatly reduced elephant herds in the aftermath. In the early part of the century, ivory still came from areas near the coast; by the mid-century, the central caravans ranged far to the interior. Caravan traders both hunted elephants and supplied hunting people with new firearms. One route stretched from Mombasa and Tanga toward Mount Kilimanjaro. It tapped ivory from the plains west of the great mountains of the Eastern Arc. In Kenya, the Kamba people became known as great suppliers of ivory until supplies became scarcer in what is now central Kenya. Further south, a route stretched from Bagamoyo through central Tanzania on to Ujiji on Lake Tanganyika. On this route, the Nyamwezi became known as the great hunters and porters. By the 1850s, the caravans had begun to cross the Great Lakes, moving over Lake Tanganyika into what is now eastern and southern Congo. In the south, Kilwa remained an important port, and routes stretched through the gap between Lake Malawi and Lake Rukwa in southern Tanzania into Zambia and Malawi. Exports seem to have held steady from the East African coast at about 400,000 pounds per year throughout most of the nineteenth century, a figure that meant roughly 4,000 elephants had to die to supply it (Steinhart 2005; Sheriff 1987).

The caravans took both ivory and captives. In the first instance, the caravans used the captives as porters to supplement the free workers hired on the coast and in the interior. Once a returning caravan reached the coast, the increasing demand for labor in the Indian Ocean fueled the expansion of commerce in the whole of eastern Africa. Politically, Oman came to dominate the Swahili

Porters with ivory tusks, Dar es Salaam, now capital of Tanzania. (Frank and Frances Carpenter Collection/Library of Congress)

towns and established a capital for the region on the island of Zanzibar. Economically, new investment in the ivory trade and in the slave trade flowed in from European, Middle Eastern, and Indian sources. Caravan trade expanded as Swahili merchants had access to new capital from these sources (and lost some of their autonomy). The increased economic activity got another boost when Omani investors began to develop clove plantations on the islands of Zanzibar and Pemba. Clove tree planting dominated the islands, and caravans brought slaves from as far as the eastern Congo to work on the plantations. Along the coast, plantations developed growing rice, maize, and sugar cane to supply both the islands and the Indian Ocean trade generally (Sheriff 1987).

Europeans also participated in this trade. Although much of the cloth that formed one of the main imports of the region came from India, a growing amount came from Britain, Germany, France, and even the United States. French and Portuguese merchants purchased slaves for shipment to sugar plantations on islands such as the Comoros, Mauritius, and Reunion and before the 1850s for smuggling to Brazil. Similarly, the island of Madagascar became a major source of captives during the nineteenth century.

The northern frontier of eastern Africa experienced less dramatic change than those areas further south. In Ethiopia, a century-long trend toward political recentralization helped spur an extension of the ox-plow complex further south in the highlands. In the Sudanic Nile region, Egyptian dominance resulted in the expansion of agricultural production along the Nile as well as an expansion of slavery and slave exports to Egypt proper. The most valuable export down the Nile seems to have been ivory. By the middle of the nineteenth century, Egyptian merchants imported around 350,000 pounds a year. Caravans made up of Egyptian and Sudanese porters and merchants traveled by boat up the Nile. They disembarked and then both traded for and raided for ivory, cattle, and slaves. By the middle of the century, they had reached the kingdom of Buganda on the shores of Lake Victoria where they faced competition from Arab-Swahili traders (Manger 1988).

SOUTHERN AFRICA IN THE NINETEENTH CENTURY

In the regions north of the southern end of the continent, increasing trade did not mean the alienation of either land or productive decisions from indigenous peoples or institutions. Before the 1870s, foreigners, Europeans on the west coast, and Asians on the east coast and along the Nile controlled external trade but otherwise laid claim to little more than enclaves on the coast or along trade routes. The trade in slaves often disrupted African societies throughout the nineteenth century, and types of plantation agriculture developed in response to new opportunities; but the vast majority of Africans lived under African rule and made their own decisions about how to gain their livelihood from their own land.

South Africa was, of course, different. During the 1800s, it saw the rise of the most powerful African state ever known in the region and then the expansion of white domination out of the enclave of the Cape Colony, despite the rivalry between Boer settlers and the British colony; the development of an extensive agriculture that would eventually drive most Africans into crowded reserves; and the discovery of the largest deposits of diamonds and gold in the world, and the creation of a capitalist economy based on their exploitation. As Mahmood Mamdani (1996) has stated, in many ways, South Africa served as the model for the later colonialism in the rest of Africa—a model white South Africans more completely developed than almost anywhere in Africa.

The nineteenth century opened two momentous political changes that would help reshape the landscape of the region. In the area that is now Kwazulu-Natal on the Indian Ocean coast, a process of political centralization began that led to the formation of the Zulu kingdom. To the west, the British occupied the Cape Colony during the Napoleonic wars and, after returning it briefly to Dutch control, retained it in the settlements after Napoleon's final fall. The result was a reconfiguration of both the political and eventually lived landscape of the region.

In the late eighteenth century, a process of political centralization began among the Nguni-speaking peoples along the Indian Ocean coast. African communities had previously lived in "chiefdoms" of several thousand people, often loosely allied with each other through links of patronage and seniority. "Chiefs" used cattle to tie commoners to them. In the late nineteenth century, Dingiswayo, the leader of the Mthethwa chiefdom, began to build a new set of alliances by using age-regiments to create stronger military forces. The leader of the Zulu chiefdom of the alliance, Shaka, took power when Dingiswayo lost his life in battle. Shaka further centralized the age-regiment system, creating a standing army and integrating young men from throughout the territory he dominated. Zulu forces waged aggressive war, and during the ten years of his rule, Shaka created a centralized state where none had existed before. He consolidated his control over cattle resources in the region, thus securing royal control over economic activities. A chain reaction of state consolidation spread across the region, partly in response to Zulu expansion and partly in response to colonial expansion from the Cape. Population shifted into the mountains of the Drakensberg and further away from the Zulu state. Several military leaders moved with their followers out of the Natal area into regions to the north and created sec-

King Shaka was the founder of the Zulu Empire in South Africa. By unifying the northern Nguni people of Natal, he built one of the most powerful states in nineteenth-century Africa and held it together through intimidation and terror. He is recognized today as the founder of the modern Zulu nation. (South Africa Library)

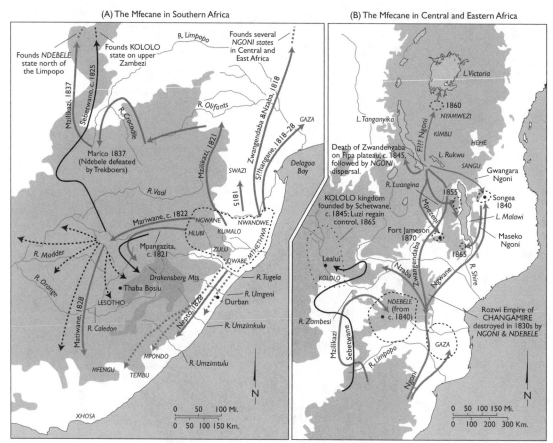

The Mfecane.

ondary states; for example, the Ndebele moved first to the Transvaal and then to Zimbabwe, the Gaza went into southern Mozambique, and the Nguni traveled all the way to what is now Malawi and Tanzania. Closer by, Lesotho in the Drakensberg and the Swazi state organized to defend themselves, adopting some of the tactics of the Zulu state. Lesotho especially also faced intense pressure from the Boers for access to land in the lowlands north of the mountains. Thousands of refugees also moved away from Natal into the Xhosa areas to the west (Peires 1989; Beinart and Coates 1995).

The rise of the Zulu state and the response to the pressures coming from the colonial expansion became known in African oral traditions as the *mfecane*, "The Crushing." It has been the subject of an intense historiographic debate (Cobbing 1988; Hamilton 1998), state consolidation took place, resulting in the rise of the Zulu state. Jeff Guy, for example, contends that the underlying cause of the consolidation of the Zulu state lay in growing population pressure on the land. Europeans had not generally victimized the region south of the Limpopo through the slave trade, and maize seems to have spread to some extent in the region during the eighteenth century. In addition, trade through the Portuguese-controlled ports of Mozambique just to the north seems to have increased both in ivory and to some extent in slaves (Guy 1987).

Other scholars have argued that Boer expansion along the frontier between the Cape Colony and the Xhosa people around the Fish River helped lead to increased conflict over land (Peires 1989). Both Boer settlers and outcast groups from the colony that collectively became known as Griqua had by the late eighteenth century begun to intrude on Xhosa to the east and Sotho lands to the north. They raided for cattle and for the captives that the Boers called "servants" instead of slaves as well as conducting trade for cattle and ivory. The invaders' advantage in these encounters lay in their greater supply of guns and horses whereas the African peoples had the advantage of numbers. Both groups used land extensively. To the west of the Fish River, the Boer and Griqua settlements generally relied more on livestock, whereas the Xhosa and Sotho communities practiced more extensive agriculture. The frontier between the colony and the Xhosa stabilized around the Fish River by the 1780s and remained there until the 1830s (Beinart and Coates 1995).

Colonial settlement did bring dramatic changes to the landscape in South Africa, however, even if settlement of the Cape Colony remained limited until the 1830s. Colonial settlers understood land as an alienable and exclusive property and regarded resources such as the skins, meat, and tusks of wild game as commodities to be harvested. They also saw trees both as commodities and as resources to be used, and they viewed the land thus deforested as fields to be farmed and pastures for cattle.

One of the most important concepts motivating settler expansion in the nineteenth century was that of permanent and inalienable property. According to African pastoralist conceptions of land use, land was a common good held by the group, and this land use was to be defended against outsiders but shared within the community. Community itself could expand or contract depending on context. African farming communities subscribed to the right of usufruct—that is the user had full rights to the land in use, with unused and fallow land controlled by a larger clan or political division. Extensive use of land meant that much land lay fallow at any particular time. Boer settlers saw this land as vacant and believed they had a moral (if not legal) right to expropriate it. They also believed that as population expanded the people had the need to acquire more land for their children rather than more intensively exploit their large landholdings.

Hunting and raiding formed the outliers of settler expansion. Merchants from Cape Town and Natal traded with settlers and Griqua for hides and tusks as well as with San hunters and with other African peoples for these goods. Behind the settlement frontier, game almost totally disappeared. The quagga, a type of zebra found only in South Africa, became extinct, and by the end of the nineteenth century only a handful of elephants remained in what had become the Cape Province.

By the 1830s, pressure had built up behind the Fish River frontier, and political conditions on both sides of the frontier became unstable. Boers, anxious to escape control from British colonial authorities and looking for new land beyond the barrier imposed by dense Xhosa settlement, began what they called the "Great Trek," an invasion of the land north of the Drakensberg and on into Natal. As they moved, they battled African communities for control over land. Their first settlements in Natal came with the permission of the Zulu state, but when the Zulu soon tried to expel them, a vicious war broke out. After the British sought to extend control over settlers near Natal, in part in response to the violence, the Boers retreated to the "highveld" beyond the Drakensberg Mountains and eventually established two independent Boer republics: the Orange Free State and the Republic of South Africa (Transvaal).

In these areas, the settlers gradually consolidated their control over land using force. They had the advantage of access to the colonial market where they purchased guns in return for livestock and their by-products. Sometimes African communities resisted, as in the case of the Ndebele, by moving away from the intruders. Boers established the same kind of extensive agropastoral system relying on "servants" as a labor force.

In the Cape Colony and Natal, the British colonial administration pursued an inconsistent set of policies. They tried to control the Boer settlements without necessarily incurring the costs of direct administration. At the same time, they sought to ensure adequate labor supplies from the surrounding and nominally independent African peoples without directly having to conquer them. These policies came to a head in the 1850s when a major disaster befell the Xhosa peoples of the Eastern Cape region. The Xhosa communities were pressed by British and Boer expansion as well as the rise of the Zulu and Sotho kingdoms. A series of conflicts in the 1850s had extended colonial control to the Kei River, leaving half of the Xhosa land under formal colonial control. In 1855, an outbreak of cattle lung sickness decimated herds owned by both settlers and Africans. In the face of the loss of herds, a millennial movement arose that claimed that if the people killed their cattle and did not cultivate their fields, the ancestors would restore the Xhosa to their lands and herds and drive the foreigners out of the country. Although by no means did all people in the region participate, enough did join the movement to create a massive disaster in the region by 1857. The colonial government took advantage of the flight of thousands of hungry refugees into the colony to both extend their control deeper into Xhosa territory and to forcibly recruit thousands of men, women, and children as inden-

Men working 2,000 feet undergound in Kimberley Diamond Mine, South Africa, ca. 1890–1905. (Frank and Frances Carpenter Collection/Library of Congress)

tured servants for settlers. While the movement of Boers into the highveld beyond the Drakensberg proved disruptive, it was the system of alienation of land and the reduction of Africans to tied labor that greatly extended in the 1850s the foundation for the development of South Africa. By the 1840s, this system had developed a sheep-rearing industry that made wool South Africa's most valuable export (Beinart and Coates 1995).

The discovery of diamonds in Kimberley gave a different impetus to this massive reorganization of space in South Africa. The Kimberley diamond fields were discovered in 1866 in a territory recognized by the British as belonging to a segment of the Girqua people. Quickly, both the British from the Cape and the Boer republic of the Orange Free State tried to claim it. The British colony eventually won out and then imposed the rule that effectively only whites could own the mines that produced the diamonds. During the first decades of the production of diamonds, Africans supplied the demand for labor willingly. They received wages that they used to purchase guns and horses, and to invest in new technology such as plows. Africans also responded to the explosion of population around the mines by increasing production of food crops for sale and by engaging in timber cutting and hunting to supply the mining sector. Whites quickly joined them, and a struggle for control over these lucrative businesses ensued. The demand for timber, meat, and leather depleted supplies throughout the area.

With the discovery of gold in the Transvaal in the 1880s, pressure for labor to work in the mines, build and run the railroads, and work the settler farms increased. Although the British and the Boers fought over the control of South Africa (in the context of the conquest of the rest of the continent), they essentially agreed on the need for cheap labor to make their investments work. The solution, implemented piecemeal before 1900 and codified after the British victory in the South African or Boer War of 1900, was to reserve 75 percent of the land of the newly christened Union of South Africa for white ownership and to confine African landownership to "native reserves." This system, developed in the late nineteenth century, became the template for colonialism in much of the rest of colonial Africa, with a division between market-oriented production and exploitation of resources reserved for expatriate capital and Africans confined to wage labor and subsistence production (Beinart and Coates 1995).

THE SCRAMBLE FOR AFRICA

In some ways, the discovery of diamonds in South Africa can be seen as the beginning of the final conquest of sub-Saharan Africa. Although many factors influenced the decisions of French, British, Portuguese, German, Spanish, and Italian policymakers to claim colonies in Africa, the discovery of diamonds in South Africa gave all the European powers something of value for which to aim. In general, the "Scramble" became a scramble for control over first, resources, and second, the people and labor to exploit those resources. Not all of Africa of course would have the kind of mineral wealth found in South Africa, but South Africa proved the possibility of gaining enormously lucrative possessions in the subcontinent.

The conquest and creation of African colonies in the years after 1870 had enormous implications. First, Europeans could conquer African societies because they finally had the ability to survive in sufficient numbers in the tropics and their control of new technologies generated by the Industrial Revolution finally gave them a decisive military advantage over African military forces. Second, the process of conquest itself was violent and caused new ecological stresses that greatly reduced Africa's population. While not on the order of the die-off that accompanied the conquest of the Americas 400 years earlier and varying greatly between regions, the 50 years between 1870 and 1920 saw significant population loss nonetheless. Third, the new European masters sought to exploit all the resources available to them in their new possessions. They generally followed a pattern of resource mining in the early decades followed by the exploitation of more labor-extensive forms of production. In some cases, such as the forest regions of West Africa, they sought to take control of already existing export-trade networks in raw materials. After World War I, the European powers that dominated Africa began to intensify their control over colonial economies. At the same time, colonialists developed a discourse of conservation that often blamed Africans for the destruction of natural resources rather than the colonial enterprises that dominated them.

The scramble for Africa could begin once advances in knowledge of tropical and infectious disease increased the Europeans' survivability in tropical Africa. Since the seventeenth century, some medical practitioners had known that the bark of the cinchona tree, originally from South America, could control the effects of fevers associated with malaria. In 1820, French chemists isolated the main effective element in the bark, which they labeled quinine. In fits and starts, French troops and officials in Algeria and Senegal began to use quinine as both a curative and a prophylaxis. After 1838, the mortality rate of French troops serving in Africa declined from an average of 163.07 per thousand per year between 1819 and 1838 to 54.05 between 1839 and 1851. These rates remained higher than those of the general population back in Europe or of soldiers stationed in Europe. Quinine's mechanism of action in treating and suppressing malaria was still not known with certainty, and treatment regimes varied greatly until about the turn of the twentieth century. British mortality rates in West Africa declined as the British more slowly adopted the quinine regime. However, British forces in West Africa demonstrated the effectiveness of both the use of quinine and of improved sanitation measures in their invasion of Asante in 1874. The British went to great lengths to prepare the 2,500-man European force for the invasion. Planning included limiting the amount of time the total force stayed in the Gold Coast to two months, but the death rate per thousand for the force for that time was only seventeen and Asante's power to challenge British dominance of the Gold Coast was decisively broken. The full effectiveness of these measures only came about in 1900; before then, application of both quinine and sanitation remained inconsistent. After 1900, knowledge of both the bacterial agent that caused malaria and the role of mosquitoes in carrying it had become generalized. Similarly, the adoption of the germ theory of disease meant that a host of other parasites and infections could be combated by testing water supplies and providing clean water (Curtin 1998).

Perhaps more important than the effect of these improvements on the ability of Europeans to fight in Africa was the improvement of their ability to explore, preach, trade, and eventually administer in Africa. Beginning in about

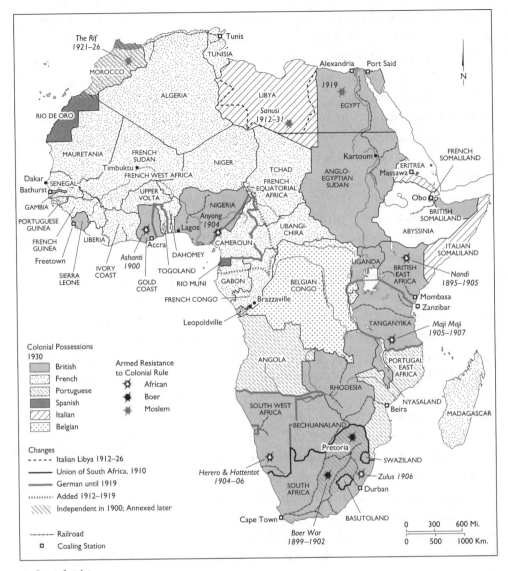

Colonial Africa.

1850, European explorers and missionaries began to be able to travel in Africa with some hope of surviving the journey there and back. Setting off to find the sources of the Nile or the route of the Niger, they also saw themselves as path breakers in the progress of Christianity. They established missions that throughout the nineteenth century had little success in winning converts but created "facts on the ground" of European presence in advance of partition and conquest.

When actually engaged in conquering African societies, Europeans minimized the potential loss of their personnel by using African troops. The French were the first to adopt this practice on a large scale, establishing first the *Tirailleurs Algeriens* and then the *Tirailleurs Senegalais*. Cases like the Asante campaign of 1874 where a predominantly European force invaded an African state became rare, and the French, British, Germans, and Portuguese all recruited colonial armies made up of African enlisted men with European officers. European conquest came not just because Europeans were better able to survive in tropical Africa, but because European states could mobilize resources to hire professional armies in Africa and equip them with better arms than African states could mobilize. The relatively small scale of African states and their large number also meant that Europeans usually had little difficulty in finding both allies and recruits in their campaigns. European-led forces did not win every battle, even when they had a technical advantage. In the Zulu War of 1879, a British force made up mostly of regular British troops was destroyed by a Zulu force that did not even use guns to any great extent at the battle of Isandlwana, and in 1896 Ethiopia defeated an Italian invasion at the battle of Adowa, retaining its independence for another forty years (Vandervort 1998).

The decades between 1870 and 1920 were portentous ones for Africa: these years encompassed not only the conquest of Africa but also, as both cause and consequence of that conquest, a series of environmental disasters that reduced African populations in many places, destroyed Africa's economic autonomy, and eventually created impoverished and dependent economies in the new colonies. In 1884–1885, the Berlin Conference met to ensure an orderly division of Africa among the European powers. The pace of conquest quickened after the conference, as Britain, France, Germany, Portugal, and King Leopold of Belgium sought to "effectively occupy" the areas they claimed at the conference. The exploits of the "men of the spot" such as Karl Peters in German East Africa, Alfred Lugard in Uganda and Nigeria, Henry Stanley in Leopold's Congo Free State, Cecil Rhodes in southern Africa, Savorgana de Brazza in Central Africa, and Faidherbe in the West African Sudan were once romanticized; however, the policies set at the Berlin conference determined the pace of conquest.

These conscious policies of conquest meant that each year new expeditions left from the port cities that came under colonial control. These expeditions followed existing trade routes and sought in many ways to control first trade in existing commodities and then people in the larger states. Wars of conquest disrupted productive systems and spread epidemic disease in their aftermath. The most intense period of conquest coincided with the spread of rinderpest in Africa for the first time. Rinderpest, a cattle disease that affects ruminants and is transmitted through spittle left in grazing areas, had been endemic in parts of Europe and Asia for many centuries, but the Mediterranean and Sahara had served as an effective barrier to its spread. In 1887, however, the Italian expedition that conquered Eritrea brought infected cattle to sub-Saharan Africa for the first time. The disease established itself in African herds and spread rapidly, with mortality rates in unexposed herds reaching up to 95 percent. The disease also struck many of the grazing game animals and spread rapidly from Ethiopia both south and west. In 1892, it reached East Africa and devastated both pastoralists' and agropastoralists' herds. It impoverished numerous groups, including the Masai, just as the British and Germans began their push to dominate those peoples. It moved further south, pausing at the Zambezi until 1897; crossed the river and spread even more rapidly through the wagon trails that carried goods and people between the European settlers of Rhodesia and South Africa; and in the west moved across the savannahs and Sahel, devastating the herds of groups like the Fulani during the last great push of colonial expansion in the Sudan (Anderson 1988; Waller 1988).

Rinderpest, coupled with the scorched earth policies of the Germans and British in East Africa when they faced resistance, made the 1890s a decade of famine in East Africa. With the movement of colonial forces and the gradual increase in trade, other diseases also spread more rapidly. A smallpox epidemic took the lives of many, while cholera spread along the trade routes. People fled the conflicts caused by colonial conquest, taking the diseases with them (Dawson 1979, 1987). In the eastern Congo Free State, fighting between the European-led *Force Publique* and the forces of Arab-Swahili merchants in 1892 decimated the entire area (Northrup 1988). In Southern Rhodesia, the *Chimurunga* uprising in 1896 not only caused great loss of life but also resulted in Cecil Rhodes' British South Africa Company expelling Africans from over half the territory of the colony (Ranger 1999). In the West African Sudan during the 1890s, the French fought for a decade with the Mossi forces of Samore Toure. When drought came, as it did in eastern Africa in 1895, famine followed. Europeans now feared that the continent was in danger of becoming depopulated.

In the long run, the rinderpest epidemic had another important consequence that permanently shaped African landscapes: the decline of stock numbers and the relocation of population emptied large portions of the landscape in the savannahs. The decline of both grazing and burning allowed the regeneration of bush. Despite the effect of periodic drought in situations where African populations already faced stress, the decade of the 1890s probably also was a wetter time than before, which aided in the reestablishment of the bush in many areas. With the bush came tsetse fly and trypansomiasis. Soon large areas of the eastern African savannahs and of the West African Sudan became almost uninhabitable (Kjekshus 1977).

The famine crises of the 1890s occurred because colonial conquest had disrupted the systems that African societies had developed for maintaining access to food in the face of climatic variability. The new stresses came in the form of a more rapid circulation of diseases, both human and animal. They also derived from the disruption and violence of conquest itself. The colonial state system added to these stresses by imposing a new series of demands on African societies. In haphazard and often initially unplanned ways, colonial states set about to remake African landscapes, and Africans, into parts of the global economy but in a subservient and dependent position. Colonial states also came to recognize the costs of converting African landscapes into scenes of commodity production; they imported ideas about conservation of natural resources and preservation of landscape with colonial rule.

REMAKING AFRICAN LANDSCAPES

There was a general consistency to the colonial enterprise across Africa: the new African colonies were to produce mostly raw materials and to consume imported manufactured goods. The resources found in the colonies belonged to the colonial state, and the state had the final say in how (or if) their exploitation occurred. But within those broad parameters, there was a great deal of variation between different colonies, depending in part on environmental conditions.

In southern Africa, the established settler communities and the expansion of diamond and gold mining dictated that almost all land and productive resources come directly under the control of either the settlers or, in some cases, the colonial state. Africans were eventually forced into the position of a dispossessed labor force. The process began with the expansion of Boer farmers after the 1830s. From then until the 1890s, they established gradual control over the areas that became the Boer Republics. Their effective control over African populations remained theoretical until the discovery of gold, but their access to guns from the British colonies at the Cape and in Natal allowed them to hold their own in the lands they settled. The British for their part gradually extended their influence in the areas to the south of the Boer Republics, making most of the Xhosa and Sotho-speaking peoples dependent on labor in the colonies.

The discovery of diamonds at Kimberley changed the situation dramatically. Demand for labor increased, which meant that even more immigrants arrived from Europe. The demand for food for the mining settlements and the demand for timber for construction and for fuel also rose dramatically. Initially, in areas within economical, ox-wagon reach of the mining areas, Africans intensified production of agricultural commodities to meet this new demand. Maize became more important in crop mixes, and farmers adopted plows and ox wagons for transport. The mines also drew labor, offering wages that young men used to pay bridewealth – fees paid to the bride's family—for wives. When gold mining developed in the Transvaal (as the South African Republic was known), the process intensified.

Historians have often considered the South African (or Boer) War of 1898–1900 as a convenient turning point in this process. The successful British effort to unify "white" South Africa marks the creation of modern South Africa. The historic compromise the British imperial government made with South Africa's settlers, both Afrikaner and English speaking, cemented racial domination and the more complete articulation of capitalism in the country. Through a series of measures, the colonial government declared African land unoccupied and then sold it or gave it away to settlers. To government officials and settlers alike, the land often looked unoccupied because African agricultural systems in the region utilized land extensively. African farmers often shifted fields, allowing them to lie fallow for up to twenty years. But the policy, similar to that of the United States toward Native Americans, also had a more direct goal. By limiting African access to land, the colonial government and settlers hoped to force Africans, especially younger males, to go to work in mines and on farms. Employers developed several tactics for getting the labor they needed as cheaply as possible. In white-owned farm areas, settlers often allowed African families to "squat" on part of their land in return for labor on the farm. Through this practice Africans were made to pay labor rent for land they or their direct ancestors had lived on for a long time. It also allowed settler farmers to abuse the relationship. After 1910, the now white-dominated independent government of the Union of South Africa moved to limit this practice in the name of both equity and efficiency (Beinart and Coates 1995).

Mining and eventually industrial employers developed a second means of ensuring a steady supply of labor by setting up labor recruiting systems. These systems limited competition among employers. They set their wages very low on the theory that only young men would become migrants and that they would return to their rural homes after a few years. The young men would supply cash for their families, while the remainder of the family continued to live on the land and farm. The reserve system in turn ensured that African households did not have enough land to produce marketable surpluses of either crops or livestock. This system reached its logical political conclusion when, after the victory of the Afrikaner Nationalist Party in the general elections of 1948, the government instituted the policy of Apartheid—"separate development"—that made black South Africans citizens of homelands occupying 13 percent of the land of South Africa while reserving the rest for white occupation or government ownership.

The result was a bifurcated landscape. Urban South Africa over the twentieth century industrialized, with Johannesburg, Durban, Port Elizabeth, East London, and Cape Town becoming centers of manufacturing surrounded by townships occupied by supposedly migrant laborers. The white-owned farms, supported by government subsidies, adopted mechanization and modern agricultural technology to become extremely productive. The government set aside large areas as national parks, most notably Kruger National Park in the northern Transvaal. The reserves and homelands suffered from overcrowding and deteriorating environmental conditions. They became eroded and less productive. Africans living on them, rather than growing their own food, became dependent on remittances from migrant workers to purchase their food. In an effort to maintain the fiction of separate development, the government tried to promote improved agricultural practices and conservation through what it called "betterment" policies after World War II, but these policies, which both previewed and mirrored policies in other colonies, often failed because they required more effort and expenditure by Africans for no measurable return (Mamdani 1996; Beinart and Coastes 1995).

In many ways, the environmental history of South Africa served as a preview for that of the rest of Africa, even those areas without large settler populations. As colonial administrations came into being, they all sought to promote the export of commodities; to develop transportation and communication infrastructures that would facilitate these exports; to promote production and reserve especially valuable resources for colonial exploitation; and to conserve and preserve resources and landscapes.

During the late nineteenth and early twentieth centuries, many colonial regimes relied on the extraction of raw natural resources found in African environments to provide their first returns. King Leopold of Belgium established the Congo Free State in the late nineteenth century without the backing of a national government. To finance his adventure, he sold concessions to private firms that had monopoly rights over exports of commodities from their territories. These firms would in turn administer the regions according to guidelines set by the agreement and develop infrastructure, including railroads. These firms concentrated on raw materials such as rubber, ivory, and timber and used highly coercive methods to force Africans to collect the commodities. An international outcry against these methods forced the Belgian government to take over the colony in 1905 (Northrup 1988).

King Leopold was not alone in using private capital to develop colonial possessions; all the major colonial powers used such companies in some cases. The Germans began their colonial adventure by chartering companies to found and administer colonies. In both East and West Africa, the British turned to such companies to lay claim to what would become Kenya and Nigeria when governments of the day feared they would not get parliamentary support for new colonies, while Cecil Rhodes' British South Africa Company founded Northern and Southern Rhodesia. The French used such companies in their possessions in Central Africa and the Portuguese in Mozambique.

Leopold II was king of Belgium from 1865 until his death in 1909. (Library of Congress)

In several cases, chartered companies served as a means of securing private capital for constructing railroads. Transportation infrastructure preoccupied early colonial administrations. In most regions of Africa, excepting only the Senegal, Niger, and Congo River basins, head portage remained the only means of moving goods around in the late nineteenth century. Even in the case of the rivers, rapids and waterfalls made water transport costly. Colonial administrations sought to build railroads that would tap potential export markets and promote effective administration. In South Africa in the late 1800s, railroad development linked the mining areas with the coast and then in the early 1900s stretched north to tap new sources of labor and food for the growing urban areas. Under British control, they reached as far north as Northern Rhodesia and into the Portuguese colony of Mozambique. Reflecting its general industrial development, South Africa, by the middle of the twentieth century, became the only region in Africa with an integrated rail network.

Elsewhere, early colonial plans for rail development concentrated on lines designed to promote exports and never expanded much beyond that before the end of colonial rule. In West Africa, the British and French built competing railroads running parallel to one another. In East Africa, the British and Germans built parallel networks in Kenya and German East Africa. The British eventually linked the two together during their invasion during World War I of German East Africa. Only in the 1920s did motorcars and trucks become relatively common in the colonies.

The railroads had dramatic effects on the landscape of African colonies. The placement of rail lines determined which regions in certain colonies became commodity-producing regions and which became suppliers of labor. The position of rail lines did not necessarily follow from economic potential. The Germans built their first line from Dar es Salaam to Lakes Tanganyika and Victoria through the arid and tsetse-infested central part of German East Africa instead of either through the more productive highlands in the northern or southern part of the territory. While they wanted to tap the potentially rich regions around the lakes, including Rwanda and Burundi, they also wanted to establish firm control of the center of the colony. The success of the Uganda Line running from Mombasa to Uganda in drawing off traffic from the northern part of the colony forced the Germans to build a competing line from Tanga to Moshi on Mount Kilimanjaro. Regions along both of these lines became commodity exporting, including sisal grown on European-owned estates and from the highlands of the region.

In Kenya, the placement of the Uganda Line was designed to tap the potential for trade with Uganda and to establish British control over both Kenya and Uganda. It ran through the cooler and fertile Kenya highlands and opened that area up to settlement by white farmers. In both colonies, until well into the

Two colonists examine the rubber collected by the workers on a plantation in French Central Africa. (Hulton-Deutsch Collection/Corbis)

twentieth century, areas away from the rail lines became labor reserves, producing migrants to the white-owned estates and in the case of Tanganyika to the mining regions to the south.

In West Africa, the French sought both to tap the existing production of groundnuts in the Sudan and to secure their control over the Senegal and Niger River Valleys by building a railroad from Dakar to the Niger at Bamako in what is now Mali. They also built railways from Abidjan to Ouagadougou and from Libreville to Brazzaville that paralleled routes built by the British and Belgians, respectively. The construction of railways in general allowed for the development of commodity production further inland and decreased the reliance on exports of high-value, low-bulk goods such as ivory. By 1910, the era of resource mining had ended. Elephant herds in East and Southern Africa had thinned tremendously and had almost disappeared from West Africa. Plantation rubber from Southeast Asia (and to some extent West Africa) replaced wild rubber from the

Central African rainforests as the wild rubber vines died out. Replacing this resource mining was commodity production. In some cases, as in much of West Africa, this meant the extension of already existing African production through opening new lands and adding new crops. In some parts of the continent it meant the introduction of new crops, often through force. In some areas, European settlers came to create estates, and in other areas, mining developed; the profitability of both depended on relatively cheap African labor. In all cases. both environmental conditions and conscious decisions by colonial officials helped determine the type of production that dominated. But African societies proved both resilient and resistant to imposed changes.

In a few regions of Africa, most notably Southern Rhodesia, Southwest Africa, and Kenya, colonial rule meant the coming of settlers. Much smaller settler sectors also developed in German East Africa, Angola, Mozambique, and the Congo. The French colonies, aside from Algeria, attracted very few settlers, partly because almost all French territory lay in malarial areas and partly because the French never encouraged many settlements. In colonies such as Côte d'Ivoire, plantations owned by French planters ran more as capitalist firms than as settler farms, and there was little segregation of landownership in the way that the British colonies practiced.

Settler colonies in sub-Saharan Africa all were inspired in one sense or another by the South African model. Southern Rhodesia, Southwest Africa, German East Africa, and Mozambique all imported settlers to a greater or lesser extent from South Africa. Settler colonies all relied on an enforced division between land open for settlement and land reserved for natives. The vision of ordered fields of grain or coffee or tea or sisal or cotton served to contrast with a vision of inefficient and unscientific agriculture practiced by Africans. Settlers, whether in Rhodesia or Kenya, saw in their place in Africa an open, indeed unsettled, landscape that they would tame. Hunting often served as the first sign of their arrival, after which came harrying the Africans on the land into submission (Steinhart 2005; Beinart and Coates 1995). But the settlers could not survive without the African communities from which they stole the land; they needed their labor and even their knowledge to survive in their new "homes."

As in South Africa, settler sectors had a hard time surviving without using the coercive power of the colonial state. They relied on the state to enforce landownership, to build the infrastructure that made their economic survival possible, and to maintain their livelihoods. In Kenya, the colonial government prohibited Africans from growing coffee in order to preserve a white monopoly. Settlers claimed that Africans did not know how to care for coffee. In neighboring Tanganyika (now Tanzania), however, African smallholders on Mount Kilimanjaro produced the highest-value Arabica coffee from East Africa throughout the colonial period, and African producers in Uganda, Burundi, Rwanda, and the Bukoba region of northwest Tanganyika dominated the world market for *robusta coffee* (used typically in instant coffee) (Iliffe 1979).

Settlers in Kenya, Tanganyika, and Rhodesia constantly fought to control the movement of African livestock. African herders had always used rangeland extensively, and the fences and claims of ownership made by settlers over grazing lands brought the two groups into persistent conflict. Settlers also claimed that African herds spread disease because of the Africans' inability to practice scientific herd management. Such claims often proved illusory. For African herders, ensuring the survival of cattle served as their primary goal. Toward this end, they practiced a variety of strategies to spread risk and share resources through reciprocal relationships with other herders. Such practices, of course, did not preclude violent conflict in the form of cattle raids and struggles over grazing between African communities, but within communities and often between communities, these relationships could mean the difference between survival and famine. Settler insistence on the sanctity of property not only destroyed these options but often served as a means for settlers to take stock from Africans. A common complaint, despite being formally outlawed by colonial governments, was that settlers confiscated any stock found on their land (Anderson 2002).

Settler sectors often survived literally by creaming off the surplus from what was essentially African productivity. In Tanganyika and Mozambique, colonial governments paid differential prices for commodities produced by settlers as opposed to Africans. Often, settlers made profits by buying African produce and then reselling at the higher settler price. Cotton, regarded as a strategic commodity during the late nineteenth and early twentieth centuries, often received this treatment, but it occurred with other commodities as well (Kitching 1980).

The net effect of settler sectors on African environments is even more readily apparent, especially when expatriate-owned estates are included in the sector. Such plantations produced coffee in Côte d'Ivoire and Congo, sisal in Tanganyika, tea in Nyasaland, and rubber in Liberia. They represented the importation of capitalist production into the colonies in the agricultural sector. They developed where profits mandated them, although they, too, relied on colonial coercion to provide a cheap and compliant labor force.

A second way in which colonial capitalism remade African landscapes was through the exploitation of African minerals. Gold had long served as one of the most important exports from Africa, and Africans had developed iron and copper working industries throughout much of the continent. From the fifteenth century on, Europeans had sought to control the supply of especially gold from Africa. However, like other merchants before them, until the late nineteenth century Europeans had to be content to let African miners supply the gold in return for imported goods. The discovery of diamonds at Kimberley changed this equation dramatically. In the late nineteenth century, foreign, mostly British, capital developed first the diamond mines and then the gold mines of the Witwatersrand in the independent Afrikaner republic of Transvaal. In both the Cape Colony and Transvaal, governments restricted ownership of the mines to whites and created the system of migrant, "temporary" labor. After the South African war of 1900 and the creation of a unified South Africa, the development of the mining industry provided the capital necessary for the beginning of industrialization in South Africa. Urbanization around Johannesburg and then in secondary centers such as East London, Durban, and Port Elizabeth proceeded as industries developed first to support the mining industry and then to meet demand for consumer goods. Coal mines also developed to provide the energy for the country. The development of an industrial economy continued to be based on labor costs held low by racial oppression and dispossession of land. A bifurcated landscape developed with white farmlands managed more or less scientifically, while urban slums (townships in South African terms) and African reserves in the rural areas became some of the most degraded landscapes in the whole of the continent by 1950.

Elsewhere, mining mostly remained a capital-intensive industry. In the copper belt that formed the border between Northern Rhodesia (Zambia) and the Belgian Congo, European capital helped create the largest copper-producing region in the world, initially following the South African model of management. By the middle of the twentieth century, however, copper belt mines had moved toward labor "stabilization" that encouraged workers to move permanently with their families to the mines. As a result, the contrast between rural and urban areas in those regions became less stark than in South Africa. Before World War II, mineral development remained small scale. In many places, the discovery of gold or precious gemstones set off gold-rush events where Africans exploited relatively easily worked deposits using limited technology. Little in the way of iron production developed on an industrial scale, and petroleum development occurred only after World War II.

Despite the development of extractive industry, urban areas, and settler colonies in Africa, the majority of Africans throughout the colonial era lived in generally self-sufficient rural communities. Colonial regimes saw these rural communities as great untapped reservoirs of labor and agricultural production that required external stimulus to develop. On terms requiring little investment from the colonial state, they sought to force Africans in these communities to produce marketable crops or provide labor for extractive industries (Iliffe 1987).

In some places, African farmers had begun to develop commodity production during the nineteenth century. In West Africa especially, Europeans had competed for control of trade routes, bringing especially palm oil and groundnuts to the coast. African farmers usually organized production on a household scale. To some extent it complemented production of food crops. In areas lacking settlers or extractive industries requiring African labor, colonial regimes sought to encourage this type of "peasant" production. They operated on the theory that African labor, particularly that of males, lay dormant in the aftermath of colonial pacification. Colonial regimes also sought to promote the production of strategic crops such as cotton. Starting in the late nineteenth century, they used a variety of means to compel production (Hill 1970).

Colonial governments imposed head or poll taxes as the most important means of compelling production of marketed crops. In many areas, Africans could initially pay such taxes in kind; however, colonial governments moved quickly to convert them to cash payments. Colonial governments also sought to direct production toward export crops such as cotton and away from food crops. In many cases, they resorted to compulsion to get Africans to produce the desired crops. In German East Africa, German efforts to compel cotton production helped lead to the Maji Maji revolt of 1905, whereas compulsion remained a mainstay of Mozambique's cotton industry throughout the colonial era (Isaacman 1996).

In the latter part of the nineteenth and early part of the twentieth centuries, these new demands often helped create subsistence crises when coupled with the normal variation in rainfall and climatic conditions. Rural African households faced demands that young men go off to work to increase production of crops or to sell more livestock. Generally, these activities produced consistently low returns, with a substantial portion going to the tax collector. In different places, household social organization changed dramatically in response to these demands. In many areas, cash crops became a man's domain, often with the encouragement of colonial regimes. Consequently, women became almost single-handedly responsible for the production of foodstuffs. These limits on economic opportunity matched a general repression of women's rights under new colonially sanctioned codes of "native law." Often up to World War I, food shortages and famines resulted from the combination of new demands, limited resources, colonial attempts to reduce mobility, and an inability to bring food imports into stricken regions (Maddox 1986; Iliffe 1987).

Colonial regimes and missionary observers often attributed the problems African farmers faced either to the conservatism of peasants or to racial characteristics. Yet across Africa, farmers responded to the new demands and opportunities of the colonial era with a variety of innovations that transformed rural landscapes. Colonial regimes sometimes encouraged such changes and at other times sought to limit their effects on social order or environment. In the Gold

Coffee and banana farm on Mount Kilimanjaro (1997). African coffee growers in the highlands of East Africa are often very prosperous. (Photo courtesy of Gregory Maddox)

Coast, African landowners invested in clearing land in the forest, using migrant labor from the north of the colony and from neighboring colonies. They dramatically expanded the production of cocoa, whose production spread into neighboring colonies, especially the Ivory Coast. In the highland regions of East Africa, farmers on the slopes of Mounts Kilimanjaro and Meru and in the Bukoba region of Tanganyika began to produce coffee by integrating its cultivation into the system they had developed of planting bananas around their homesteads. The bananas shaded the coffee, and its leaves provided mulch. Stall-fed cattle provided manure, and channels brought water from the mountain streams. In some areas on the plains throughout Africa, cotton became a staple crop. Successful adoption of cash crops depended on several factors beyond climatic and soil compatibility. In particular, many crops required the development of a transportation infrastructure in order to bring decent returns. African farmers in the Mwanza area of Tanganyika located south of Lake Victoria became major producers of cotton, whereas farmers in Uganda produced large amounts of cotton and coffee because railroads and lake transports made moving the commodities to international markets relatively inexpensive.

Colonial fiscal policy and control over marketing often played a critical role in the success or failure of cash cropping. Colonial states often tried to limit the buying of commodities to licensed merchants. These merchants often were from immigrant communities, mostly from the Indian subcontinent in eastern Africa and from the Middle East or Mediterranean in western Africa. These controls, though not totally effective in preventing Africans from engaging in trade, depressed prices paid to farmers. In response, African farmers supported the creation of cooperatives as a means of gaining more control over the marketing of their crops. In French and British Africa, governments sought to co-opt these organizations. They also created "marketing boards" that could set maximum and minimum prices for crops, and pay subsidies during years when prices fell too low. Throughout British Africa in particular, these organizations were managed conservatively and collected relatively large surpluses by the 1950s.

The new demands on rural households also brought dramatic changes in agricultural and pastoral activities tied to food production. The first half of the twentieth century saw an intensification of production and the adoption of new crops and techniques. Both food and domestic animals acquired a market value that was sometimes at variance with the reproductive value to households. As such, rural households began to shape their decisions about what to produce, how to produce it, and how much to sell based on a new set of factors. Their decisions often frustrated colonial administrators who insisted on a sharp division between subsistence and cash crops. Depending on circumstances, African farmers began using plows, built irrigation works, adopted maize and cassava as food staples, and diversified production. Often included in their calculus was an effort to economize on labor requirements in the face of labor shortages caused by the necessity of wage labor.

Perhaps the most generalized effect of the creation of colonial economies on the rural landscapes of Africa was the adoption of two New World crops on a large scale during the twentieth century. In better watered areas, maize became an increasingly preferred crop over traditional African grains such as sorghums and millets. Maize gave higher yields under optimum conditions, matured faster, and offered better protection against birds, but it also carried a greater risk of failure if rainfall proved inadequate. Highly capitalized settler farmers in South Africa and Kenya adopted maize early in the twentieth century. By the middle of the century, hybrids developed in South Africa spread throughout the subcontinent. Smallholders switched to maize because of its potential for higher yields and because maize brought a far higher price than sorghums and millets. In many areas of unpredictable rainfall, it remained a gambler's crop (McCann 2005).

Cassava also spread in many parts of Africa during the twentieth century, but the dynamics of its adoption turned on the impoverishment of both African rural peoples and African landscapes. Cassava fields require the building of mounds in which to plant the cuttings, but only minimal weeding is needed afterward. The cassava tuber stores in the ground for long periods. It became popular in both drought-prone areas and in forested regions with relatively poor soils. Its adoption represented both labor constraints in particular regions and declining soil fertility due to shortened fallow.

Throughout Africa, rural people responded to the growth of urban areas by increasing the production of vegetables and other crops. Such production often flew under the radar of colonial officials. Similarly, urban dwellers developed techniques of urban farming that supplied a substantial portion of their subsistence. Such changes reshaped African landscapes and lives, but officialdom often missed their significance. Instead, starting in the 1930s, colonial regimes began to focus on degradation that was perceived to be caused by African misuse of resources. The result was three decades of struggle over control over natural resources.

6

AFRICAN ENVIRONMENTS IN THE AGE OF CONSERVATION AND DEVELOPMENT

Paradoxically, the second half of the twentieth century was a period when environmental change accelerated in Africa at the same time that conservation efforts increased dramatically. Both resulted from the same process: the incorporation of African societies into the global economy as primary product producers. The dramatic expansion of production during and after the colonial era consumed more natural resources than African societies had before. The demand for increased production, in a context where capital for intensification remained in extremely short supply, led to the intensification of production through the use of more labor and land. This intensification of demand for labor helped set the conditions for rapid population growth. In turn, the expansion of population and production put more stress on environments and taxed their ability to recover.

Recognizing this contradiction, colonial regimes introduced the twin goals of conservation and development. Conservation could mean both the preservation of natural landscapes and resources and the conservation of productive resources, especially agricultural and pastoral productivity in land. In the 1930s, the many colonies began to develop forest reserves, game reserves, and national parks as a means of preserving natural landscapes and protecting valuable resources. In the 1930s, colonial officials also began to express concern about the degradation of soil productivity through overuse and damaging use. Governments, often influenced by American and European models, began to implement a variety of programs designed to promote soil conservation.

After 1950, colonial governments began to promote "development" as a goal for their policies. Development meant using scientific knowledge to better manage the use of resources as well as investment in new and appropriate technologies. Throughout Africa, development schemes sought to irrigate new lands, increase mechanization in agriculture, halt soil erosion, and increase the amount of land under government management in the form of reserved forests and parks. These top-down schemes proved singularly unsuccessful in achieving their aims.

The fault did not lie with African resistance to change, as many colonial and development officials have claimed, for African producers adopted new crops and techniques and sought technical assistance throughout the continent when innovation benefited them. The fault, rather, lay in the fact that these programs consistently meant more labor from Africans for little return.

In many parts of Africa, resistance to colonial conservation programs ultimately led to resistance to colonial rule itself. In some cases, nationalist political movements adopted populist critiques of conservation measures. However, at independence most African states continued the same policies (sometimes with the same personnel). The logic of "seeing like a state," in James Scott's phrase (1998), saw control over resources as a critical element of state power. In some parts of Africa, this continuation played a role in the decline of state legitimacy. In others, it created an ongoing set of conflicts with African states often taking direction from international agencies and donors at the expense of their own people. In several African nations, conflict led to civil strife and the destruction of any efforts at conservation.

For African communities, these twin drives often played out as alienation and exploitation. Forest and game reserves generally required that communities vacate land they had controlled for generations. Just as in the rest of the world, the creation of nature preserves meant the creation of a new landscape, not the preservation of wilderness. Soil conservation programs usually required extensive labor inputs from communities already short of labor. They often became the target of protests and in many cases helped rally support for nationalist independence movements in the 1950s.

The African elites that took over the newly independent states from the late 1950s also inherited many of the same conceptions as the former colonial governments. They often pursued policies that removed control over land and local resources from local communities. Deepening economic crisis starting in the 1970s forced many into the arms of the International Monetary Fund and World Bank, which likewise retained a strong bias toward external control as a means of conservation.

Starting with the Sahelian drought of the 1970s, a series of disasters rocked the continent and led many to believe that sub-Saharan Africa had become a lost cause. These "Afro-pessimists" pointed to rapidly rising populations, failed states, civil wars, the collapse of economies, conflict in southern Africa, and resource mining as leading to the irreversible degradation of African environments. Just as many African states lost control over their own economies because of internationally imposed financial plans, so several innovated with turning over conservation programs to aid agencies and Western-funded nongovernmental organizations. Many of the conflicts of the colonial era replayed themselves.

The rapid spread of HIV/AIDS in sub-Saharan Africa has proved only the most recent of the iniquities visited upon the people of the continent. By the turn of the century, AIDS mortality had dramatically reduced life expectancy in many African countries. While fertility rates for women remained high compared to much of the rest of the world, the pandemic sparked fears of population decline and stagnant economies because the working-age population has been struck hardest by the disease. Although African peoples and African environments have proved more resilient than the degradation narratives of the recent past have claimed, serious threats to the long-term health of African environments and people remain. In the 1990s, agricultural production and productivity in much of the continent increased, and much of the population growth occurred in urban areas. Most African countries continue to be extremely poor and as a result have weak state institutions. In these contexts, protecting the environment can be perceived as an unaffordable luxury. Threats range from exploitation of resources such as oil or minerals by multinational firms with little regard to environmental impact to uncontrolled clearing of forests for timber and/or agricultural land to cultivating or grazing in national parks by impoverished farmers with no other resources.

FROM HUNTING TO WILDLIFE CONSERVATION

One of the most abiding images of Africa in non-African perception remains that of wild Africa, teeming with wildlife. The image dates back to antiquity with the use of African animals in Roman circuses and Hannibal's use of elephants in his attack on Rome. It stretches across the globe through the Muslim world and on to China. In the nineteenth century, Europeans avidly read the works of explorers like David Livingstone and Henry Stanley and of hunters like F. C. Selous and R. G. G. Cumming. In the twentieth century, Theodore Roosevelt's African safari in Kenya and German East Africa in 1909 (Roosevelt 1910; Steinhart 2005; MacKenzie 1989) sparked interest in an imaginative genre of literature that included the literary works of Ryder Haggard and Edgar Rice Burroughs. From Tarzan to Disney's animated feature and television series *Lion King,* the image of wild Africa has struck a responsive cord in non-Africans.

In the last several decades, the image of Africa as a hunting ground has declined in favor of an image of wild Africa at risk. While conservation efforts in parts of Africa date to the late nineteenth century, in the 1950s concern over the threat to wildlife in Africa became an international movement. The 1959 film, *Serengeti Must Not Die,* made by the head of the Frankfurt Zoo, Dr. Bernhard Grzimek, made wildlife conservation a part of the growing environmental move-

Theodore Roosevelt standing next to an elephant killed during his safari to Kenya, ca. 1909. (Library of Congress)

ment in the West just as much of Africa became independent. Since then, a string of popular entertainment works, drawing in some cases on the work of widely renowned scientists such as Diane Fossey and Jane Goodall, have fronted a concentrated effort by nongovernmental organizations to promote wildlife conservation. The pressures these groups have brought to bear, and the money they have raised and that is raised from tourism, have convinced many African governments to continue to expand colonial-era efforts at conservation (Adams and McShane 1996).

The results have been spectacular in some areas. Several African nations have set large amounts of their territory off limits to economic activities outside of tourism. Tanzania, Kenya, Togo, South Africa, Zimbabwe, Zambia, Botswana, and several other countries all have allocated over 10 percent of their total area to national parks or reserves. In parts of eastern and especially southern Africa, wildlife populations of many species are larger and more stable than they were 100 or 50 years ago. The growth of population and the spread of cultivation to feed that population has restricted range in some areas and created what some scholars call "fortress conservation," which uses military-style force to protect wildlife. On the negative side, some species, especially rhinoceroses and in some areas elephants, subject to hunting for their horns and tusks, have come under threat. In Central and West Africa, animal populations face increasing stress as a result of forest clearance for agriculture and hunting. "Bush meat," especially in forested regions, has become a major commodity available in markets. In general, however, conservation is an established practice, and the amount of land in reserves of various kinds is expanding.

Even so, conservation remains a contested idea within African nations. In times of crisis, such as the recurrent civil wars in the Democratic Republic of the Congo (formerly Zaire) or in Uganda before 1987, conservation generally becomes one of the first casualties of war. Part of the reason is that in many cases local populations view wildlife conservation in the same way that they saw colonial rule: as an oppressive denial of resources, rights, and heritage. The total segregation between people and wilderness imposed by modern conservation denies local people access to land and to important religious and historical sites. The beneficiaries are seen to be governments and their officials as well as the foreigners who visit the parks. The roots of this conflict lie in the origins of the colonial era's conservation efforts (Brockington 2002; Neumann 1998; Carruthers 1995).

Before the twentieth century, African societies lived side by side with game. As noted in earlier chapters, disease, especially trypanosomiasis, carried from wild animals to humans and livestock by the tsetse fly, had helped limit the areas available to dense human settlement and had given Africa the largest population of big game animals in the world. Human societies lived in a patchwork with areas infested by tsetse and harboring large animal populations. The animals could make forays into areas free of tsetse as humans could go into tsetse areas to hunt. Humans could and did push back this frontier through intensive clearing, burning, and grazing of domestic animals. The boundaries changed as human societies expanded or diminished the area they could manage to keep clear.

Two factors in the nineteenth century served to upset this "mobile environmental equilibrium." First, extensive hunting driven by European settlement in southern Africa cleared large areas of the southern part of the continent of its animal populations. Second, at the end of the century, the great rinderpest epidemic that killed a substantial proportion of the cattle in the continent forced the abandonment of large areas by herders and farmers and the rapid generation to tsetse-bearing bush. This expansion of tsetse-bearing bush coming at the beginning of the colonial era then allowed colonial regimes to develop the idea of segregated game reserves that contained no permanent human population (Adams and McShane 1996; McCann 1999b).

Colonial concern over the destruction of wildlife in Africa developed by 1900 just as colonial expansion put wildlife populations at risk. Support for conservation came first from European hunters. Africans had always hunted available game. In some areas, specialists like the BaAka of the Central African rainforest, the San of southern Africa, and the Dorobo of East Africa lived among agricultural and pastoral peoples and traded the fruits of the hunt with them. In many societies, hunting remained an important source of meat as well as other goods such as hides and bone. Ivory served as a major export for centuries. Hunting methods varied and included traps, pitfalls (deep pits filled with sharp stakes), driving game into enclosed killing areas, and poison as well as bow and arrows and spears. However, clearing for agriculture and burning pastures, especially in arid regions like the Sudan and the East African savannahs, represented the most important human check on animal populations (MacKenzie 1989).

In the nineteenth century, as trade between Africa and the rest of the world intensified and diversified, demand for ivory especially expanded. In East and West Africa, this demand led to more intensive hunting and the decline of elephant populations. In South Africa, this demand, coupled with an expanding area under white domination, almost totally destroyed animal populations. Boers and Griqua (Afrikaner-speaking "mixed-race" peoples) hunted extensively using horses. They not only exported ivory but also used hides for a variety of items. After the Great Trek of the 1830s, hunting provided a subsidy to white expansion. It provided meat as well as income, and Boers used the meat to attract Africans to their farms in the Orange Free State and Transvaal as laborers (Beinart and Coates 1995).

Hunting, by both settlers and their offshoots and by Africans, became more efficient as the century wore on. In much of South Africa, whites, Griqua, and Africans all used horses in hunting. The development of breech-loading and repeating rifles greatly increased the firepower in the hands of hunters. Similarly, improvements in range and accuracy increased the kill ratio in hunting. Rifles made their way into African hands relatively easily, although generally they were older models than were available to Europeans. Traders offered rifles in return for tusks and hides. After about 1870, the deadliness of guns had become one of the main causes of the decline of wildlife.

Although the Dutch East India Company government had provided some rudimentary regulation of hunting as early as the seventeenth century in the Cape Colony, by the 1820s the decline of game in the Colony led the British colonial government to issue the first game law in 1822. This law provided for a closed season for hunting, protected some animals by requiring a special license,

Springbok in the Kalahari Gemsbok National Park, South Africa. (Michael and Patricia Fogden/Corbis)

and provided rewards for the killing of vermin such as leopards and wild pigs because of the damage they caused to crops and domestic lifestock. Africans, servants, and slaves could not hunt except under the direction of whites. Landowners retained the right to shoot any animals on their land at any time in the name of protecting their property. The Boer republics enacted similar laws in the 1850s. The Cape law had little effect. By the 1880s, hardly any game remained in the Colony proper (MacKenzie 1989).

A new round of legislation followed the enactment of a new law in the Cape in 1886 to protect wildlife. The colonial government then extended game regulation to African areas such as Griqualand and the Transkei. Natal adopted regulations starting in 1890. The Transvaal created its first reserves in 1889, and Natal followed shortly thereafter, but in many respects these laws proved too late. The year 1896 marked the last great migration of springbok from the highveld in the north down toward the Cape. The migration had always been a time for farmers and hunters to raise their kill counts. That year, the springbok moved south around the area of Kimberley in a mass about 3 miles wide and 15 miles long.

Hunters lined the way in an effort to prevent the mass from reaching cultivated areas. Hundreds of thousands were killed, and only a few made it across the Orange River. None made it all the way back to the highveld, killed by continued hunting and the great rinderpest epidemic. Springbok remained in the Cape only on farms as almost tamed animals, and sometimes they were raised for sport hunting (MacKenzie 1989).

North of the Limpopo, the hunting "frontier" of Europeans overlapped with that which formed in response to the demand for ivory from both the west coast and the east coast. Intensive hunting followed the Pioneer Column that the British South Africa Company sent north to establish Rhodesia as early settlers practiced "asset stripping" without concern for sustainability to subsidize settlement. The same intensification of hunting occurred in the German and British areas of East Africa following the establishment of colonial claims. In these areas, European hunters/traders competed with hunters and traders tied to older networks. However, the presence of tsetse and trypanosomiasis limited the use of horses and provided game populations with refuges from the slaughter (Steinhart 2005).

By the 1890s, the obvious decline of game in places like South Africa had led sport hunters to spearhead a drive to protect African wildlife. The movement had important supporters among the elites of Britain and Germany who came from a sport-hunting tradition at home. On the ground in eastern and southern Africa, calls for conservation usually included demands for an outright ban on hunting by Africans. Sportsmen decried African hunting techniques as cruel and wasteful, and they claimed that hunting by Africans had led to the rapid disappearance of game. Some European observers, such as Hermann von Wissmann, a longtime official in German East Africa and the main force behind conservation measures in that colony, noted that European hunters and traders played a large role in the decline of African wildlife. The main thrust behind their stance was Europe's aristocratic hunting traditions, which reserved hunting for the elite and prevented the lower orders from exploiting the resources around them. Increasingly, calls were heard to establish reserves that would exclude human habitation or exploitation except for licensed hunting. German East Africa, Natal, and the Transvaal established the first reserves in the 1890s. These reserves did not necessarily mean the immediate removal of humans, but efforts to prevent hunting by unlicensed persons gradually made them effective. In German East Africa, the rinderpest epidemic emptied the Serengeti of Masai herders, and in its aftermath the German government declared the first large game reserve in East Africa. All these efforts, however, maintained the right of white landowners to kill animals on their property. In theory, Africans had to rely on colonial officials to protect their herds and fields from animals, although in practice in many areas

Africans continued to hunt for these same reasons with official approval (MacKenzie 1989; Neumann 1998).

Other elements of the emerging colonial regimes opposed game protection and especially reserves. Settlers (and in some places missionaries claiming to speak for African interests) pointed out that animals carried sleeping sickness and that reserves served as special foci for the disease. Critics also noted that game could destroy both fields and herds of both settlers and Africans. A few even noted the importance of hunting in the subsistence strategies of many African communities. However, conservation had powerful allies in the elites of Britain and Germany (MacKenzie 1989).

German and British officials called a joint conference in 1900 in London dedicated to preserving African wildlife. The convention arising from the conference called on colonial powers to enact legislation to protect wildlife. Among the measures suggested were bans on the killing of young females and closed seasons to allow populations to grow, hunting by license only, bans on the use of traps, snares, and poison, prohibiting Africans from easy access to firearms, and the creation of reserves or refuges for wildlife. In the interest of protecting young elephants, the convention mandated that tusks smaller than 5 kilograms (about 11 pounds) could not be sold. The convention also included a schedule of animals to be accorded different levels of protection, extending from the highest for animals deemed endangered to the lowest for animals deemed "vermin" and a threat to humans. Listed as vermin were lions, hyenas, baboons, crocodiles, and poisonous snakes. Although most countries never ratified the convention, the British and Germans in particular incorporated many of its goals in the legislation for their colonies. In West Africa, British colonies enacted game controls, even though they had little in the way of settlers or safari hunting and continued to allow hunting for subsistence and protection by Africans. Other colonial powers moved more slowly. The French generally did not enact game reserves until the 1920s and 1930s and then usually as an adjunct to forest preserves. The Portuguese and Italians enacted some legislation before World War I; in fact, the Italians cooperated with Britain and the independent government of Ethiopia to control ivory smuggling in the Horn of Africa. The Belgians also lagged behind but in the 1920s began to pursue game protection energetically (MacKenzie 1989).

In 1903, the Society for the Preservation of the Fauna of the Empire (SPFE; also occasionally called the Society for the Preservation of Wild Fauna of the Empire) organized as a lobby both for protection of wildlife and for preservation of hunting rights in the British Empire. Throughout the colonial era, this organization played a critical role in developing the drive first for reserves and then for national parks on the American model. It was a powerful lobby group, including such luminaries as the colonial governor Sir Harry Johnston and famed hunter and writer F. C. Selous as well as a number of members of Parliament and the aristocracy.

Up to World War I, the main effect of the conservation action was the gradual establishment of reserves. German East Africa had eleven reserves by 1914, for example, whereas neighboring Kenya had greatly extended its reserves. These continued to meet resistance from both local populations and settlers. Calls to rescind the reserves were frequently sounded, and Africans continued to hunt when and where they could. The war brought to a halt much conservation effort, but it was renewed after the war, with other colonial nations joining in the effort. The Belgians created the first national park in Africa in the Albert (later renamed Virungu) National Park developed to protect mountain gorillas in the eastern Congo bordering Uganda in 1928. The French in West Africa passed their first law allowing for the creation of game reserves in 1925 (MacKenzie 1989).

Colonial regimes also pursued policies designed to protect and promote forests; however, from the first, colonial regimes were motivated by openly economic as well as conservation concerns. In forest areas of West and Central Africa, the period from the 1890s up to about World War I saw the exploitation of forest resources such as rubber and timber on an extensive scale. Forestry policy concentrated on trying to make these systems more productive. Foresters experimented with the introduction of exotic species, including eucalyptus and black wattle, both of which became widespread because of their rapid rate of growth. After the collapse of rubber prices following the development of rubber plantations in Southeast Asia (and to a lesser extent Liberia), forestry policy shifted toward a conflict between the promotion of agriculture and the protection of valuable wood species. By the 1920s, most colonies had adopted forest regulations that sought to control access to forests. The goals of these policies included protection of forests from burning in forest/savannah regions, protection of forests on hill and mountain slopes, preservation of economic forests, and the granting of concessions to foreign firms to exploit forests. Most moved to systems of licensing for the harvesting of trees even by local populations. For example, in southern Tanganyika the British in the 1920s waged a long campaign against the cutting of large trees for canoe building along the Rufiji River. Eventually, forest reserves followed the policies of game reserves and sought to promote conservation through the total segregation of local populations from forests. This policy denied local populations access to fuel wood, hunting and fishing grounds, and new agricultural grounds. The hardships this policy caused led to both resistance and to circumvention of the rules (Fairhead and Leach 1996a, 1998).

Despite the economic crisis of the 1930s, conservation efforts of all sorts increased dramatically. Colonial concern extended from conservation of forests

Aerial view of eucalyptus forest in Karkloof Valley, KwaZulu/Natal, Republic of South Africa. (Peter Johnson/Corbis)

and wildlife habitat to conservation of the productive capacity of fields and pastures. Forest conservation expanded from West Africa, whereas wildlife conservation moved from East Africa. South Africa pioneered efforts to promote soil and pasture conservation. Gradually, a new rhetoric took hold among colonial administrators that emphasized "development" as the goal of policy. Implicit in this new vision was the idea of sustainability as a goal (Anderson 1984; Anderson and Grove 1987).

Colonial planning for development and conservation grew out of an explicit commitment to the superiority of Western science compared with the experiential knowledge of Africans (and even of settlers). Much of the effort came from the British Imperial College of Tropical Agriculture founded in Trinidad and Tobago in 1928. It trained much of the higher staff of the agricultural departments of the British colonies. The first full expression of this conceit came in South Africa where agricultural scientists were influenced by the development of scientific agricultural and pastoral research in the United States and Australia. In the 1930s, South Africans visited the United States and toured the Dust Bowl and agricultural research universities. Agricultural officials from other colonies visited South Africa as well as the United States (Tilley 2004; Conte 2004).

In the 1930s, cooperation between the colonial powers with regards to conservation increased. The wildlife conservation lobby, still dominated in Britain by hunters but increasingly drawing support in scientific circles, helped lead to the calling of a series of conventions to strengthen wildlife protection. In 1931, an International Conference for the Protection of Nature was held in Paris, and two years later, the conference reconvened to consider a British-proposed Convention for the Protection of African Fauna and Flora. The resulting convention called for strengthening the provisions of the 1900 convention and in particular creating "national parks" as permanent refuges for wildlife (MacKenzie 1989).

The general effect was the development of segregation as the major means of wildlife and forest protection. Coupled with the creation of reserves and from the 1930s national parks, many African colonies also created game departments whose jobs consisted mainly of hunting game in the name of protecting crops and animals. In Tanganyika and Uganda, for example, the game departments maintained an avid shooting culture even as reserves and national parks became established. Similarly, enforcement of game laws occurred mainly in reserves and national parks. Outside those areas, settlers retained the right to shoot almost any animal that crossed onto their land. Limited budgets meant that enforcement of hunting restrictions on both Africans and Europeans in remote areas, or even in areas bordering reserves, remained haphazard. As a result, game populations expanded inside the reserves while they declined outside them (Adams and McShane 1996).

Colonial efforts to promote development had the same contradictory impact. These efforts in the 1930s reflected growing concern over the sustainability of African agricultural systems. By this decade, population had begun to grow in almost all parts of the continent. In the late nineteenth and early twentieth centuries, outside and colonial observers often remarked on the crisis African populations seemed to face. They ascribed the causes to the slave trade in eastern Africa, the violence of conflict and resistance throughout the continent, and the spread of diseases, including smallpox, cholera, malaria, and sleeping sickness during the 1890s. In southern and eastern Africa, settlers and mine operators feared for labor supplies in the face of declining populations. The West African population seems to have faced less of a crisis around the beginning of the colonial era, although in general no observers described it as expanding rapidly.

By the 1930s, populations had begun to recover. Colonial observers ascribed the recovery to the imperial peace, the expansion of markets, and the introduction of scientific medicine and public health measures. While all three factors played a role in the rise of African populations, they do not totally explain the

Troops in German Southwest Africa (now Namibia) at the time of the Herero Rebellion of 1904. (Getty Images)

rapid population growth. The reduction in violence only became generalized across Africa after the end of World War I. In the first decade of the twentieth century, several major uprisings took place against colonial rule such as the Maji Maji revolt in German East Africa and the Herero Rebellion in Southwest Africa. The Germans suppressed the Herero Rebellion ruthlessly. In fact, the depopulation of large parts of southern Tanzania dating from this period helped ease the establishment of the Ruaha game reserve, which has become the Selous National Park, the largest in the world. World War I also brought fighting to several parts of the continent, including especially the German East Africa Campaign, which left famine in its aftermath. In the central region of that colony, as much as a third of the population died in the period between 1916 and 1920. Similarly, the Chilembwe uprising in Nyasaland led to severe repression by the British (Cordell and Gregory 1994).

The extension of markets under colonialism was especially driven by the development of a transportation infrastructure and brought both benefits and costs. The availability of railroads and steamboats and, after 1920, motorcars all meant that food could be moved around efficiently and that African producers potentially had markets for their produce that would give them the cash income to pay for food imports. These improvements made the movement of people to find employment much easier. Scholars have noted that the intensity of food shortages and famines declined after 1920 (Iliffe 1987). At the same time, the greater circulation of people meant a more rapid spread of infectious disease. African migrant laborers proved particularly at risk. They often moved into new disease environments and so suffered from diseases, including malaria. Living in close quarters resulted in higher rates of infectious diseases such as tuberculosis. The expansion of colonial towns and cities saw similar results. Like pre-industrial cities the world over, they became *consumers*, with their populations maintained only by continuing immigration (McCann 1999b).

Western medicine and public health measures took hold very slowly and in many cases with little effect before World War II. Governments provided only limited access to health care. Christian missions offered significant amounts of health care in some localities, but lack of funds limited their reach. Non-Christians also found it difficult to receive care from some missions. Some scholars argue that population began to grow in Africa after World War I not because of a decline in mortality rates by itself, but because of an increase in fertility rates (Cordell and Gregory 1994). African demographic regimes before the colonial era varied according to disease and environmental conditions. In areas of dense population or subject to subsistence crisis caused by regular drought, African populations seem to have moderated fertility through a variety of measures, including delayed marriage and birth spacing through extended breast feeding. Although mortality declined in some regions of Africa, the main cause of increased population growth seems to have been increased demand for African labor. African families eventually responded by increasing fertility. They began to produce enough of a surplus to pay taxes and to cover the cost of an increasing range of consumer goods and services, including education and health care. Given very low use of capital goods, growth had to come through increasing labor and land usage. As a result, population growth rates accelerated beginning in the 1920s and continued to grow into the postcolonial era (Koponen 1986).

Increased population, coupled in many cases by reduced access to land because of settler alienation and the creation of reserves, caused African farmers to intensify production and reduce fallow. By the 1920s, these issues came to dominate the thinking about African agriculture throughout the colonial empires. Colonial scientific establishments saw in these efforts the threat of degradation through overexploitation. In the forested regions of West Africa, they became increasingly concerned with forest loss and the expansion of what they perceived to be derived savannah. They responded with the first attempts to promote conservation measures (Tilley 2003; Fairhead and Leach 1996a, 1998; Conte 2004; Beinart 1984).

These measures met some success in settler/agricultural areas. In those areas, especially South Africa, Southern Rhodesia, and Kenya, the state not only provided expertise to increase conservation efforts but also directly and indirectly subsidized these measures. The metropolitan government set up a Colonial Development Fund in 1929 to provide grants to colonial governments. African agricultural areas, however, received less in the way of investment in conservation. In those areas, colonial officials expected Africans to pay both directly and indirectly for the implementation of conservation. Both the British and the French began to develop such plans in the 1930s. In French West Africa, the French government established the *Office du Niger* in 1932 to develop irrigation along the Niger River in what is today Mali to produce cotton and rice. In this case, the French sought to recruit African families to take up irrigated plots under tightly controlled terms of production. Despite significant financing, the scheme relied on forced labor and eventually only resulted in the irrigation of one-tenth of its projected total (van Beusekom 2002).

In British Africa, a mania for conservation emerged during the 1930s. By the 1930s, the British had developed a number of institutions designed to promote efficient use of resources in its imperial possessions. Some of these included institutes designed to train foresters, agricultural instructors and researchers, doctors, and a host of other scientists. These institutes began to create a cadre of officials with knowledge about the latest advances in their fields and supposedly trained to apply this knowledge in a practical manner (Anderson 1984). This initiative created an interesting effect on policy in the British Empire. Through the Africa Survey project an attempt was made to create a systematic account of the state of their African colonies. The project, which led to the publication of *An African Survey* by William (Lord) Hailey in 1938, attempted to gather together information on many aspects of the African colonial world, including environmental conditions. The process of producing the volume drew on the expertise of hundreds of scientists, scholars, and colonial officials. It and other publications from the project provided an institutionalized base for development policy after the war (Tilley 2003).

The systematic approach to knowledge moved policy on a variety of issues out of the range of knee-jerk superiority toward Africans common among an older generation of colonial officials. Researchers discovered the effectiveness of African agricultural practices and the limits imposed by environmental conditions to change. For example, researchers in Tanganyika in the 1930s finally discovered that acid soils limited the ability of farmers in some areas such as the Usambaras and the Southern Highlands to grow coffee despite a climate well suited for the crop and similar to that of the major coffee-producing region of Kilimanjaro (Conte 2004). This kind of attitude had always marked some Europeans in Africa even in areas such as South Africa and Rhodesia, but it took hold most strongly in colonies without large settler populations, including all of the British colonies in West Africa and in parts of East and Central Africa.

At the same time, this systematic approach to generating knowledge created a different type of hubris towards Africans and African environments. Particularly as the French and Belgians created their own institutions of colonial knowledge, they generated a belief in the efficacy of scientific and technical solutions to wildlife and forest conservation, soil conservation, disease control, and a host of other issues. This culture of expertise would lead to the creation of a different kind of hierarchical approach to the new development initiatives (Van Beusekom 2002; Fairhead and Leach 1998).

World War II put a hold on development initiatives. The British passed the Colonial Development Act in 1940, which required development plans for each colony and promised a degree of metropolitan support for those plans, but these plans had to await the end of the war for implementation. The war years showed how fragile development remained in African colonies and the costs of growing dependence for Africans within colonial economies. Not as much fighting occurred in sub-Saharan Africa during World War II as in World War I, but the effects of the war on African peoples and landscapes were significant all the same. The West African Sudan suffered a series of food shortages as imports of both food and equipment were cut off. East Africa's massive food shortages during the war turned into famines in Tanganyika and Kenya when lack of transportation and wartime demand limited the ability of food imports to make up for production shortfalls.

AFRICAN ENVIRONMENTS AND THE CRISIS OF COLONIALISM

The end of World War II heralded a "second colonial occupation" in the words of John Iliffe (1979) just as African peoples learned to organize themselves to fight against colonial rule. Much of the new energy that Britain and France in particular brought to their colonial empires took the form of efforts to create an environmental policy. Efforts to protect natural resources, to expand production, to promote soil conservation, and to control disease all intensified with new investment from the British, French, Belgians, Portuguese, and even in settler-dominated South Africa. In the immediate aftermath of the war, the colonial regimes also needed to increase production to help support the costs of domestic reconstruction. Many colonial economies grew rapidly in the two decades after the war as demand for their agricultural and mineral production remained strong. Yet the colonial governments' increased interference in the day-to-day life of Africans became the immediate cause for the rejection of colonial rule. Whether in the creation of Apartheid in South Africa with its more oppressive segregation or new rules requiring farmers to build conservation works in their fields or the expulsion of herders or farmers from an area newly declared a game reserve, all helped fuel anticolonial nationalism.

In addition, urbanization and even some forms of industrialization began to accelerate during the late colonial era. South Africa diversified its economy into manufacturing rapidly after World War II, and it remains the most industrialized economy on the continent. Mining expanded greatly in Central Africa and consumer-goods processing increased in many colonies. This development drew increasing proportions of the population of African colonies to cities and led to increased pressures to increase productivity in agriculture to feed this growing population.

In South Africa, the institution of Apartheid developed during the postwar era. An Afrikaner nationalist government was elected in 1947 and sought to divide South Africa into homelands for each of its constituent peoples. Whites—whether Afrikaner, British, or other European immigrant—would control over 80 percent of the territory of the country, including almost all its valuable assets. Africans lost South African citizenship and became citizens of "tribal homelands." Theoretically, they had to have a job to live in "white" areas; otherwise, they had to return to the homelands. "Black spots," black-owned land in white areas, were to be removed. This policy of removal, even if never completely enforced, created even more overcrowding in the reserves. The Apartheid government responded to the potential for severe environmental degradation in the reserves with a program it called "betterment." These policies provided a modicum of investment in equipment and expertise to help develop conservation efforts. However, in this case, the efforts tended to support the efforts of local authorities cooperating with the white government and often led to the increasing alienation of rural people from the state. The degradation resulting from this policy was extreme. The homelands became some of the most degraded landscapes in all of Africa, serving basically as rural slums. The white vision of self-sufficient peasants supporting themselves while a few young men became migrant laborers no longer had a basis in reality. Whereas some people continued to farm and raise stock in the reserves, many could not raise substantial amounts of food and remained dependent on wages earned in white South Africa by migrants. From the 1940s the settler-dominated government of Southern Rhodesia followed a similar policy, although the percentage of land in the country retained by Africans as "tribal lands" made up about 50 percent of the country (Beinart and Coates 1995; Jacobs 2003).

The environmental damage done to the reserves of South Africa (and to those of Rhodesia and Kenya) by the policy of land alienation was much like that done by the policy of reserving land from human use through game parks in East Africa and forest reserves throughout the continent. In the aftermath of World War II, a growing population and a growing demand for African commodities put the land under increasing pressure. Land outside the reserves filled up. In Kenya, African herders constantly took their cattle into land reserved for white farms but were often unoccupied. In South Africa, despite efforts to end the practice in the name of enforcing wage labor, Africans continued to live as "squatters" on white-owned land, paying rent through labor. Throughout the continent, people sought to graze animals, to collect firewood, and even to farm land to which colonial states declared they had no right (Adams and McShane 1996).

This mentality that only whites could productively develop natural resources did not extend to the entire continent, but the notion that African producers needed to undertake a radical transformation of their farming and herding practices was pervasive. In large parts of West and Central Africa, the focus remained on the development of smallholder production and natural resource extraction. Even before World War II, the British government turned to a developmentalist approach toward its colonies. To replace and improve on the 1929 Colonial Development Fund, in 1940 the British imperial government created a new Colonial Welfare and Development Fund to promote economic growth and social welfare; implementation had to await the end of the war. The French government created its own development agencies in the *Fonds d'investissement pour le développement économique et social* (FIDES) in 1946 and the *Fonds d'équipement et de recherché pour le développement et social* (FERDES) in 1948. The institutions of both empires had a wide charge to promote positive change, and so both invested in social welfare and in economic development programs. Both empires sought to utilize scientific knowledge and expertise to promote sustainable development. They took different approaches, however, as did the other colonial powers, the Belgians, Portuguese, and Italians. The French promoted centralized control over their colonies in particular, whereas the British developed a more decentralized structure for their development programs. However, both worked within an evolutionary framework that saw African societies as potentially advancing to a more modern state of affairs. This modernizing vision shaped policies toward all aspects of colonial development, including environmental policy (Fairhead and Leach 1998). The nationalist opponents of colonial rule that emerged to contest continued domination adopted much of its rhetoric and policies by the 1960s.

The dramatic rise in demand for African commodities, whether mineral or agricultural, in the aftermath of World War II provided the background to the development programs of the late empire. Prices for many commodities rose sharply. African producers responded by increasing production of crops like coffee, tea, cocoa, cotton, and tobacco. In West Africa, African farmers expanded their production. In Kenya, Rhodesia, Nyasaland, Mozambique, and Angola, European immigrants took up newly alienated land to convert to highly capitalized farming. The rapid growth in the production of copper in the Belgian Congo and Northern Rhodesia, and eventually oil in Nigeria, Angola, and Gabon, meant increasing urbanization and greater demands for food. As this increased production occurred, concern over the sustainability of the increase in production grew.

Late colonial development projects took three important forms. Colonial regimes invested in infrastructure, especially roads. Although the transportation networks remained mostly export oriented, road construction made commodity production more feasible in many parts of the continent. The colonial governments also invested, with very little success, in large-scale production. Finally, they created development programs focused on specific regions that included efforts to promote sustainable resource use. These programs, which brought colonial intervention down to the level of the village and farmer, generated a great deal of resistance, even if, as in some cases, they promoted technology transfer.

By the 1950s, almost every region in Africa saw some level of government effort to promote soil conservation. Most British colonies appointed a "soil conservation officer" to oversee the development of conservation plans and their implementation by district officials, agricultural officers, and "native authority" chiefs. These efforts included the promotion of terraces of various sorts on sloped land, extension of forest reservations on hilltops, rotational cropping and grazing schemes, and in many areas, resettlement out of overcrowded areas into less densely populated areas. These efforts resulted from a perceived decline in forest cover, soil productivity, and grazing facing African rural communities. In many cases, they represented a strikingly myopic vision by colonial observers. James McCann (1997) has recounted for Ethiopia (although only under colonial rule for a few years) how European observers from the mid-nineteenth century on commented both on the lack of tree cover in the most densely settled highlands and then proceeded to claim that the losses had occurred only in the last few decades. They only briefly commented on the development of stands of trees developed by indigenous institutions within Ethiopia to provide fuel and building materials. In short, observers projected an "Edenic" past onto the landscape and ignored evidence of landscape manipulation by local populations that were of long standing. Colonial officials, officials of the technical services, and colonial scientists repeated this reasoning across the continent.

The result from the 1930s on was a mania for conservation driven by claims of desertification and permanent degradation. The rapid increase in population with its accompanying increase in cultivation and grazing provided what seemed the final evidence for the "degradation narrative" outside experts developed. The colonial state's authoritative and top-down approach for the most part failed miserably, even though throughout the continent African farmers and herders adopted a huge number of new crops and techniques throughout the colonial era—sometimes with the assistance of colonial agents and sometimes totally outside the sight of those agents. Colonial efforts generally took the form of "rules" that local courts had to enforce. These courts, of course, were staffed by local "chiefs" in the British case and appointees in the French case, and agricultural instructors were almost all African. The procedures required often brought little direct benefit to farmers. They often necessitated huge amounts of extra labor and sometimes interfered with long-established productive systems. In many areas, colonial governments sought to ban or sharply control burning in the name of preventing deforestation. Perhaps the most intensively developed case study of these efforts comes from Malawi (colonial Nyasaland). In that colony, farmers practiced a form of burning for cultivation called *citimene* in the southern Shire Valley and *visoso* in the north. Farmers cut bush and made piles in the fields just before the rains began. They burned the dried piles of brush, and the rains washed nutrients back into the mounds left by the brush. Colonial officials saw these practices as leading to deforestation and desertification and sought to ban or limit them. They also tried to impose "strip" cultivation where farmers planted in alternate strips of land every year. The British required that the strips be marked by rows of grass. A number of tests carried out by local officials showed that these methods led to less production and much more labor. Yet the colonial government continued to enforce them in the name of conserving trees and soil. They used the local courts to impose fines, require supervised labor on the workers, and even jail resisters. By the 1950s, opposition to these rules had spread throughout the country, and the nationalist movement led by Dr. Hastings Banda used this discontent as a major organizing tool in the struggle for independence (Mandala 1990; McCracken 2003; Vaughan 1987).

Such resistance occurred throughout Africa. During the MauMau uprising in colonial Kenya in the early 1950s, conservation rules requiring the building of terraces in the increasingly overcrowded "Kikuyu reserves" helped build support for the uprising. In Buganda in Uganda, soil conservation rules also sparked resistance. Throughout the West African forests, efforts to prevent people from cutting wood drove them to increasingly ignore local African authorities and spurred corruption.

Perhaps no set of colonial conservation policies created more resistance than those designed to protect grazing lands. From South Africa to West Africa, colonial officials contended that African herders kept too many cattle and caused degradation and desertification through overgrazing. As happened with so many colonial concerns, the issue first surfaced in South Africa. In the aftermath of the rinderpest outbreak of the 1890s, both colonial officials and African herders had sought to rebuild herds rapidly. In areas that kept stock, colonial governments sought to provide veterinary services to prevent diseases. Colonial officials, however, viewed these herds as potential sources of cash for their owners and revenue for the colonial state. Herders often looked at them very differently; for them, herds represented subsistence. Herders sought to hold as many cattle as possible and to share the risk in cattle keeping from disease and drought by reciprocal cattle pawning. Herders generally took advantage of markets for both stock and for hides when they needed cash, but they did not raise cattle as a commodity to be sold every year. Hence, colonial officials saw overgrazed land and too many head of livestock as one of the great causes of degradation in the drylands of Africa (Mortimore 1988).

Throughout Africa, from the 1920s on, colonial governments sought to impose rotational grazing and stock reduction programs in many parts of Africa. In the late 1930s, the South African government killed thousands of head of cattle in the Transvaal in an effort to control disease. In the 1940s, such efforts spread throughout Africa. In Kenya and Tanganyika, colonial governments tried to force Africans to sell stock starting in the 1940s. In the 1950s, the government of Tanganyika tried to impose a mandatory grazing plan on the Kondoa Highlands in the central part of the country. This highland "island" suffered from what officials called severe erosion, and they attempted to almost completely remove stock from the area (Östburg 1986). As happened in many cases, the effort in the 1950s led to rapidly rising support for anticolonial political movements.

Imperial and colonial governments also attempted to create major development projects after World War II. These efforts generally resulted in monumental failures despite the large amounts of capital invested in them. The French, for example, made a major effort after World War II to complete the *Office du Niger* project, which would provide irrigation along the Niger River. The British created the "Groundnut Scheme" in several of their colonies. In the French case, the plan failed because it did not take into account the difficulty associated with the hydrology of the Niger River (van Beusekom 2002). In the British case, the imperial government planned to use surplus military equipment to begin large-scale farming of groundnuts (peanuts) to meet a perceived shortage in edible oils. They picked several areas in Tanganyika and Nigeria, chosen mostly because they lacked dense populations in areas where local farmers produced some groundnuts. In Tanganyika, the result was spectacular failure. They chose two areas on which to concentrate, one in the south of the territory and one along the Central Railway. In both cases, the machinery proved totally inappropriate for the conditions. In the south, the lack of transportation meant they tried to build a railway that took up most of the funds for the project. In Kongwa in central Tanzania the soil dried so hard when it was cleared that plows could not break it and harvesters could not harvest the crop. The managers resorted to "hiring" labor, often under conditions that bordered on coercion. The collapse of both of these schemes led to the fall of governments in France and Britain (Hogendorn and Scott 1983).

Almost all the African colonies in the 1950s were subjected to a myriad of schemes designed to promote soil conservation and bring "development" to rural areas. Often these schemes involved resettlement away from fertile areas by population deemed surplus in order to promote soil conservation (Hoppe 2003). In some cases, these lower-capital intensive programs achieved a certain measure of success. Two reasons can be identified for the limited successes. First, in some cases, "resettlement" plans resulted in opening up "new" lands that proved reasonably fertile. In East Africa, several resettlement schemes, most notably the Mbulu resettlement plan in northern Tanganyika, resulted in the retreat of tsetse-bearing bush, thereby making lands habitable once again for humans and livestock. In the Mbulu case, the result was a gradual recolonization of the plains around the Mbulu Highlands by agriculturalists (Lawi 1999). In other cases, the expansion of markets or of cash crops brought cooperation with such schemes. In Kenya, farmers in the Machakos and Nyeri Districts gradually adopted the building of bench terraces in order to grow vegetables for the

Nairobi and coffee for export (Tiffen 1996). In Tanganyika, the Uluguru Land Usage Scheme tried to introduce terraces in the Uluguru Mountains around Morogoro. It failed completely and led to a major riot. But in one area, Mgeta, farmers adopted the terraces because they proved successful at producing vegetables for the Dar es Salaam market (Maack 1996). Similarly, across Africa, farmers adopted new breeds of crops that occasionally were developed by colonial agricultural services. In many areas, farmers integrated exotic wood species such as eucalyptus and wattles into their agroforestry practices. In short, development and conservation efforts worked best when they made incremental changes that promoted the well-being of African populations. When they required "more work for little to no benefit" as one official described a district-level development plan, they found very little support (Maddox 1991). When they threatened to cause impoverishment, as de-stocking campaigns did, they brought active resistance.

In the face of the rapidly expanding populations in much of Africa after World War II, the concerns over conservation become obvious. Population in Africa began to grow rapidly because of several factors. Demographers, studying the history of population growth across the globe, developed a theory labeled "demographic transition" that charts the effects of economic growth on population growth. They suggest that as economic development proceeds and as people and societies gain the ability to produce more, whether food or goods to trade for food, population grows. Population growth proceeds until a point is reached where the costs for households of having additional children outweigh the benefits that the children bring. That point is usually reached when schooling becomes both universal and extended. Then population growth rates begin to fall toward replacement levels.

Demographic transition theory starts with observed phenomena: the rapid increase in population growth rates experienced throughout the globe over the last 300 years and the decline in population growth in wealthier countries over the last century. But behind those observed phenomena the reasons for rapid population growth are murkier. Population growth rates are determined by both mortality rates and fertility rates. In Africa in the twentieth century, both changed dramatically. Medical advances played an important role in reducing mortality, especially after World War II. Gradually at least rudimentary elements of modern medicine reached most parts of Africa, with antibiotics and widespread quinine treatment for malaria playing perhaps the biggest roles. Public health measures, particularly in urban areas, also played a part. Population growth and the increase in area cleared for cultivation or grazing reduced the incidence of sleeping sickness. Improved transportation that allowed the movement of food into regions with food production shortfalls played the biggest role in reducing mortality in some areas. These changes would prove reversible in some cases and did not reach every part of the continent evenly. The uneven impact of these improvements points to the major underlying reason for population growth: an increase in fertility. The increase in fertility seems almost universal in Africa and derives from the direct relationship between labor and production. It explains why population has continued to grow even in countries where the other elements, such as health care systems and transportation networks that could bring food to food shortage areas, have collapsed. It explains why population has continued to grow despite the reduction in life expectancy caused by the spread of HIV/AIDS since 1980 (Cordell and Gregory 1994; Iliffe 1987).

This rapid population growth has led to tremendous changes in African landscapes since the 1940s. Africans have put more land under cultivation in one way or another than ever before. They have found ways to increase production and productivity in agriculture that have generally kept pace with population growth, despite the reduction in fallow times for land under cultivation. Observers and officials both noted with alarm the increasing "deforestation" of Africa throughout the continent by the 1950s. It led to increasing efforts to promote nature or wildlife conservation through the total segregation of populations from reserves just as in much of Africa colonial powers were reluctantly ceding political power to African leaders. The creation of "fortress conservation," in Daniel Brockington's phrase, at the very end of the colonial era symbolized the continuing dependence of the newly independent African states on the wealthy nations of the world (Brockington 2002).

AFRICAN ENVIRONMENTS IN THE POST-COLONIAL ERA

Julius Nyerere, the leader of the struggle for independence in Tanganyika and longtime president of Tanzania (created by the union of Tanganyika with Zanzibar in 1963), announced at the independence celebrations for Tanganyika that in combating a food shortage in many parts of the country at the time, the new nation was now fighting "not man but nature" (Iliffe 1979). The image of fighting nature, of trying to ensure survival for African peoples, captures the perception of many people in Africa in facing the string of disasters that have befallen parts of the continent since the 1960s. Famine has reappeared at times in several parts of the continent. Malarial parasites acquired resistance to the most commonly used (and least expensive) form of treatment for the disease. In some parts of the continent, the expansion of reserves seemed to indicate that the world placed a greater value on nature than on the people who depended on African environments for the survival. Droughts, storms, and floods wreaked havoc on

Tanzania president Julius Nyerere, at the African Heads of State meeting in Addis Ababa, Ethiopia, 1963. (Bettmann/Corbis)

communities and infrastructure alike as climate change made the weather even less predictable. HIV/AIDS became an epidemic and reduced life expectancy across the continent. It is understandable why many Africans like Nyerere saw their relationship with nature as a war.

At the same time, Nyerere's comment proved optimistic. Human agency continued to cause many of the problems perceived as environmental in Africa. While drought, floods, and crop failures continued at about the same pace as before 1960, famines occurred not just because of crop failure, but because of human action. Famine, the generalized rise in mortality in a region because of a lack of food, became associated with war. In Nigeria, the civil war of 1967–1970 caused a famine in Biafra, the southeastern region of that country that sought independence. In Ethiopia in the late 1970s, in Sudan in the 1980s, and again in Ethiopia in the early 1990s famine resulted when civil strife combined with a climatic event.

Newly independent African nations tended to maintain similar policies toward environmental resources as colonial governments, despite the occasional populist action in loosening some conservation regulations. This continuity occurred for several reasons. First, the newly independent governments were made up of people who were both trained under colonial rule and accepted the generalized development paradigm that had taken hold in the late colonial period. They believed that the way for Africa to "catch up" was to practice scientific management of resources. Second, large-scale development projects legitimized the new governments in their own eyes and in the eyes of the people. They gave the new rulers patronage to pass around. Third, international aid agencies generally all followed the same scientific orthodoxy. Whether multilateral like the World Bank or United Nations, or bilateral, whether Western or Soviet bloc in the period up to 1989, most promoted the same general policies. Hence, after protests against de-stocking in Kondoa helped mobilize support for independence in Tanzania, the Tanzanian government turned around and in the 1970s required the complete de-stocking of the Kondoa Highlands, and then followed with de-stocking in two other areas of central Tanzania. The colonial state had never been able to muster the authority to carry out such plans. The de-stocking resulted in a dramatic reconfiguration of local communities as their wealth now had to be kept miles away, even as it led to the regrowth of thicker vegetation in the de-stocked areas (Östberg 1986).

Many regions of Africa faced food shortages throughout the period since independence without widespread increase in mortality. The transportation and marketing systems developed during the colonial era continued to be able to bring food into food-shortage regions. The most well-studied drought of the post-colonial era, the mid-1970s drought in the Sahel (as well as in eastern and southern Africa), caused relatively little increased mortality. A series of years with rainfall below the century-long norms brought crop failure, the reduction of cattle herds, and the spread southward of desert conditions throughout the Sahel. It also brought crop failures to many other countries in the continent. Yet it only coincided with famine in those countries undergoing civil conflict.

That drought caused by a generalized, though temporary, reduction in rainfall, became central in an important debate about environmental change in postcolonial Africa. Many experts charged that rapid population growth had caused degradation on fragile environments such as the Sahel. They argued that the re-

Military training recruits during the civil war in Biafra, Nigeria, 1968. (Bettmann/Corbis).

sult was "desertification," the permanent spread of desert, in many parts of the continent. Further south in West Africa as well as in Central and eastern Africa, they saw the spread of savannahs at the expense of forests as a result. However, it turned out that drought occurred not because of the actions of Africans, but because of people outside of Africa. It marked a phase in the rapidly accelerating trend of climate change due to global warming. In general in the past, global warming would bring greater rainfall to much of Africa, and in parts of Africa, the twenty years before 2000 saw an increase in rainfall. However, global warming has also led to an intensification of periodic climatic events such as El Niño, and hence to more extreme swings in climate. As the rains returned (and even in some areas of the Sahel increased above century-long averages) the "desert" retreated and people shifted their fields and cattle northward in the Sahel again (Mortimore and Adams 1999).

More important for the health of Africa's environments as well as its peoples was the link of Africa's economies to the global economy. During the 1960s, the post–World War II economic boom continued, and markets for African commodities remained strong. The worldwide recession that began in the early 1970s struck Africa particularly hard. African economies faced falling prices for commodities and rising prices for imported energy. A few countries, such as Nigeria, Angola, and Gabon, developed oil export sectors, but more than in most cases, these industries developed few linkages with the local economy. The crisis that gripped Africa for over a decade saw dramatic impoverishment, declining investments in environmental protection, and widespread civil strife.

Nigeria became the first major oil-exporting nation in sub-Saharan Africa. As oil prices rose throughout the period after 1973, multinational oil firms signed agreements to explore for oil and gas deposits throughout the continent, especially in offshore areas. By the 1990s, major producers included Angola, Gabon, and Equatorial Guinea. Oil production provided relatively few linkages to other sectors in African economies. Oil revenues were siphoned off into the military, wasteful public spending, and corruption. Oil money (along with proceeds from illicit diamond sales) helped fuel a two-decade-long civil war in Angola and propped up repressive governments in Nigeria and Equatorial Guinea. In addition, oil companies stood accused of causing severe environmental damage in parts of Africa. They generally sought to ensure that environmental protection regulations were weak and that enforcement was minimal. In Nigeria, environmental destruction in the Niger Delta region became one of the major issues surrounding protests against the then military dictatorship led by General Sani Abacha. A writer and activist, Ken Saro-Wiwa of the Ogoni people of the area organized a movement that called for greater local control over the proceeds of oil production, greater local political autonomy, and an end to the environmental destruction. The movement he led—the Movement for the Survival of the Ogoni People (MOSOP)—launched a campaign against the largest producer in the region, Shell Oil, which forced it to abandon production temporarily. The Nigerian government, anxious to keep the spigot of oil money open, arrested Saro-Wiwa on what were generally regarded as false charges and had him executed in 1995. The execution mobilized opposition to the military regime, which finally left power after Abacha's death in 1999. However, the aftermath of the Saro-Wiwo execution also proved the influence of oil interests, both multinationals using money to buy influence and consuming nations putting the security of supplies above other interests. Everyone, including elites in some African governments, put money above the interests of African environments or African economies (Saro-Wiwaand 1995).

The economic crisis of the 1970s and 1980s led to resource mining of sorts in Africa by people and governments desperate to get cash. The "poaching" of wildlife, especially elephants and rhinoceroses, presents perhaps the most extreme version of resource mining. Rhinoceroses have almost totally disappeared from the wild, and for a period in the 1980s, some thought elephants would face the same fate. Poachers used automatic weapons readily available in eastern and southern Africa to decimate herds. The international ban on ivory sale did much to reduce the demand for tusks, although the protection may have come too late for rhinoceroses. Elephant populations, as in the early twentieth century, recovered fairly rapidly, and by the late 1990s, southern African nations were again complaining about having an overpopulation of elephants (Adams and McShane 1996).

An added crisis was the outbreak of the HIV/AIDS epidemic in the early 1980s. Scientists now think that the Human Immunodeficiency Virus spread from chimpanzees in the Central African rainforest to humans. Viruses that make such a jump often prove quite lethal to their new hosts, and as Simian Immune Virus (SIV), transformed itself into HIV, this proved doubly the case. The exact trajectory of the disease is still unclear. The virus may have moved into human populations many times, but the one that "took" because a mutation in the virus allowed it to reproduce in humans probably occurred about seventy years ago. Since then at least three main strains and eleven subtypes of HIV have developed. The virus requires basically a blood to blood transmission. The original jumps came because people ate and continue to eat chimpanzees. In humans, transmission can occur via sexual intercourse, blood transfusions, incompletely cleaned medical equipment, and syringes.

By the time the disease had been identified in Europe and the United States, it had already become an epidemic in Africa. Infection rates seem to have spiked first in Uganda, fueled by a decade of war after the fall of Idi Amin. High rates of infection then came to be recognized in the rest of East and Central Africa. The highest rates of infection have been recorded in southern Africa. Botswana had an estimated infection rate of 38.8 percent. Between one-third and one-quarter of all adults are infected in Lesotho, Swaziland, Zimbabwe, and South Africa (which has the largest number of HIV positive individuals in the world). Kenya, Namibia, and Zambia all have infection rates of between 15 percent and 20 percent. Most other sub-Saharan African countries have infection rates between 5 percent and 10 percent, including some such as Uganda and Tanzania, which have seen significant declines from rates well over 10 percent to under that figure in the last decade (Hunter 2003).

The result has been dramatic, with life expectancy falling in most African countries. Of the 28 million AIDS deaths up to 2003, 26 million were in Africa. Five countries may even see population decline as a result of the epidemic. The effects are compounded by the increase in the incidence of tuberculosis in part because HIV positive individuals are more likely to contract and then spread the disease. Although these figures do not mean that Africa will become depopulated, they do mean that the continent will be poorer and people will suffer more. The resources necessary to create sustainable development for African countries will be harder to come by because so much will be taken up in combating the spread of the disease. Resource mining to cover the costs of the epidemic will increase. This includes the exploitation of minerals, petroleum, and timber as well as, for example, the expansion of the "bush meat" market in Central Africa (Hunter 2003).

The 1990s were years of both hope and despair for Africa's environments and peoples. The promise made by South Africa's relatively peaceful transition to majority rule helped bring hope to a continent wracked by war in Sudan, large parts of West Africa, Burundi, and the Democratic Republic of the Congo and by genocide in Rwanda. Africa, the mother of all humanity, at times seemed particularly harsh to her offspring that remained close to home. But she continued to offer them the promise of a decent life if they cared for her properly. Overall, agricultural production increased during the 1990s in Africa, fueled by a rise in land and labor productivity. African farmers in many parts of the continent adopted "Green Revolution" technologies and increased yields dramatically. These technologies of course meant dependence on the use of chemical fertilizers and even dependence on multinationals for hybrid seeds in some cases. Although population continued to grow, most of the growth took place in urban areas. Fertility rates in parts of Africa also began to fall, indicating that some parts of Africa were fully in the process of demographic transition. These indicators often flew under the radar. The massive disasters such as the genocide in Rwanda and the civil wars of West Africa and Sudan created too much chafe.

Kenyan ecologist Wangari Maathai poses after receiving the 2004 Nobel Peace Prize at a ceremony December 10, 2004, in Oslo. Maathai became the first African woman and first environmentalist to receive the prestigious award. Tree planter Maathai received the prize for her campaign to save Africa's forests and for being a "leading spokeswoman for democracy and human rights, and especially women's rights." (AFP/Getty Images)

CONCLUSION

In 2004, Wangari Muta Maathai received the Nobel Peace Prize for her work as an environmental activist and political dissident in Kenya. Dr. Maathai, the first woman in Kenya to earn a Ph.D. (in veterinary medicine), rose to become a dean at the University of Nairobi. In the 1970s, she became active in the Green Belt movement in Kenya, which promoted tree planting as a means of combating soil erosion around Kenya. Maathai argued that planting trees improved the life of rural and poor women because they could then have easier and more secure access to firewood. This benefit came on top of the struggle to combat deforestation. Her grassroots environmental activism reached a peak in 1989 when she led the movement to save Nairobi's Uhuru Park from a plan supported by the government of Daniel Arap Moi to build a large office building on it. From this point she became active in pro-democracy politics and was arrested several times; she received serious injuries in 1999 when she was beaten by police as she led a group trying to plant trees in a national forest to protest continuing deforestation. After running briefly in the 1997 presidential election, she was elected to the Kenyan Parliament in 2003 as the leader of the small Mazingira Green Party and became assistant minister in the Ministry of Environment, Natural Resources, and Wildlife in the new government that had defeated the candidate of the previous regime. She had won several major prizes for her work before being named a Nobel laureate in 2004.

Dr. Maathai's long struggle illustrates the promise for Africa's environments. Her fame and even her effectiveness derived in part from her position in postcolonial Kenyan society. As a member of the postcolonial elite, albeit a pathbreaking one as the first female holder of the doctorate and head of a university department, her voice carried weight when she decided to speak. Yet her success in creating a movement that has planted over 30 million trees over two decades in Kenya came not because of her degrees, but because of her understanding of the costs of environmental degradation for poor people in Kenya. It is in such action that the hope for Africa's future lies.

7

CASE STUDIES

CASE STUDY 1:
THE FIRST FRONTIER: THE SAHARA DESERT

The Sahara Desert is the largest desert in the world and from the beginnings of humanity has played a critical role in human history. This fact may seem contradictory since as an arid desert few people live in it today or have lived in it in the recent historical past. Yet the Sahara's environments have helped form the contours of human development. The Great Desert served as a biological barrier during the formative eons of human development that shaped the environments within which humanity developed. Nonetheless, the Sahara's variability as the Earth's climate changed made it less an impenetrable obstacle and more a filter. Humanity's ancestors took advantage of the changing Sahara to both cross it and live in it. Brian Fagan, the noted archaeologist, has described the effect of the Sahara's variations as the "great Saharan pump." Over the millennia as the Sahara dried during colder phases of the Earth's climate and became moister during warmer phases, different kinds of life—plant, animal, and eventually human—moved into the Sahara and were forced out of it in all directions (1999).

The effect of the Sahara on human history has been large. It generally isolated the environments of sub-Saharan Africa, giving the region distinctive flora and fauna. North of the Sahara, the Mediterranean shores of Africa participated in the biological exchange of the Old World. However, during warmer, wetter phases of the Earth's climates, the Sahara shrank and filled with life. During one of these phases about 1.8 million years ago, part of the population of *H. erectus* ancestral to modern humans crossed the Sahara, probably using the Nile Valley as the passage. While *H. erectus* populations (and their progeny) spread across much of the Old World, those in Africa, still buffered by the fluctuations of the Sahara, gradually produced modern humans. Only as recently as 50,000 years ago did modern humans successfully cross the Sahara and begin the permanent colonization of the rest of the world. The Saharan pump then served alternatively to isolate and reconnect human populations. During wetter

169

The Sahara Desert is the largest desert in the world. (Corel)

phases, humans followed vegetation and game into the Sahara and fled its aridity during drier ones. At the beginning of the Holocene about 13,000 years ago, humans once again moved into the Sahara, but this time, when it began to dry, they domesticated one of the species of wild oxen that had moved into the desert. While their kinsmen to the east, in western Asia, began to experiment with cultivation, the Saharan peoples developed the first animal husbandry. When the Sahara began to dry again, the two lifeways came together, and the movement of people out of the Sahara may have set off the chain of events that led to both the spread of agriculture in the rest of Africa and the development of urban civilization in the Nile Valley. After the "final" drying of the Sahara about 5,000 years ago, it continued to serve as a filter for contact between the human (and other biological) populations around. Humans learned to live in it and cross it with the assistance of domesticated animals, especially the camel that made its way from Central Asia by about the beginning of the common era. The Sahara served as the southern boundary of the great empires of the classical Mediterranean world from Egypt to Carthage to Rome to the Arab Empire, but the human populations in that area never lost total contact with each other.

Two men plowing with oxen during the five-year drought in the Sahelian Zone of West Africa, ca. 1973. (Library of Congress)

Ideas and goods flowed through the Sahara, right down to the age of European conquest in the nineteenth century.

Since the twentieth century, the Sahara has become an important element in the debate about humanity's effects on the global environment. Throughout the last century, observers claimed that the Sahara had begun to advance, especially to the south. They charged that overexploitation of the fragile lands of the Sahel, the southern shore of the Sahara, had caused land degradation and that deforestation in the broader regions had reduced the amount of rainfall. These concerns peaked in the 1970s when a several-year-long drought caused massive suffering. The United Nations created a special program to combat "desertification." The sustained warming trend after 1980, however, brought a reversal of the expansion of the Sahara. While rainfall has increased and vegetation returned to many areas thought permanently degraded, concerns still remain about the sustainability of economic practices in the region. Degradation has replaced desertification as the main concern (Mortimore and Adams 1999).

The Sahara Desert occupies 8 percent of the world's total land area. It is around 8,400 kilometers (about 5,200 miles) long and occupies more than 9 million square kilometers (about 3.5 million square miles). The desert is bordered on the north by the Atlas Mountains of Morocco and the Mediterranean Sea, on the east by the Red Sea, on the west by the Atlantic Ocean, and on the south by a strip of arid land running from the Atlantic to the Ethiopian Highlands known as the Sahel. It is mostly a high plateau, between 400 and 500 meters (from about 1,300 to 1,500 feet) in elevation. In a few depressions, it sinks lower, including the Qattara Depression, which lies below sea level in western Egypt. The Sahara is divided into the western Sahara, the central Ahaggar Mountains, the Tibesti Massif (a region of desert mountains and high plateaus), and the Libyan Desert in Libya and Egypt.

The Sahara is a true desert in most areas, with annual average rainfall of less than 150 millimeter (about 6 inches) per year. It is also subject to high and variable winds. As a result, its sandy landscape is always on the move. It contains both *regs*, plains filled with stones, and *ergs*, fields of dunes that move with the winds. More moisture reaches the mountain ranges of Air and Tibesti. During the moister phases of the Earth's climate, more regular rains carved out riverbeds that now run dry. In some places, these *wadis* cut low enough to put them close to the water table and oases occur. Humans have long supplemented the moisture in these areas with wells and have practiced at least limited agriculture in some of them.

The basic outline of the Sahara seems to have formed about 5 million years ago. Since that time, its climate has been governed by the Intrertropical Convergence Zone (ITCZ), which also governs the monsoons. As the ITCZ moves north, it brings rain with it during the Northern Hemisphere summer. When it returns south, high pressure builds up over the Sahara, and arid conditions return. This shift is generally caused by the Hadley Circulation of warm, moist air rising at the equator, cooling, and then falling back to the Earth away from the equator coupled with the Earth tilting on its axis while rotating around the Sun. Annual and short-term shifts in the northern extent of the monsoon result from changes in sea surface temperature in the surrounding oceans. Global temperature changes determine longer-term shifts. During warmer epochs, the Hadley Circulation takes warmer air farther north and south from the equator. The increase in the amount of moisture in the air means that more rain reaches farther north in the Sahara. During cooler epochs, the circulation is more constricted, the air dries, and the Sahara expands (Grove 1978).

Both long and short variations in the Sahara's climate have had important consequences for human history. The changing size of the Sahara has been behind some of the most important developments in the invention of agriculture and animal husbandry. It has perhaps helped set the conditions for the rise of some of the earliest civilizations on the globe. It has also caused death and hardship as drier conditions took hold after times of plenty and humans had to adapt to new conditions. The area of the Sahel (from the word for shore in Arabic) and Sudan to the south of the Sahara suffers from perhaps the most variable climate in the world, and the human societies that developed there had to create innovative ways to survive in that landscape. However, by the twentieth century, with increasing numbers of humans and livestock trying to survive in that landscape, it was feared that these lands had become too stressed to recover from the periodic incidences of drought (Nana-Sinkam 1995).

During the last glacial age, between about 30,000 and 14,000 years ago, the Earth's climate cooled by about 5 degrees centigrade. Glaciers existed not only in the Northern Hemisphere but in Lesotho and South Africa. The border of the Sahara shifted south, and dunes extended in places along the West African coast. The great rivers of Africa that touch the borders of the Sahara ceased to flow, and many of the lakes of Africa disappeared. The warming that marks the end of the glacial age and the beginning of the Holocene pushed the monsoon much farther north. A dry millennium starting about 10,500 years ago slowed the progress, but after about 9,500 years the Sahara (and the Earth) reached what is called the Holocene Optimum. Grasslands stretched far to the north of the desert border today. Lake Chad, a damp depression during the glacial age overflowed its banks and drained through the Beneu River into the Atlantic. Strong rivers flowed through the desert to reach the Niger and the Senegal rivers (Grove 1978, 1993).

Plants and animals (including humans) followed the rains into what had been the desert. From about 9,500 years ago until about 4,500 years ago, the Sahel extended about 700 kilometers (about 450 miles) into what is now the desert. Africa's large fauna roamed throughout the area, and predators, including humans, followed them. In this semiarid space, however, some of the human communities began a series of experiments that would lead to the development of both animal husbandry and agriculture. In addition, around the expanded bodies and courses of water in the broad area, communities developed settled lifeways based on the exploitation of fish and shellfish as well as regular harvesting of plants.

A gradual drying of the Sahara began about 6,000 years ago and resulted in the movement of the southern frontier of the desert south. It coincided with the development of animal husbandry in the eastern Sahara. Animals and plants retreated out of the region, and humans followed. By about 5,000 years ago, the Sahara had reached its same general extent that it holds today. However, fluctuations in climate have continued; these fluctuations have occurred over time scales ranging from single years to centuries. The annual and short-term variations result in drought that affects vegetation and hence both animal and human food sources. Drought means that vegetation is in short supply, while in wetter-than-average years pasture and grazing for both wild and domestic animals extend farther north. Shifts on the order of decades and centuries cause more shifts in the boundaries of the Sahel and the Sahara.

Perhaps as a result of the drying trend in the Sahara, peoples on all sides of the expanding desert gradually began to adopt technological and social innovations. In northern Africa, especially in the lower Nile Valley, communities utilized crops such as wheat and barley to create true agricultural societies. These communities remained linked to the cattle-herding societies that inhabited the more arid lands away from the river. State formation in the Nile Valley coincided with the final drying of the Sahara. Some scholars have suggested that an influx of people from the desert helped create the conditions that required more intensive agriculture and the social controls necessary to manage it (Ehret 1998; Hassan 1997a).

To the south, the compacting human population seems to have encouraged experimentation with more intensive forms of agriculture to couple with the gradual spread of animal husbandry. The archaeologists Roderick McIntosh, Susan Keech McIntosh, and their colleagues have developed a model of the effects of this process in the Niger Bend region of West Africa. They suggest that each dry-period population collapsed back onto the Niger River (McIntosh and McIntosh 1984; McIntosh 1993, 1998). This process created a layering of settlement and lifeways across the West African Sudan as communities not so much merged together but remained separate but interconnected. When the rains stretched north again, communities moved north while maintaining connections to the south with its more reliable rains.

Roderick McIntosh (1998) has compared the archaeological evidence with evidence about climate change to suggest the following chronology for the Niger Bend region:

Climate in the Niger Bend Area

Time Period	Conditions	Comments
4500–4100 B.C.E.	Very dry	
4100–2500 B.C.E.	Rapid oscillations	Current desert/Sahel boundary formed
300 B.C.E.–300 C.E.	Extremely arid	"The Big Dry"
300–700 C.E.	Increasing moisture	
700–1000	Recent optimum	
1000–1200	Variable	
1200–1550	Very dry	
1550–1630	Wet	
1630–1860	Generally dry	"Little Ice Age" in Europe
1860–1985	Variable	Severe droughts in 1910s and 1968–1985; wetter conditions 1860–1900 and 1950–1960
1985–?	Wetter	

Source: McIntosh, 1998.

Historians have long speculated that during the era that McIntosh called "The Big Dry" between 300 B.C.E. and 300 C.E. contact between North Africa and West Africa was completely lost. Prior to 300 B.C.E., indirect evidence, including trade goods found in West Africa and rock paintings in the Sahara, indicates that some degree of contact was maintained. After 300 B.C.E., little evidence of contact in West Africa can be found. After 300 C.E., however, there is increasing evidence that trade resumed across the desert. Camels gradually spread among the peoples of the desert edge and allowed for more efficient transport across the Sahara (McIntosh 1998).

Camels changed the way people lived in the Sahara. Camel herders, who in the beginning were Berber speakers, could take camels deeper into the arid lands at the border of the Sahara on both the north and south sides in search of grazing. In the Sahel, camel herders added another tier of resource exploitation on top of cattle and goat herders. Across the Sahel a banded ethnic distribution developed, with camel herders following the scant rains far to the north, cattle herders taking their herds into the Sahel during the rains, and farming communities staying in the Sudan where rainfall permitted the cultivation of millet and sorghum. During the dry season, both sets of nomads moved back south, engaging in trade and sometimes conflicting with each other and with the settled communities (Brooks 1993; Lovejoy and Baier 1975).

After about 800 C.E., a well-developed trading system using camels as the main means of transportation emerged in the Sahara. Berber speakers moved through the desert from oasis to oasis. In the highlands, which collected slightly more moisture such as Air and Tibesti, they constructed semipermanent bases. They set up salt mines at places like Bilma and Fachi north of Lake Chad, Taghaza north of Timbuktu, and Awlil and Ijil further to the west. They constructed towns such as Awdaghust in what is today southern Mauritania and Sijilmasa in southern Morocco. In some of the oases, nomads ruled over "slave" populations that farmed and collected salt. A tenth-century account describes Awdaghust:

> [It is a] large town . . . there is one cathedral mosque and many smaller ones . . . Around the town are gardens with date palms. Wheat is grown there by digging with hoes, and it is watered buckets . . . Excellent cucumbers grow there, and there are a few small fig trees and some vines, as well as plantations of henna which produce a large crop . . . [There are] wells with sweet water. Cattle and sheep are so numerous . . . honey is abundant, brought from the land of Sudan. (Al Bakri from Nehemia Levtzion and J. F. P. Hopkins, *Corpus of Early Arabic Souces for West African History* [Cambridge: CUP, 1981])

By the eighth century, the Sahara had become crisscrossed with trade routes. The nomads of the desert relied on the sedentary peoples of both sides for access to grain, metal, cloth, and leather goods. Cloth manufacture became widespread

A worker harvests salt from evaporative ponds in Fachi, Niger, 1999. (Michael S. Lewis/Corbis)

across the Sudan. In return, the nomads brought salt and moved goods such as textiles and metal goods both ways across the Sahara. They also brought their weapons, raided for slaves, and established dominance over desert-edge communities. Although these everyday commodities made up the bulk of the trade conducted across the Sahara, the most well-known resource that moved from the Sudan to North Africa was gold. Gold came from West Africa as early as 300 C.E.; by the eighth century, the Empire of Ghana emerged as the dominant Sudanese state controlling the trade. But the Saharan peoples also contested the control of that trade. In the eleventh century, the Almoravid movement emerged among the Zanga (better known as Sanhaja to Europeans) Berbers of southern Morocco. They swept both south to Awdaghust (perhaps even defeating Ghana) and north to conquer the Arab-dominated lands of North Africa and Spain. Their reign lasted only a century before another Berber movement, the Almohads, swept out of the central Atlas Mountains to replace them. The North African historian, Ibn Khaldhun, saw in the constant interplay of conflict and trade between nomads and sedentary peoples the great engine of history, and his theory finds plenty of evidence on both sides of the Sahara.

This interaction between desert and sedentary peoples continued to turn the wheels of Saharan and Sudanic history until the twentieth century. The rise and fall of empires such as Ghana, Mali, and Songhai in the Sudan followed the rhythm of the annual movement of camel and cattle nomads south during the dry season. In the desert itself, Arabs moved south and west, displacing and absorbing Berber communities after the thirteenth century. The final element of this movement came with the Moroccan invasion of the last, great Sudanese empire, Songhai, at the end of the sixteenth century. An army equipped with guns marched across the Sahara and destroyed the unity of the Songhai Empire without effectively replacing it with Moroccan control.

The opening of the Atlantic trade did not destroy trans-Saharan trade, however; it coincided with the beginning of a long-term shift toward aridity in the whole of the Sahelian region, and this environmental change helped produce dramatic change in the region after 1600. The boundary of the Sahel may have shifted south by about 250 kilometers (about 150 miles) between 1600 and 1850. According to James L. A. Webb, by the beginning of the twentieth century, the areas around Adrar in the western Sahara, where cattle had once grazed, could now only support camel nomads. The area around the Senegal River was heavily wooded; today it flows through an almost treeless Sahelian environment (Webb 1995).

With this shift of the "desert frontier" southward came a reconfiguration of societies on both sides of the frontier. Arab-speaking camel nomads had moved south out of the North African Maghreb into the southern Sahara as early as the fourteenth century. By 1600, a new "white" African community had begun to form in the southern Sahara. Arabic in language and Muslim in religion, it also included substantial elements of Berber and other African communities that had lived in the area during more humid times. Agriculturalist communities retreated south, and cattle pastoralists found their movements squeezed. The desert pastoralists' control of both camels for use in the desert and horses gave them a distinct military advantage in the struggle between nomads and sedentaries. Thousands of agriculturalists were enslaved and carried to the desert oases to work in the salt mines and on plantations, growing dates and food (Lovejoy 1986).

The Atlantic slave trade intersected with this frontier several times after 1600, but with its generally low population density the western part of the Sahel was only a minor source of captives for the trade. After 1700, the French became the most important European presence in the western Sahel and Sahara with their base at St. Louis on the mouth of the Senegal River. A substantial number of captives in the late eighteenth and early nineteenth centuries flowed through their base from deeper in the West African Sudan. However, the end of the legal slave trade early in the nineteenth century did not end the French presence in the region; instead, it led to the eventual French conquest of most of the West African Sahel and Sudan all the way to what became Nigeria (Lovejoy 2000).

In the Nilotic Sudan, the same process of drying coincided in the nineteenth century with the expansion of Egyptian control and intensified "Arabization" of what is now the northern part of Sudan. As happened further west, camel nomads became increasingly Arab in language and culture and distinguished themselves more sharply from the cattle nomads and agricultural peoples of the region.

The Sudan, Sahelian, and Saharan zones were some of the last areas of Africa conquered by Europeans. The major European powers established control along the borders of the region; areas such as Algeria and Senegal came under the French in the first half of the nineteenth century. However, the French were not able to establish effective control over the Sahel or desert until the beginning of the next century. In the Nilotic Sudan, an Islamic reform movement drove out the Egyptians in the 1880s and held off the British-backed counterattack for ten years. In the West African Sudan, the French fought almost fifty years of campaigns to subdue the region. In what became northern Nigeria, the Sokoto Caliphate did not surrender until after 1900.

Once they had established control over the region, the British and French struggled to learn the hard lessons the African peoples had long taken to heart. The extreme variability of climate made mobility the key to survival. The movement of nomads following the rains and even the movement of agricultural peoples in search of food supplies in dry times interfered with what the Europeans regarded as the proper management of a landscape. Instead of a finely tuned system of risk minimization, European observers (with a few notable exceptions) saw a wasteful use of land and its resources. Early in the colonial era, a series of droughts gave European observers and administrators the evidence to create the concept of desertification as an explanation for the variability.

The 1890s marked the onset of a long dry period, with a major famine occurring in the Sahel and Sudan between 1900 and 1903. The drought of 1910–1915 in the Sahel was perhaps the worst in the recorded history of the region. The drought began in 1913 and with some variation lasted to about 1920. When rain did not reach as far north as before, cattle nomads were forced to keep their cattle further south longer and to deplete the available grazing much faster than usual. Rinderpest, which spread into the region for the first time in the 1890s, again decimated herds. Similarly, grain yields declined sharply (Nicholson 1980).

But colonial rule presented a new factor in the equation of climate change and human mobility that Africans had established. The French controlled most of the West African Sudan, whereas the British took what is now northern Nigeria and dominated the Anglo-Egyptian Sudan along the upper Nile. These new colonial regimes brought with them both borders that cut across lines of mobility and a general belief in the superiority of permanent settlement and cultivation. Although officials at the local level often recognized the importance of mobility and extensive land use by both agriculturalists and cattle keepers, policy generally promoted solutions to the problem of food shortage that encouraged a reduction of mobility. The most extensive case of this type of policy came in the French effort to promote irrigation from the Niger and its tributaries in the *Office du Niger* scheme (van Beusekom 2002).

A second important change during the colonial era came with the ability to accurately measure climate change and compare records from across the region and even the continent. By 1950, colonial regimes recognized that the climate of the Sahara and Sahel had become much drier than it had been before the 1890s. Colonial scientists invented the term *desertification* to describe the process they observed. They claimed that the desert frontier southward expanded because of the actions of African farmers and herders. They blamed the burning of pasture for the encroachment of the desert and the burning of the forest farther south for the expansion of the savannah. They blamed overgrazing for the degradation of grasslands and the loss of wildlife. They blamed desertification on population growth. They believed this growth was caused by the imposition of colonial peace, the spread of scientific health care, and improvements in transportation that allowed the importation of food during shortages, for a reduction of fallow and declining yields.

A wetter period during the 1960s coincided with independence for the nations of West Africa. During this period, debates about desiccation and desertification receded as governments and international bodies promoted agricultural expansion in the Sahel. In the 1970s, after the construction of the high dam at Aswan, the Egyptian and Sudanese governments began to construct the great Jonglei Canal project in Sudan to drain the swamps of the Sud and establish irrigated agriculture in the Sahel.

Disaster struck in the mid-1970s. From 1975 to 1985, rainfall decreased dramatically, and the Sahel moved south by about 100 kilometers (about 60 miles). Concern about the toll from decreased food production and the movement of people southward created an air of international crisis. Led by the United Nations, the international community responded with a massive effort to prevent starvation in the Sahel. These efforts proved reasonably successful at providing access to food for millions in the affected areas (Mortimore 1989).

Again led by the United Nations, the international community recognized the need to find a long-term solution to the problem of desertification. In 1977, the Organization for Economic Co-operation and Development created the *Club du Sahel* (later expanded to the Sahel and West Africa Club) to coordinate both emergency assistance and development aid to the semiarid regions of Africa. After much debate, the United Nations in 1994 proposed the United Nations Convention to Combat Desertification. The convention pledged to "combat desertification and mitigate the effects of drought in countries experiencing serious drought and/or desertification, particularly in Africa" (Nana-Sinkam 1995).

Although these efforts did result in more resources becoming available for monitoring climatic conditions in order to more accurately predict rainfall levels and provide support for infrastructure development, they had at their base a set of Malthusian presumptions about the path of environmental degradation. One influential report put it baldly:

> *The rural poor, many of whom live in environmentally fragile areas, are both the main victims and the unwilling architects of soil degradation.* Nomadic herders in the Sahel region, increasingly impoverished as a result of drought and the expansion of arable agriculture, have been forced to graze their herds on fragile grasslands. *Similarly, staple-food producers working on marginal soils have little choice but to sacrifice the future for the present, clearing trees and mining soils in an unsustainable manner to provide a livelihood.* They are often unable to invest in soil and water conservation. Loss of tree cover contributes to erosion by exposing soils to wind and rain. It also results in women being forced to walk increasingly long distances to collect and carry home bundles of fuelwood—a process that we have witnessed right across the region. Apart from its implications for their health, this diverts their labour from food production and household activities. (Nana-Sinkam 1995. Emphasis in the original)

Such dire statements run through the literature on desertification from the first French colonial administrators who identified the problem. It links impoverishment to environmental degradation in a causal way, rather than linking climate and environmental change to impoverishment.

Even before the hue and cry of disaster regarding the expansion of the Sahara had reached its peak in the later 1980s and early 1990s, a strange set of phenomena became apparent. Rainfall levels in the Sahel began to rise, and satellite imagery began to show the desert retreating. Droughts still occurred, but scholars began to recognize that drought was a relative term. Some began to suggest that rather than "soil mining" African herders and farmers utilized opportunistic strategies that maximized production in years of more abundant rainfall because of the certain knowledge that drier years would follow (Mortimore and Adams 1998).

The general (but variable) increase in rainfall since 1985 has raised other questions, however. First, some evidence indicates that lands, which in one way or another had seen efforts at conservation measures during the dry years, regenerated faster and with a more sustainable mixture of plants than land that had been truly overgrazed by cattle or left unprotected subject to erosion. Although the degree of "green" visible from a satellite may have been dependent on rainfall rather than on human use, the quality of green could vary greatly. This concern over "degradation" as opposed to permanent "desertification" reflected the replacement of grasses with shrubs in overgrazed lands and the decline of nutrients in soil farmed too long without fallow. In short, increasing resource use by expanding populations may limit the option for mobility that people in the Sahel had in the past.

The role of climate change in variability makes this question critical for the future. Increased moisture in the Sahel may be associated with a warmer globe, but current projections are uncertain as to the exact effects of continued global warming on tropical climates. It is quite possible the current wetter period may be like that of the 1950s and early 1960s—a prelude to a greater disaster as people find their mobility reduced.

CASE STUDY 2: DEATH AND LIFE ON THE SERENGETI: WILDERNESS IN AFRICA AS MANAGED LANDSCAPE

The Serengeti ecosystem occupies a large space in outsiders' imaginations about Africa. Straddling the border between Kenya and Tanzania, the Serengeti is defined by ecologists as the entire territory over which the great Serengeti wildebeest population ranges during its annual migration. The area is large, over

The Serengeti Plain in the Serengeti National Park, Tanzania. (Tim Graham/Getty Images)

25,000 square kilometers (about 10,000 square miles), and encompasses a number of distinct environments. It includes volcanic highlands and dense woods as well as broad, dry plains. It is home to the largest concentration of wild ungulates on Earth. Today most of the ecosystem has protected status of one form or another. It is the very picture of "wild Africa," untouched by human hands (Adams and McShane 1996).

Like wilderness landscapes in much of the world, the Serengeti is a managed wilderness. A variety of institutions participate in efforts to conserve the area. These include the Serengeti National Park in Tanzania and Masai Mara National Reserve (in Kenya); the Maswa Game Reserve, the Grumeti Game Reserve, the Ikorongo Game Reserve, the Ngorongoro Conservation Area, the Loliondo Game Controlled Area (all in Tanzania); and the Mara Ranches (in Kenya). In the Serengeti National Park and Masai Mara National Reserve, only tourism is allowed. The game reserves permit some settlement and grazing of cattle but no hunting. In the Loliondo Game Controlled Area, land use is not restricted but hunting is. The Ngorongoro Conservation Area is a joint-use area with settlement but sharp restrictions on land use. The Mara Ranches are private property

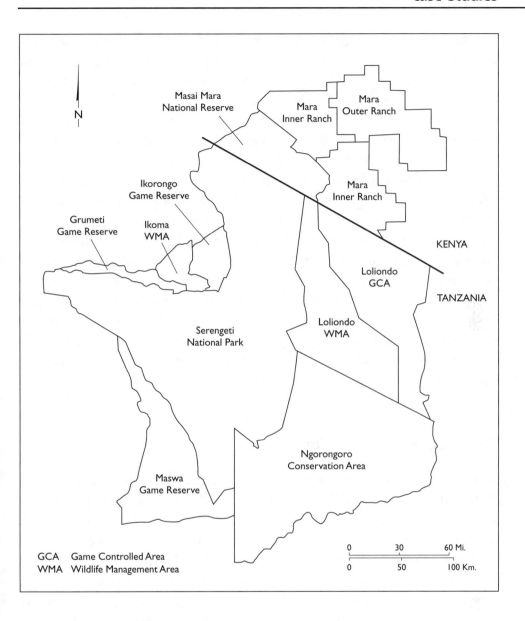

N

Masai Mara
National Reserve

Mara
Inner Ranch

Mara
Outer Ranch

Ikorongo
Game Reserve

Mara
Inner Ranch

Grumeti
Game Reserve

Ikoma
WMA

KENYA

Loliondo
GCA

TANZANIA

Serengeti
National Park

Loliondo
WMA

Ngorongoro
Conservation Area

Maswa
Game Reserve

GCA   Game Controlled Area
WMA   Wildlife Management Area

0   30   60 Mi.

0   50   100 Km.

under varying types of tenures, but they have voluntary game management (see map above).

The Serengeti ecosystem is vast, complex, and ancient. Conservationist literature often portrays the region as both relatively untouched by human action and at the same time threatened by increasing human populations in the area. For example, the official Web site of the Serengeti National Park, maintained by Tanzania National Parks and the Frankfurt Zoological Society, claims that only scattered foragers occupied the area before the coming of the Masai, which they date to about 200 years ago (www.serengeti.org). Such views are misleading to say the least. The Serengeti has been "managed" for centuries, but to different ends from that of conservation. It is a unique ecosystem, home to one of the last great animal migration cycles in Africa, but it was not unoccupied by humans before 100 or 200 years ago as the Frankfurt Zoological Society maintains. Human communities lived in it and sought to manage its bounty in order to produce human society. Today, it is still managed in order both to conserve wildlife and habitat and to support human society.

These goals—conservation and supporting human society—often come into conflict, but not just in terms of preserving land for wildlife versus using land for agriculture or grazing. Tourism in the Serengeti ecosystem contributes a substantial amount to the economies of both Kenya and Tanzania. In addition, the contribution is in a form that both governments find easy to tax and regulate. Finally, outside organizations, charities, and multilateral and bilateral aid organizations underwrite much of the financial support for conservation. Hence, these struggles have multiple actors, each with different sets of resources at their disposal. The result is a managed landscape created by human action. It is an artificial landscape, but like national parks throughout the world, it has created its own rationale for existence (McCabe 2002).

The Geography of Paradise

The Serengeti ecosystem has always been home to large numbers of wildlife. Within the boundaries of this system specific climatological and geomorphic conditions determine types of vegetation and hence types of animal life that it can support. To the south is Lake Eyasi; north of the lake, the Crater Highlands (including Ngorongoro Crater) are separated from the Gol Mountains by the Salei Plains, which the wildebeest use. The Gol Mountains run north across the Kenyan border to the Loita Hills. The border turns west along a variable frontier in the arid Loita Plains that changes with rainfall every year. To the northwest, the Isuria Scrape of the Rift Valley marks the boundary that turns south along a heavily cultivated area that has long been the home of agricultural peoples. In the southwest, the boundary runs along an arid area toward the Speke Gulf of Lake Victoria Nyanza before turning south along the Mbalageti River toward Lake Eyasi. Rain falls in the typical East African monsoon pattern of short rains in November and December and long rains from March through May. The rainy seasons may fuse. Rainfall increases from the southeastern corner (averaging about 500 millimeters (about 20 inches a year) of the system to the northwest in Kenya (averaging 1,200 millimeters (almost 50 inches a year). Generally, the south and southeast are grass plains with few trees except along the Olduvai

Elephant in Ngorongoro Crater, 1997. (Photo courtesy of Gregory Maddox)

River. In the far south, shorter grasses dominate because volcanic soils create a hardpan where only short-rooted grasses survive. Moving north, the grass becomes taller, and eventually at about the Orangi River in the middle of Serengeti National Park today gives way to woodlands dominated by acacia species (Sinclair 1979b).

Some of the areas neighboring the Serengeti ecosystem are themselves home to large wildlife populations and are protected as reserves of one form or another. The Crater Highlands contain the Ngorongoro Conservation Area. Ngorongoro Crater itself is a large caldera with a substantial resident wildlife population that generally does not leave the crater. Lake Manyara National Park and Tangire National Park lie nearby to the south and east in Tanzania. In addition, much of the Serengeti ecosystem does not lie in the two totally protected areas of Serengeti National Park and Masai Mara National Reserve. Besides Ngorongoro in Tanzania, much of the ecosystem is covered by the other game control areas, while the northern extremities in Kenya lie in the Mara Group Ranches (Sinclair 1979b).

Wildebeest herd at watering hole in Masai Mara, Kenya. (Norman Reid/iStockPhoto.com)

Wildlife, especially the wildebeest, defines the Serengeti ecosystem. The Serengeti and its neighboring areas, despite their vastness, are still just "a dot on the African continent" according to Adams and McShane in *The Myth of Wild Africa* (1996), but they are still host to about one-fifth of the continent's large wildlife. The annual migration of the wildebeest is one of the last two great animal migrations in Africa and certainly the largest. Many migratory systems, such as that of the springbok in South Africa, have been destroyed, and only that of the white-eared kob in Sudan survives today. Between 1 million and 1.5 million wildebeest take part in the migration (other large populations stay within more limited areas like Ngorongoro Crater). The wildebeest spend the rainy season on the short grass plains in the south of Serengeti National Park and Ngorongoro Conservation Area. As the rains end in May, they move northwest into the long grass plains and woodlands. Around October they turn north and cross Kenya's border in the Masai Mara. Around 200,000 zebra accompany the wildebeest, along with 350,000 to 400,000 Thomson's gazelles. Other species, such as buffalo, giraffe, topi, impala, hartebeest, and dikdik, are

Wildebeests, zebras, and flamingoes at a watering hole near the Ngorongoro Crater in the Ngorongoro Conservation Area, Tanzania. (Galen Rowell/Corbis)

more territorial and generally stay in the better watered areas of the system. Elephant range in the northern reaches (and the Crater Highlands); whereas hippo are found along the rivers. Black rhinos, once fairly widespread in the ecosystem, live in small numbers only in Ngorongoro Crater. Carnivores prey on this assemblage, but most are territorial. Hyenas (the most numerous), lions, cheetahs, leopards, and wild dogs (which also are almost extinct) feast as the migration passes their territory, but they live most of the year on the less mobile species resident in their domains. Scientists consider the wildebeest the "dominant" species not just because they are most numerous, but because the availability of forage determines their numbers and their numbers determine the availability of forage for other grazing species. These in turn determine the number of predators (Campbell and Borner 1995).

The Serengeti ecosystem probably took its current form about 1 million years ago, with wetter and drier epochs altering the ratio between woodland and grassland in the system (Adams and McShane 1996). Even before the ecosystem and migration stabilized, the wildlife population seems to have followed similar patterns back to 2.5 million years ago. Of course, as seen in Chapter 1, ancestral humans made up part of that population. Olduvai Gorge lies in the southeastern corner of the Serengeti, which today is part of the Ngorongoro Conservation Area. Australopithecines scavenged after predator kills, and *Homo habilis* learned to make tools to crack wildebeest bones. *H. erectus* may have evolved to hunt in the short grass plains.

Although foragers hunted and gathered in the region from the beginning of human existence, food-producing populations have lived in the Serengeti for at least the last 2,500 years (Sutton 1990). Foraging populations probably began the management of Serengeti's ecosystem through burning. Much of the northern and western reaches of the ecosystem receive enough rainfall to harbor tsetse and hence trypanosomiasis. Foragers may have selectively set fires to drive game toward killing zones; these fires would have kept the bush down and improved survival for human populations. Food production may have come later to the Serengeti than to the more favored surrounding areas because of both trypanoso-miasis and the hardpan soils of much of the region. Hence, a cumulative process of burning may have slowly opened the area to grazing and in some cases to agri-culture (McCann 1999b).

Evidence for cattle throughout the Serengeti dates to about 2,500 years ago. Agricultural settlements have been located in the northern reaches to about the same time. These communities survived in part by burning to increase the amount of safe pasture. Wildlife continued to live in these pastures, but the burning may have created more habitat for ruminants such as wildebeest and less for browsers such as giraffe (Gifford-Gonzales 1998).

About 400 years ago, a general drying in the environment took place con-nected to the cooler period known as the "Little Ice Age." This event coin-cided with the decline of agricultural communities in the broader region and the development of the Masai pastoral tradition. Agriculture retreated into the better watered lands around the Serengeti, while Masai pastoralists dominated a broad swath of land running from central Tanzania to northern Kenya. Masai traditions claim that the Masai never practiced agriculture and never ate wild game. However, Maa (the name of their language) speaking communities grew up that hunted (called Dorobo) and farmed (most notably the Arusha of Mount Meru and Il Chamus of Lake Baringo). Maa speakers constantly interacted with hunters and farmers, trading cattle, exchanging wives, and raiding. Up until the late nineteenth century, a system that included Masai pastoralists domi-nating the Serengeti stretched across much of East Africa (Spear and Waller 1992; Spear 1997).

In the Serengeti (and surrounding areas) Masai pastoralism and even agricul-ture coexisted with large wildlife populations. The people of the Serengeti used

Lake Kivu, Great Lakes region. (Yves Grau/iStockPhoto.com)

land extensively, although they remained linked to the more densely populated, intensive agriculture areas to the west along Lake Victoria Nyanza, to the north in the Kenya Highlands, and to the east in the areas around Mounts Meru and Kilimanjaro. As seen in Chapter 5, the late nineteenth century brought a series of severe disruptions that dramatically changed the human landscape of East Africa and the Serengeti.

The first major disruption to hit the Serengeti ecosystem came with the dramatic increase in the demand for ivory during the nineteenth century. In general, the short grass plains do not serve as an ideal habitat for elephants. Hence, their populations in the broader Serengeti have generally remained low and were limited to the more wooded areas to the north and west of the ecosystem. In those areas, they played an important role in keeping bush density low because they consumed trees before they reached too large a size. However, Ndorobo and immigrant hunters began to move into the area in the nineteenth century as the large elephant populations in the better watered areas to the east declined in the wake of increased hunting. The intensity of hunting increased as Arab/Swahili caravans began to reach the area after 1850. When the first Europeans reached the area in the last part of the century, they thought elephant could not survive in the area because they found none. However, during the early twentieth century, elephant returned to the area, moving in mostly from the north (McCann 1999b; Dublin 1991, 1995).

As important for the ecosystem as the loss of elephant population was, a more important disaster occurred in the 1880s and 1890s. First, bovine pleuropneumonia and then rinderpest struck the herds of the Masai in the area very hard. The bovine pleuropneumonia epidemic stuck first in 1883. It reduced herds sharply and resulted in the Masai's increased raiding both of surrounding peoples and among themselves. Rinderpest first appeared in 1891. The people lost almost all their cattle and so retreated from the plains of the Serengeti to the agricultural settlements that surrounded it. This retreat emptied the region of its population and greatly reduced the incidence of burning. Very quickly, bush began to regrow and tsetse recolonized much of the area (Waller 1988).

In addition to cattle, rinderpest also decimated the herds of ungulates in the Serengeti. Wildebeest, buffalo, and giraffe populations all fell dramatically in the aftermath of the 1890s epidemic. While nonungulates such as zebra did not suffer from rinderpest, the decline in wildlife populations further reduced the ability of the human populations to survive in the Serengeti. The disease became endemic among wildlife, spreading back and forth between domestic cattle and wildlife for the next seventy years. It kept populations of ungulates down by breaking out as an epidemic every few years or so. Wildebeest populations in the Serengeti ecosystem stabilized at about 300,000, with heavy mortality among yearlings. The disease repeatedly also spread among domestic cattle as populations recovered (Dobson 1995).

For the Masai communities of all of East Africa, the aftermath of the rinderpest became the central event in their historical memory. Smallpox struck shortly after the cattle epidemics. The Masai called the time *Emutai*, meaning everything was finished. Smallpox had commonly struck people throughout East Africa, but in the aftermath of the collapse of the cattle herds and the disruptions of colonial conquest in Kenya and German East Africa, it became particularly deadly. Population fell dramatically. By 1900, however, both human and animal populations had begun to recover. The Masai gradually regrouped into their sections, moving away from their hosts in surrounding communities, and sought to reoccupy their lands. However, in Kenya, the new British government, despite relying on the Masai for armed men during their conquest of agricultural groups in the colony, confined the Masai to reserves much smaller than they had occupied previously. In German East Africa, the spread of tsetse effectively accomplished the same result, although the Germans did allow the establishment of a farm on the floor of the Ngorongoro Crater. They also sought to encourage European settlement along the Kenya border, but they failed to attract many (Waller 1990; Århem 1985).

After World War I and the British takeover of what became Tanganyika, a twenty-year period of recovery followed. Pastoralists increased their herds to a stable level and reoccupied part of the Serengeti in the west around permanent sources of water. The British also moved to protect wildlife in the broader region. By 1930, both the Tanganyikan and Kenyan governments had declared their portions of the Serengeti "closed reserves." These acts made hunting by Africans illegal and prohibited any agriculture in the area but did allow Masai to continue to graze their cattle throughout the area (MacKenzie 1989).

Gradually, in thrall to a preservationist view of conservation, government administration tightened in the area. Although regulations continued to allow hunting by Europeans (who paid sizable fees for the privilege), they banned all hunting by Africans. In 1940, the Tanganyikan government declared the Serengeti a national park. For the next ten years, war and the relative inaccessibility of the region meant that the new situation meant little on the ground. However, around 1950 a series of factors began to come together to create increasing conflict over the Serengeti (MacKenzie 1989).

First, both colonial governments began to alienate increasing amounts of land for agriculture in the areas traditionally dominated by the Masai. In Tanganyika, large-scale, mechanized wheat farming began to spread. In addition, agricultural peoples in Arusha and Mbulu began to spread their fields into the lands the Masai had used for grazing. In Kenya, the area around Mara also came under increasing pressure from both white farmers and African agriculturalists. The Masai responded in part by moving more cattle into the Serengeti (Broten and Said 1995; Perkin 1995; Galvin et al. 2002).

This movement of cattle into the Serengeti ecosystem coincided with increasing efforts to enforce the wildlife preservation regulations. Park ordinances preserved the resident Masai's right of access to pastures in the new park, but park authorities had wide authority to regulate park usage. They began to enforce the ban of agriculture in the Serengeti and to put restrictions on the movement of livestock and the building of settlements in the park. In particular, the new park authorities began to enforce the restriction on the use of fire to maintain pastures. The Masai of Tanganyika responded with political agitation and won the support of the district administration against the park authorities. In 1956, a commission recommended reducing the amount of land under park regulation and opening the rest to normal grazing and agriculture. Only the far west of the Serengeti, Ngorongoro Crater, and Empakaai Crater would remain as park land (Århem 1985).

The international outcry was immense. The highly public struggle over the Serengeti was both a sign of a growing conservationist movement in international affairs and a spur for the further expansion of that movement. A number of individuals and organizations began to mobilize to oppose the dismembering of the Serengeti. American conservationist organizations sent teams to evaluate the proposals. The Fauna Preservation Society of London (formerly the SPFE) sent Professor W. H. Pearsall to report on the situation, and it was his report that became the basis for the eventual compromise. Publicly, however, Professor Bernhard Grzimek became the key figure in the battle to "save" the Serengeti (MacKenzie 1989).

Grzimek was head of the Frankfurt Zoo, and he and his son Michael had begun to travel to Africa to collect specimens in the early 1950s. In the middle of the controversy, the Grzimeks made a documentary and wrote a book arguing that the proposal would destroy the Serengeti ecosystem by breaking up the wildebeest migration route. The documentary and book, *Serengeti Shall Not Die*, became an international sensation and created pressure that neither the Tanganyikan colonial government nor the British imperial government could resist (Grzimek and Grzimek 1961; Grzimek 1970).

The original proposal to break up the park reflects the orientation of both the pastoralists who lived in the area and the colonial administrators who governed them. Extensive pastoralism was in fact consistent with wildlife. Preservation through a national park program meant the eviction of people whose ancestors had long used the land. It meant moving people into areas where they had no claims to resources and potentially creating even more environmental damage in those areas. At its base, preservation through eviction sought to make the livelihoods of the Masai of Serengeti someone else's problem, and at its worst, it argued that animals were more important than people (MacKenzie 1989).

Conservationists based their arguments on three factors that did not necessarily apply to the Serengeti or were not even generally applicable in the 1950s. First, they argued that any human activity in the Serengeti disrupted a natural environment that had been unoccupied until recent years. This argument ignored the long history of human occupation of the Serengeti in favor of taking the post-rinderpest retreat of people from the Serengeti as the baseline. Second, they contended that wildlife populations were dangerously low, partially because of competition for resources and partially because of hunting. In fact, endemic rinderpest kept wildlife populations low. The removal of people and their herds did not drive out rinderpest; only the development of an effective vaccine for livestock that greatly reduced the incidence of disease in domesticated animals in the 1960s did. Hunting by Africans drew off relatively few animals, most being taken not by the resident pastoralists but by people in agricultural communities to the west. In fact, European hunting may have been a bigger threat to some species such as elephants and lions. Finally, the conservationists projected more intensive land use than pastoralists actually practiced. Continued extensive pastoralism in the Serengeti would not have threatened wildlife any more than it had up to the 1950s. In this projection, they were wrong in the 1950s but perhaps correct in the long run. The expansion of mechanized agriculture in regions in Kenya and Tanzania surrounding the Serengeti demonstrates that at least parts of the region could support some form of more intensive exploitation. If this intensification had occurred in the Serengeti Plain, it undoubtedly would have resulted in the breakup of the migratory system in the Serengeti (Norton-Griffiths 1995; McCann 1999b).

The colonial government negotiated a compromise in 1959, two years before Tanganyika declared its independence. According to this compromise, Masai living in the western Serengeti would abandon their lands, while the Crater Highlands would be separated from the National Park and made into the Ngorongoro Conservation Area. In this area, Masai could continue to live and practice pastoralism, while a new body, the Ngorongoro Conservation Authority continued to administer wildlife conservation measures in the area. In Kenya, the Masai Mara National Reserve was declared and was administered similar to a national park (McCann 1999b, Århem 1985).

This new territorial division left some parts of the Serengeti ecosystem outside both national parks and controlled areas; hence, it created the conditions for continued conflict between conservation authorities and local populations. Independence for Tanganyika in 1961 and Kenya in 1963 did little to change the situation. The new governments adopted conservationist positions for several reasons. First, for the elites that governed the new states, wildlife became part of the iconography of their new nations. Just as American national parks for Americans and Kruger National Park for white South Africans have become symbols of nationality, so the Serengeti (and other national parks like Kilimanjaro in Tanganyika) became symbolic for Kenyan and Tanganyikan nationalists. Second, both countries remained dependent on varying degrees of support from international organizations and wealthy nations, who continued to insist on conservation policies. In particular, the national parks and reserves attracted substantial support, and conservation almost paid for itself.

Perhaps most importantly, the first decade of independence in the 1960s witnessed a dramatic increase in tourism and revenues from the game parks. Tourism brought in hard currency and attracted direct foreign investment. These revenues flowed in part directly to government accounts, in a way that an increase in livestock herds or even cash income to pastoralists would not have.

Given these incentives, local communities dependent on access to the resources in the national parks and reserves were playing a weak hand.

After the people were removed from the Serengeti National Park, the landscape changed dramatically, although not just because people left. First, beginning in the 1950s, the end of extensive burning ironically created conditions ripe for the conversion of much of the woodland of the western Serengeti back to grassland. Pastoral practice had created a patchwork landscape of woods and areas used as grazing. Wildlife had generally thrived in this context. The end of burning allowed a buildup of vegetation, which resulted in even more intense fires that in the long run helped reduce wood cover. Second, as people moved out of the park to surrounding areas, animals, especially elephants, moved in. Elephants also did damage to the wood cover. Finally, the development of an effective vaccine for rinderpest for cattle reduced the incidence of the disease (Norton-Griffiths 1979; McCann 1999b; Dublin 1991, 1995).

As a result, wildlife population in the 1960s boomed, especially the ungulates. The wildebeest population increased from around 300,000 to as many as 1.5 million in some years. Other animals also increased in number. In the Serengeti ecosystem as a whole, animal populations probably exceeded populations not only of the twentieth century but also much of the nineteenth as well. Hunting also increased in areas around the park as African populations took advantage of increased animal numbers to supply themselves with meat (Runyoro et al. 1995).

Throughout the 1960s and 1970s, the Tanzanian and Kenyan authorities faced two types of conflicts with local communities over the Serengeti. The first was with the resident pastoralists. The Masai who lived in the Ngorongoro Conservation Area experienced severe restrictions of movement and ability to exploit their land. For much of the time, they could not legally farm. They could not move their herds into Ngorongoro Crater itself or over the unmarked border of the Serengeti National Park. They protested such restrictions at every opportunity. Finally, in 1972, the local representative to the Tanzanian Parliament persuaded the Tanzanian government to intervene to allow agriculture in the Ngorongoro Conservation Area. The Conservation Authority reluctantly agreed but enforced a rule dictating that only hoe agriculture would be allowed (Århem 1985; McCabe 2002; Galvin et al. 2002).

The second conflict erupted over local peoples' harvesting game for meat, particularly in the western reaches of the Serengeti. The wildebeest migration often left the confines of the park in the west, and local populations hunted, often using traps and snares. They also sought to protect their fields from the animals and their stock from predators that followed the migration. Technically, this hunting was "poaching," and park authorities tried all types of schemes to limit it. Finally, they proposed extending the boundaries of the protected areas in the west as game reserves with limited human use for grazing. These proposals were debated within the Tanzanian government throughout the 1970s and 1980s and finally implemented in the 1990s (Sinclair 1995b).

Throughout the 1970s, economic conditions worsened throughout East Africa and had a dramatic impact on the Serengeti ecosystem. As a result of political differences between Tanzania and Kenya, Tanzania closed the border to ground traffic, an action that dramatically reduced the amount of tourism in the Tanzanian side of the border. Most visitors to the Serengeti had previously arrived overland via Nairobi. Tanzania's economy worsened considerably during the late 1970s and early 1980s, which meant its infrastructure, including its tourist infrastructure, suffered, and tourist visits declined even further. As state salaries lost ground to inflation, corruption among park staff increased.

In particular after 1979, poaching of elephants and rhinoceroses increased dramatically. Both ivory and rhino horn continued to be in high demand, especially in Asia. The collapsing economy of Tanzania and the less severely affected economy in Kenya meant that both hunters and park staff could not resist the temptation to provide for their families through poaching. Modern arms, available more widely in the region due to conflicts in the Congo, Uganda, Sudan, Burundi, Rwanda, and Somalia, made elephant and rhino hunting even more efficient. Elephants disappeared from the Tanzanian side of the Serengeti and came under stress throughout Kenya. Rhino were effectively exterminated in the Serengeti, with only a small population surviving in the enclosed space of the Ngorongoro Crater (Sinclair 1995a).

In 1989, under immense public pressure and with the support of the governments of Kenya and Tanzania, the Convention on International Trade in Endangered Species of Wild Flora and Fauna banned all sales of ivory. The proposals, backed by an array of conservation groups, faced opposition from the governments of South Africa, Zimbabwe, and Botswana where elephant populations continued to grow and poaching barely existed. The East African governments basically argued that they could not effectively police their own parks. Conservationists argued that banning the trade in East, Central, and West Africa while allowing it in southern Africa would only encourage smuggling (Adams and McShane 1996).

The ban worked after a fashion, in part because of the publicity surrounding the rapid decline in elephant numbers. The ivory trade, legal and illegal, declined sharply, and poaching decreased dramatically. Southern African nations continue to call for limited trade in ivory, but fear of renewed poaching prevents it from gaining much support. This publicity came too late for most rhino populations, however. Although their numbers have increased somewhat in East Africa, they remain highly endangered. In contrast, in the Serengeti, by the mid-1990s, elephant numbers had recovered to about their level at the beginning of the 1960s (Adams and McShane 1996).

Major issues concerning the Serengeti ecosystem are still unresolved. Park officials in both Kenya and Tanzania, as well as conservationists, believe that hunting by surrounding peoples threatens the herbivore populations as well as their predators. Such concerns seem overblown at present, given the stability of most herbivore populations, especially the wildebeest. In addition, Tanzania in particular has increased the number of tourist hunters it allows into the country (although hunters are supposed to stay out of Serengeti National Park itself). Of greater concern is the potential loss of habitat through increased intensive farming. In Kenya, the conversion of rangeland surrounding the park to farmland increased throughout the 1990s. Gradually, the land around Masai Mara became privately owned group ranches, and then in many cases individual freehold land. Masai landowners right around the park converted their lands to tourist attractions consistent with wildlife conservation, but those further away often leased their land to farmers who began mechanized farming and fenced the land. This conversion closed it to wildlife. Conflict within local communities between those who had invested in wildlife, those who had invested in farming, and those squeezed out of either option became intense (Broten and Said 1995).

In both countries, some Masai have continued to agitate for increased access to park land for cattle. They have made claims using the language of dispossessed indigenous people that they are the rightful owners of the land, in part basing their arguments on the example of the recognition of land rights in Australia for Aboriginal peoples. Although they have generated sympathy and support for their cause, and both governments are pledged to ensure some degree of continuation for their lifeways, conservation considerations still trump claims for right of access.

So, the Serengeti has not died and will not. The Serengeti ecosystem may not be what it was in the past, but it still exists as an ecosystem. Support for conservation remains widespread in East Africa; indeed, the idea of conservation, of preservation, has become part of the reality of the region and is part of the very identity of Tanzanians and Kenyans. It is perhaps greater among urban elites than among rural peoples affected by conservation restrictions, but it is real. On a less romantic note, conservation pays: it brings in both tourist dollars and outside support for conservation efforts. But the benefits to the nations of East Africa and to the world have been borne by the people of the Serengeti.

CASE STUDY 3: TRANSFORMING THE LANDSCAPE: FOOD PRODUCTION AND AGRICULTURE IN EASTERN, CENTRAL, AND SOUTHERN AFRICA

This case study examines the momentous transformation in landscapes undertaken by human societies in Central, eastern, and southern Africa between about 1500 B.C.E and 1500 C.E. During this broad sweep of time, societies in these regions dramatically intensified food production, and their populations increased substantially. Before about 3,500 years ago, communities practiced agriculture and animal husbandry in areas north of the great Central African rainforest and the Great Lakes region of East Africa. South of this rough line, human communities still relied on various foraging strategies. In the centuries that followed, agriculture and stock keeping spread rapidly south, iron working diffused throughout the region, and as a result, the landscapes became transformed. This broad transformation took many local forms across the varied environments of the region. Human societies expanded their numbers in rainforests, open plains, and highland regions. In some places, they used fire extensively to clear bush and make the land habitable for themselves and their domestic animals. In other areas, they consumed massive amounts of wood in the process of making iron to clear even more forests. By about 1,000 years ago, food-producing societies occupied all but the most difficult environments for agriculture and pastoralism. This dramatic transformation changed African landscapes throughout the region.

This case study also addresses a major theme of African historiography from an environmental perspective. This same period saw the expansion of Bantu languages and associated material cultures as revealed by archaeological discoveries. The transformation of African landscapes from the northern Central African rainforest toward the east and south has often been explained in terms of the spread of Bantu languages. Historians such as Jan Vansina (1990, 1995), Christopher Ehret (1998, 2001), David Schoenbrun (1998), and Kairn Klieman (2003) have debated the changes in terms of the reconstruction of language history and its conjunction with archaeological dates. In this case study, we will examine the same phenomena through the lens of transforming environments. In short, rather than seeing the spread of Bantu languages as a product of "migration" or diffusion by "waves," we will view it as part of a technological revolution that gave human societies a measure of power to control their environments. For the first time, this revolution made humans the "dominant" species in this region.

Before about 3,500 years ago, two different food-producing systems had emerged in Africa north of the equator. In the West African rainforest, communities had gradually domesticated several plants that grew in the forest, especially in clearings made either by natural action such as storms, elephant foraging, or fire. These plants included African yams and the oil palm. People supplemented their diets with hunting and fishing since the presence of tsetse fly and trypanosomiasis prevented the keeping of livestock. Given the difficulty in clearing land in the rainforest, population densities remained relatively low. Stone tools could not clear the larger trees very effectively, and the dampness of the forest and lack of a long dry season prevented extensive burning. Population seems to have been higher in the drier reaches of the forest (Ehret 1998).

North of the forest, agropastoralism had spread across the Sudanic regions from Ethiopia to the Atlantic. Cattle keeping had existed in the area for several thousand years, and goats and sheep had spread from southwest Asia. By this time, communities had domesticated several grain crops suited to the tropical climates and could pick crops based on climatic conditions (Ehret 1998).

This system of production could not expand into the wetter environment of the forest or easily into a belt of wetter territory running roughly south of the Great Lakes across East Africa primarily because of tsetse and trypanosomiasis. In the forest and in the more open woodlands of East Africa, trypanosomiasis circulated between flies and larger ungulates such as zebra and wildebeest. In some regions of northern East Africa, people began to use fire to clear fields and maintain pasture. By burning during the dry season, people kept the bush down, permitting grass to grow for livestock. Human diseases also prevented the expansion of plains dwellers into wetter areas. People born in wetter areas acquired a basic resistance to malaria that people from drier plains lacked.

In the forests of Central Africa and across the plains and woodlands of eastern and southern Africa, communities still survived by foraging. They hunted using bows and arrows tipped with poison; they set snares and pitfalls; they created game drives that forced larger animals into killing fields; and they hunted larger game in the plains and woodlands, and smaller game in the forests. They also foraged extensively and fished in the rivers. Their population density remained relatively low, and they probably caused little environmental change. In a sense, humans were not the "dominant" species in this region yet (Vansina 1990).

Starting about 3,500 years ago, a series of transformations occurred that allowed agriculture and stock keeping to expand into this area. The first change was the expansion of forest agriculture across the northern reaches of the Central African rainforest, followed by a gradually filtering down of agriculture through the rainforest. The second was the creation of a new synthesis between West African agriculture and agropastoralism in the Great Lakes region of eastern Africa. Once created, versions of this system would spread very rapidly east and south. The final transformation was the widespread adoption of iron working. Iron gave people new tools for clearing and tilling the land. It also required large amounts of fuel for its production in the form of wood. The end result was a large expansion in human population and a reduction in the animal population. Many landscapes became much less wooded as people burned to create fields and used wood as fuel (Ehret 2001).

The first transformation process began in the northwest corner of the Central African rainforest in what is today eastern Nigeria and Cameroon. The original "proto-Bantu" language seems to have emerged about 6,000 years ago in this area. Planting agriculture began to expand along the northern fringe of the rainforest, as evidenced by the beginnings of small settlements about 3,500 years ago. This expansion of agricultural communities is associated with the spread of Bantu languages. These farmers cultivated yams and oil palms, and they also brought small stock, particularly goats, that could survive in the rainforest. It probably entailed a gradual expansion of population and a process of splitting off settlements into new land. These new settlements had higher densities of population than the existing forager communities and over generations moved in slow waves south through the forest and east toward the Great Lakes. Although the establishment of these agricultural settlements did require the opening of the forest, the population remained relatively low and the extent of deforestation remained limited (Ehret 1998).

The forager communities gradually became absorbed into the agricultural communities until the remaining foragers became cultural and political dependents on the agricultural communities. Agricultural communities, though, acknowledged the precedence of the foraging communities by often maintaining a special status for foragers as "owners of the land" in their oral traditions. They sometimes used the metaphor of foragers "teaching" them how to live in the forests. Historic foraging communities like the Aka and Twa of the forest regions developed over centuries of interaction with agricultural communities (Klieman 2003).

The cause for this initial expansion of Bantu speakers remains obscure, although the general advantage of agriculture as practiced by the first Bantu speakers is of course part of the explanation. Agricultural techniques could easily have diffused without the expansion of the languages, as did stock keeping in southern and eastern Africa a little bit later. Little evidence exists of a dramatic technological change in the practice of agriculture by Bantu speakers at this point. Historians have therefore concentrated on social factors reconstructed from the history of language change. They have suggested that a flexible system of social organization that focused on "houses" led by senior males who could attract followers gave Bantu speakers a means both to assimilate outsiders and to organize expansion (Vansina 1990, 1995; Ehret 1998, 2001; Nurse 1997a, 1997b; Phillipson and Bahuchet 1995).

By about 1000 B.C.E., some Bantu-speaking communities had established themselves in the northeastern corner of the rainforest around the Great Lakes

Ruwenzori mountain range, Uganda, ca. 1936. (Library of Congress)

region. Here the second and third major transformations began, even while other agricultural Bantu-speaking communities continued to expand southward through the rainforest. In the Great Lakes region, these agricultural communities encountered a different set of environments, some of which were already occupied by food-producing peoples (Ehret 1998).

The Great Lakes region can be described as a circle formed by Lakes Rwitanzige, Ruiru, Kivu, and Nyanza, along with several smaller lakes. The gentle highlands of the Ruwenzori provide a well-watered environment around the lakes that encircles a drier area of lowlands in between. In the lower lands between and to the east of the Great Lakes, communities practicing agropastoralism had long been established. They kept cattle and grew sorghum and millet, depending on local conditions. They generally avoided the forested highlands around them, leaving those areas to foraging populations. Bantu speakers moved

The Ngwenya iron ore mine during blasting operations in the 1970s. This deposit of valuable iron ore supplied raw material to the steel mills of Japan, acquiring valuable foreign exchange. (Bettmann/Corbis)

first into the highlands where their particular form of agricultural technology based on yam cultivation thrived. Population in the highlands expanded rapidly, and local foraging populations, like those to the west, began to be assimilated into the agricultural communities, with a few foragers becoming specialist producers (Schoenbrun 1998).

In this region, a fusion of agricultural and cultural practices took place that transformed the practices of Bantu speakers and led to even greater and more rapid expansion. In addition, iron working became widespread. Ironically, Bantu speakers, though newcomers to the region and adopters, not inventors, of the new technology, became the dominant linguistic tradition in the area. Great Lakes Bantu speakers began to adopt both cattle keeping and grain agriculture from their Cushitic- and Sudanic-speaking neighbors. They also began to engage in iron working and expanded into the plains with variable seasonal rainfall. They absorbed much of the preexisting agropastoral as well as foraging populations (Ehret 1998).

The development of iron working in Africa had far-reaching consequences for African environments. Its origin in Africa remains a topic of hot debate. Iron working was first invented in Anatolia about 1500 B.C.E. and from there spread throughout the Mediterranean, extending east towards India and China. Although many societies throughout the world before that time had learned to work metals such as bronze, gold, and silver, iron working represented a striking technological innovation. Whereas the other metals are worked by melting their ores to purify them and then shaping them, iron requires a chemical change in the ore created by a precise mixture of ore, heat, and oxygen. Early iron working required larger amounts of fuel in the form of charcoal made from wood. All Old World iron working outside of sub-Saharan Africa diffused from its Anatolian origins, and American societies never developed iron working before the arrival of Europeans (Diamond 1997).

Scholars long believed that African iron working represented a straightforward diffusion from the Mediterranean, perhaps through two routes. Early iron-working sites have been found in the Nubian Nile around the town of Meroë dating to shortly after 500 B.C.E. Discoveries of ironworking sites in northern Nigeria dated to earlier, perhaps around 800 B.C.E., and scholars have speculated that these sites resulted from trade contacts with Carthage in North Africa. However, over the last two decades, ironworking sites in the Great Lakes region have been discovered from the period well before 500 B.C.E. These dates remain debatable: the sites are few, and some scholars have suggested that the earliest dates result from older organic material corrupting the test samples (Holl 1993; de Maret and Thiry 1996). Other scholars, notably Christopher Ehret and Peter Schmidt, have argued that these dates prove an independent invention of iron working in Africa. Ehret maintains that it was probably discovered someplace between Lake Chad and the Great Lakes and spread slowly both east and west (1998).

Whatever its source, after about 500 B.C.E., iron working began to become part of the Bantu-speaking agriculturalists' way of life. Bantu-speaking communities adopted a series of innovations in their forest-based way of life, which propelled the expansion of their languages and culture into new environments. As a result, these environments were altered considerably (Schmidt 1997b).

As iron working spread to the older, more established Bantu-speaking communities in the rainforest, the pace of their expansion into the rainforest and then on to the southern savannahs quickened. The two streams of expansion, west and east, moved in rough time with each other. Linguistic historians have debated whether these streams should be seen as equal "branches" of the Bantu languages' family tree or as a series of branching movements (Vansina 1990; Ehret 2001; Nurse 1997a). In terms of lifeway, as opposed to linguistic relation, they represent two streams. In the savannah, by sometime after 500 B.C.E., grain cultivation and stock keeping also spread from the east. Over the next 1,000 years, Bantu-speaking communities expanded south, reaching the Zambezi by circa 300 C.E.

As their population expanded, Bantu speakers began to move into the less-well-watered lowlands around the Great Lakes and undertook more intensive grain cultivation. They adopted cattlekeeping in areas free of tsetse, and they also began to make and use iron tools. All of these changes resulted in more extensive clearing of the land. Iron tools cut bush and trees much more effectively than stone ones. Iron hoes tilled the ground more thoroughly, allowing for more productivity in grain agriculture. The need for fuel for iron furnaces precipitated the beginnings of extensive clearing of woodlands (Ehret 1998).

As Bantu-speaking communities grew in population and then in territory to the east and south of the Great Lakes, they absorbed much of the previous population in the area. Some of the agropastoral peoples living in what is today northern Tanzania, Kenya, and Uganda maintained their languages in pockets scattered across the area. The new Bantu-speaking communities not only absorbed much of the population, but also adopted a variety of these populations' cultural practices, words, and material goods as well as productive techniques (Ehret 1998).

In the earliest expansions out of the Great Lakes, Bantu-speaking communities sought out better watered environments in which to practice their new lifeway. They moved across a chain of highlands from the Great Lakes to the mountain ranges of northeast Tanzania, to the highlands of central Kenya, and to the well-watered coastal belt of East Africa, reaching the coast in about 300 C.E. Other groups moved to the highlands of what is now southern Tanzania at about the beginning of the common era. In these areas, their more extensive use of the land almost totally displaced the foraging communities that had lived there before they arrived (C. Kusimba, 1999).

During the first centuries C.E., the ramifications of reorganizing land use in northern East Africa began to be felt further south. Stock keeping, both of cattle and especially sheep, began to spread in southern Africa among the foraging communities that had long lived there, well in advance of the coming of either agriculture or Bantu speakers. Shortly thereafter, Bantu-speaking communities began to follow into the region. They brought iron working as well as grain agriculture. By the middle of the first millennium, agricultural communities lived as far south as the northern reaches of what is today South Africa. When they reached what is now Zambia, they met not only foragers but communities of Bantu speakers that had spread from the rainforest (Hall 1997).

The rapid expansion of these agricultural communities initially represented a sort of leapfrog movement across different environmental zones as Bantu speakers sought out environments similar to those in which they had lived. The addition of grain cultivation and cattle keeping gave them the ability to exploit different types of environments from the moist woodlands where their communities had originally developed. As they occupied the better watered open forests and grasslands around them, the population continued to grow both because of the communities' greater capacity for food production and their absorption of surrounding populations.

This continued growth soon put pressure on their resources, but fortunately, iron working and fire gave them the ability to transform landscapes that would support their methods of food production. Clearing land both for agriculture and by burning for stock keeping opened up new areas for expansion, driving back into smaller areas both the tsetse fly and the game animals that harbored trypanosomiasis. Pollen cores taken from the Great Lakes clearly show a gradual decline in tree pollen in the cores as this process continued over the centuries after the beginning of the common era (Schmidt 1997a).

There then occurred a process of filling in, as Bantu-speaking communities expanded into environments that had earlier been left to foragers or other agropastoral communities. Bantu speakers moved into the drier lowlands of East Africa where the new communities sometimes continued to coexist with previous agropastoral peoples. The new communities often adopted many elements of the culture of their predecessors in these regions. They also faced the difficulties associated with climatic variation, especially periodic drought as in the lowlands of East Africa. To counter the variations in rainfall, they created networks of social and economic ties.

Although iron ore is common across Africa, many areas had none, and as a result, a trade in iron products developed. Other specialized agricultural products also became trade goods, in particular *eluesine* millet, which was used for beer brewing. The most important trade good, however, was livestock. If the rains failed, cattle, goats, and sheep could always walk themselves to areas where they could be purchased in return for other forms of foodstuff. Systems of stock exchange, including pawning, loans, kinship ties, and political allegiance, developed that crossed both environmental and "ethnic" boundaries. After 500 C.E., the landscapes of eastern and southern Africa were dramatically transformed by the coming of ironworking agricultural communities. The number of people in these communities remained small by modern standards, but all the same they were able to change the landscapes. In most areas, these communities still relied on agricultural methods that required long or "bush" fallow to restore fertility after several years of farming. However, the initial clearing of the land resulted in a more open landscape in many areas, and the continued use of land as pasture for stock prevented the complete regeneration of bush or forest. Wild animal populations became more confined to particular parts of the landscape, often areas that continued to harbor tsetse or that lacked permanent water sources necessary for human habitation (Schoenbrun 1998; Schmidt 1997b).

In particularly well-watered areas, such as the Great Lakes and the highlands of East Africa, the population continued to grow, and a more intensive means of production emerged. Eventually, the spread of banana cultivation after about 1000 C.E created the conditions for permanent habitation of these areas, and they became the most densely population in the region. By 1000 C.E., deforestation in some regions, especially the highlands around the Great Lakes, seemed almost complete (Schoenbrun 1998).

This case study has tried to highlight some of the most profound results of the last few decades of research to counter some of the mythology that continues to cling to the past of the peoples and landscapes of the continent. African communities have long acted to transform their landscapes in the interest of making them productive. Indeed, the scale of human-made environmental change has increased with the dramatic population growth of the last fifty years, but the process began several millennia ago, not at the beginning of the twentieth century.

DOCUMENTS

C ollected below are a series of documents that survey different ways of viewing African environments. Each has been chosen to represent a slightly different viewpoint on African landscapes. Given the relative paucity of documentation by Africans (or even about Africa) before the nineteenth century, most of the documents come from the last two centuries. I have avoided using scholarly writings in favor of documents produced directly by participants or observers. Hence, there are no writings on the "prehistory" of Africa in this section. Each selection begins with a brief description of its importance and a full bibliographic citation.

MUSA KONGOLA'S LIFE

The first selection comes from an unpublished autobiography by Musa Kongola of Dodoma Region, Tanzania. In this selection, the very beginning of Kongola's narrative, Kongola recounts the history of his patrilineal clan, Mbukwa Muhindi Wevunjiliza of the Gogo people who live in Dodoma Region. The story printed here follows the format of the recitation of the clan history done in public at certain occasions such as funerals. The story is an account of the migration of the clan founder from the south in Uhehe (the country of the Hehe people. Languages in the Bantu family make nouns by adding a prefix to a root. In this case, U- denotes country and Wa- people. Hence, U-hehe is the land of the Hehe people and U-gogo the land of the Gogo people.) They crossed the Ruaha River into Dodoma Region at some point in the past. The account demonstrates the way that many African traditions portray environment as a social landscape, with geographic features serving to reinforce social relationships. The story also relates a process of social transformation by a group leaving one environment where they were one type of people and entering another and becoming another type of people. Kongola wrote this account in 1981 when he was almost 100 years old. He became one of the first teachers for the Anglican Church Missionary Society in the region at some point after 1900. The autobiography is held by his son, the noted local historian Ernest Kongola, and appears here with

*his permission. The translation from Kiswahili is by the author. For further de-
scription of this account and of the work of the Kongolas, see Gregory H. Maddox with Ernest M. Kongola,* Practicing History in Central Tanzania: Writing,
Memory, and Performance *(Portsmouth, NH: Heinemann, 2005).*

Mbukwa Muhindi

Musa Kongola

This book is written about all the events concerning the origin of my tribe, and
myself, the writer, and my life from my birth up to today as well.

My tribe is originally Wahehe from a place called Imaje which is the name of
the country which I was told by my father's sister Maloji, sister of Munyambwa,
my birth father, who told me much before she died. I will write much that I was
told as well as what I have seen myself in my life from my childhood until I became an adult. Now I return to the explanation of the origins of my tribe.

The first ancestors were wandering Wahehe who did not have the customs
of the people of the country called Ugogo.

They did not live that type of life at Imaje. They left Imaje which is south
of Ugogo.

After the departure of our father and mother from the country where they
were born and raised, they went to other countries to the north looking for a
good place to live. They passed many countries after they departed from Imaje.
First they arrived at a place which had a big river filled with much water and it
was difficult to cross because the people were too afraid to enter. They were
very afraid; they looked for a place to pass across.

The way the ancestors got help to cross the river was this. They cut grass
which is called *madete lulenga,* they wove like grain storage baskets which
called by the name of vitundu, then when they finished weaving they entered
them and crossed to the other side. Many did this, entering on one side and
crossing to the other.

The name of the people who crossed to the other side in baskets became
Wanyagowe. It means people who cross in a basket and who all did not have
fear and those who turned back were called Wanyarudi because they returned to
their country which was called Rudi. Those who went by baskets stayed in
country called Gulwe after a mountain called Gulwe. The country has the same
name today.

It can be written that the people who left Imaje by this manner were called
all by one tribal name, Wazungwa and they did not practice circumcision.

After they crossed the big river which was filled with water they first scattered to their homes when they arrived at the place from which their scatter-
ing occurred.

Those who were afraid and returned back were called Wanyarudi and those who used the vessels for crossing were called Wanyagone.

Afterwards those who remained increased in number and continued their journey. They arrived at a place called Ibeleje and when they seized this country they began to use iron to pierce their ears so that they were called Walimba Mwanya because they did this to themselves making a space to place a fruit in their ears. They were called Walimbamwanya and this country is called Ibeleje up to today.

Those that remained and their offspring continued until they arrived at a country called Chimambi which is where our grandfather lived and is the origin of our lineage. I write the events from the beginning to the end and I hope I will do my best to write as I was told by my grandfathers and grandmothers who I met when they were still alive like my father's sister, the sister of my birth father all the way through my own life.

When they arrived inside this country they met the natives of the country. The local people did not have the ability of people today; fighting occurred time and again, and these foreigners had greater strength. They still did not have a new tribal name so they continued to be called by others Wazungwa. They kept this name, those who came to Imaje with our grandfathers, until they got a new name.

What were they called and what was the origin of their name? First, in another country there was a girl called Nyalaga who was not married then and this girl was carried away by a boy of the Wazungwa tribe.

When they were married they called her inside the house and covered her with a cloth on the face with the intention of her not to be seen by people. Because they began to cover their wives' faces in cloth so we who had been Wazungwa became known by the name Wevunjiliza, which means to close. The name Wazungwa has died out so we are called Wevunjiliza until today.

And this country was called Chimambi and the people who lived there were taken by the hands of our grandfathers because our grandfathers defeated the tribes. The strangers time after time attacked and captured many while the local people sat on their hands. Our tribe which was called Wazungwa moved to the country of Nyalago, from whom they took a wife. She was the daughter of the chief. With her knowledge of the country the elders pacified it and claimed their right to rule.

Events continued from generation to generation. The Wevunjiliza became the natives of the country which is called Chimambi. From then until now, the country is called Chimambi, and the people who rule it are called Wevunjiliza the same up to today.

Now I start to write like I said I write that which I was told. Many nearby peoples left Uhehe and entered Ugogo. They became part of the Gogo people.

There are many names of peoples, and I cannot write all the different ones. I will write names like those below moving from grandfather to father. Others will know those names and remember how they are related. They will read the history of the origin of our tribe and the lives of our elders, one by one, and will understand the way it is meant to be read.

In our first section I explained the history of the marriage of a girl and a boy concerned with using a cloth to cover the face which prevents one seeing her face. This is the origin of our name Wevunjiliza.

Now I explain the story of how we came not to eat intestines. When they killed a goat, cow, or sheep it was thrown away. It is said it is the taboo for the intestine of any animals to be eaten, this belief was held strongly and continues until today. Today it is said a person should not eat the intestines of an animal including the intestines of cows, goats, or sheep. It is a true taboo. From this is when we were called those people the Wejunviliza who did not eat intestines of animals so that we were called Mbukwa Muhindi. We are the people who do not eat intestines of animals, so from that we follow our practice of not eating intestines.

Now we enter the names of nearby peoples who were born here and whose families continue until today. They did not live together all the time. They moved to other places because of war and lack of food.

    . . .

The lives of our grandfathers and fathers and others depended on tradition; they did not have religion which came from foreigners. They worshipped at the graves of the dead of their lineage who helped them when they were attacked by disease and/or any problem. They came at their grandfather's or their father's graves where they performed rituals. Sometimes they carried a sheep and beer. They had faith that they would get help from the person who had died. But I and others have not found this the way of the modern world.

The lives of their children are described below.

The subjects we were taught by the ancestors are these: cultivation, herding, and weaving with different things like baobabs and inka. The baobab was carried into the corner so that it would dry during the day. When it was dry it was prepared so that it would separate beer by squeezing it through and it would be sold for the price of one for one chicken. It was the time before we had money; when money was brought the price was 25 hellars which was equal to 50 cents. And the work of weaving sieves is in Cigogo now called *kuhuziza ujimbi*.

And they who were taught this skill were Kongola and Manginido, the children of Munyamgwila, Makwala and Boyi who were the children of Yobwa the brother of Manyangwila.

The progress of we children was very different. Some of us got religion. Others followed to the present the beliefs of the elders while some followed the religion of Christianity. I was the first to follow the religion of Christianity. I write the news from the beginning which I have seen and others which I have heard.

I came to hear first from my mother who sent me to listen to the English men who arrived to preach the word of God.

I was herding livestock, cattle and goats, in the camp in the bush when the servants arrived to call people to go hear and when they went they heard the style of playing the piano they listened to the voice only meaning the voice which was playing was explaining its meaning this is the first day of being called to hear the word of God.

MUNGO PARK
THE SAHEL AND THE NIGER RIVER

This selection is taken from Mungo Park's account of his first journey to the Niger River from Senegal, which took place between 1795 and 1797 at the behest of the African Association of London. At this time, North European interest in the interior of Africa had begun to increase, in part because of the development of a strong antislavery movement. Although Park's journey for the African Association was by no means the first by a European (Park was from Great Britain) to the Niger River, it became the most famous such trip. Portuguese officials and representatives had traveled in the interior of West Africa in the sixteenth century. However, the relative decline of Portuguese power meant that direct contact between African peoples away from the coast and Europeans declined as the slave trade increased as African rulers and merchants sought to keep Europeans confined to the coast of West Africa. Park traveled along the Sahel from the Senegal Valley to the Niger Valley. The excerpts here are Park's account of the Sahel (the land of the "Moors" or Berber-speaking peoples) and of the far western reaches of the Niger River in the Bambara state that succeeded the old kingdom of Mali. Park's account was shaped by two factors. First, he traveled in Africa at the peak of the trans-Atlantic slave trade, and he constantly refers to the violence that the traffic sparked. Second, Park, though not nearly as blinded by racism as some later European observers of Africa, is keen to mark the distinction between "Moors" and "negroes" as a racial distinction. Park never traveled the entire length of the Niger. He returned to West Africa in 1805, determined to complete his journey, but was killed in a conflict with local people. The excerpts are from Mungo Park, Travels in the Interior Districts of

Africa: Performed under the Direction and Patronage of the African Association, in the years 1795, 1796, and 1797 *(London: 1799 [1st ed.]). Chapter 12 is from Vol. 1 and Chapter 16 from Vol. 2.*

Mungo Park
Travels in the Interior Districts of Africa:
Performed under the Direction and Patronage of the African Association, in the years 1795, 1796, and 1797
CHAPTER XII—OBSERVATIONS ON THE CHARACTER AND COUNTRY OF THE MOORS

The Moors of this part of Africa are divided into many separate tribes, of which the most formidable, according to what was reported to me, are those of Trasart and Il Braken, which inhabit the northern bank of the Senegal river. The tribes of Gedumah, Jaffnoo, and Ludamar, though not so numerous as the former, are nevertheless very powerful and warlike, and are each governed by a chief, or king, who exercises absolute jurisdiction over his own horde, without acknowledging allegiance to a common sovereign. In time of peace the employment of the people is pasturage. The Moors, indeed, subsist chiefly on the flesh of their cattle, and are always in the extreme of either gluttony or abstinence. In consequence of the frequent and severe fasts which their religion enjoins, and the toilsome journeys which they sometimes undertake across the desert, they are enabled to bear both hunger and thirst with surprising fortitude; but whenever opportunities occur of satisfying their appetite they generally devour more at one meal than would serve a European for three. They pay but little attention to agriculture, purchasing their corn, cotton, cloth, and other necessaries from the negroes, in exchange for salt, which they dig from the pits in the Great Desert.

The natural barrenness of the country is such that it furnishes but few materials for manufacture. The Moors, however, contrive to weave a strong cloth, with which they cover their tents; the thread is spun by their women from the hair of goats, and they prepare the hides of their cattle so as to furnish saddles, bridles, pouches, and other articles of leather. They are likewise sufficiently skillful to convert the native iron, which they procure from the negroes, into spears and knives, and also into pots for boiling their food; but their sabres, and other weapons, as well as their firearms and ammunition, they purchase from the Europeans, in exchange for the negro slaves which they obtain in their predatory excursions. Their chief commerce of this kind is with the French traders on the Senegal river.

The Moors are rigid Mohammedans, and possess, with the bigotry and superstition, all the intolerance of their sect. They have no mosques at Benowm, but perform their devotions in a sort of open shed, or enclosure, made of mats.

The priest is, at the same time, schoolmaster to the juniors. His pupils assemble every evening before his tent; where, by the light of a large fire, made of brushwood and cow's dung, they are taught a few sentences from the Koran, and are initiated into the principles of their creed. Their alphabet differs but little from that in Richardson's Arabic Grammar. They always write with the vowel points. Their priests even affect to know something of foreign literature. The priest of Benowm assured me that he could read the writings of the Christians: he showed me a number of barbarous characters, which he asserted were the Roman alphabet; and he produced another specimen, equally unintelligible, which he declared to be the Kallam il Indi, or Persian. His library consisted of nine volumes in quarto; most of them, I believe, were books of religion—for the name of Mohammed appeared in red letters in almost every page of each. His scholars wrote their lessons upon thin boards, paper being too expensive for general use. The boys were diligent enough, and appeared to possess a considerable share of emulation—carrying their boards slung over their shoulders when about their common employments. When a boy has committed to memory a few of their prayers, and can read and write certain parts of the Koran, he is reckoned sufficiently instructed; and with this slender stock of learning commences his career of life. Proud of his acquirements, he surveys with contempt the unlettered negro; and embraces every opportunity of displaying his superiority over such of his countrymen as are not distinguished by the same accomplishments.

The education of the girls is neglected altogether: mental accomplishments are but little attended to by the women; nor is the want of them considered by the men as a defect in the female character. They are regarded, I believe, as an inferior species of animals; and seem to be brought up for no other purpose than that of administering to the sensual pleasures of their imperious masters. Voluptuousness is therefore considered as their chief accomplishment, and slavish submission as their indispensable duty.

The Moors have singular ideas of feminine perfection. The gracefulness of figure and motion, and a countenance enlivened by expression, are by no means essential points in their standard. With them corpulence and beauty appear to be terms nearly synonymous. A woman of even moderate pretensions must be one who cannot walk without a slave under each arm to support her; and a perfect beauty is a load for a camel. In consequence of this prevalent taste for unwieldiness of bulk, the Moorish ladies take great pains to acquire it early in life; and for this purpose many of the young girls are compelled by their mothers to devour a great quantity of kouskous, and drink a large bowl of camel's milk every morning. It is of no importance whether the girl has an appetite or not; the kouskous and milk must be swallowed, and obedience is frequently enforced by blows. I have seen a poor girl sit crying, with the bowl at her lips, for more than an hour, and her mother, with a stick in her hand, watching her all the while, and using the stick without mercy whenever she observed that her daughter was not swallowing. This singular practice, instead of producing indigestion and disease, soon covers the young lady with that degree of plumpness which, in the eye of a Moor, is perfection itself.

As the Moors purchase all their clothing from the negroes, the women are forced to be very economical in the article of dress. In general they content themselves with a broad piece of cotton cloth, which is wrapped round the middle, and hangs down like a petticoat almost to the ground. To the upper part of this are sewed two square pieces, one before, and the other behind, which are fastened together over the shoulders. The head-dress is commonly a bandage of cotton cloth, with some parts of it broader than others, which serve to conceal the face when they walk in the sun. Frequently, however, when they go abroad, they veil themselves from head to foot.

The employment of the women varies according to their degrees of opulence. Queen Fatima, and a few others of high rank, like the great ladies in some parts of Europe, pass their time chiefly in conversing with their visitors, performing their devotions, or admiring their charms in a looking-glass. The women of inferior class employ themselves in different domestic duties. They are very vain and talkative; and when anything puts them out of humor they commonly vent their anger upon their female slaves, over whom they rule with severe and despotic authority, which leads me to observe that the condition of these poor captives is deplorably wretched. At daybreak they are compelled to fetch water from the wells in large skins, called girbas; and as soon as they have brought water enough to serve the family for the day, as well as the horses (for the Moors seldom give their horses the trouble of going to the wells), they are then employed in pounding the corn and dressing the victuals. This being always done in the open air, the slaves are exposed to the combined heat of the sun, the sand, and the fire. In the intervals it is their business to sweep the tent, churn the milk, and perform other domestic offices. With all this they are badly fed, and oftentimes cruelly punished.

The men's dress, among the Moors of Ludamar, differs but little from that of the negroes, which has been already described, except that they have all adopted that characteristic of the Mohammedan sect, the turban, which is here universally made of white cotton cloth. Such of the Moors as have long beards display them with a mixture of pride and satisfaction, as denoting an Arab ancestry. Of this number was Ali himself; but among the generality of the people the hair is short and busy, and universally black. And here I may be permitted to observe, that if any one circumstance excited among them favorable thoughts towards my own person, it was my beard, which was now grown to an enormous length, and was always beheld with approbation or envy. I believe, in my conscience, they thought it too good a beard for a Christian.

The only diseases which I observed to prevail among the Moors were the intermittent fever and dysentery—for the cure of which nostrums are sometimes administered by their old women, but in general nature is left to her own operations. Mention was made to me of the small-pox as being sometimes very destructive; but it had not, to my knowledge, made its appearance in Ludamar while I was in captivity. That it prevails, however, among some tribes of the Moors, and that it is frequently conveyed by them to the negroes in the southern states, I was assured on the authority of Dr. Laidley, who also informed me that the negroes on the Gambia practice inoculation.

The administration of criminal justice, as far as I had opportunities of observing, was prompt and decisive: for although civil rights were but little regarded in Ludamar, it was necessary when crimes were committed that examples should sometimes be made. On such occasions the offender was brought before Ali, who pronounced, of his sole authority, what judgment he thought proper. But I understood that capital punishment was seldom or never inflicted, except on the negroes.

Although the wealth of the Moors consists chiefly in their numerous herds of cattle, yet, as the pastoral life does not afford full employment, the majority of the people are perfectly idle, and spend the day in trifling conversation about their horses, or in laying schemes of depredation on the negro villages.

Of the number of Ali's Moorish subjects I had no means of forming a correct estimate. The military strength of Ludamar consists in cavalry. They are well mounted, and appear to be very expert in skirmishing and attacking by surprise. Every soldier furnishes his own horse, and finds his accoutrements, consisting of a large sabre, a double-barreled gun, a small red leather bag for holding his balls, and a powder bag slung over the shoulder. He has no pay, nor any remuneration but what arises from plunder. This body is not very numerous; for when Ali made war upon Bambarra I was informed that his whole force did not exceed two thousand cavalry. They constitute, however, by what I could learn, but a very small proportion of his Moorish subjects. The horses are very beautiful, and so highly esteemed that the negro princes will sometimes give from twelve to fourteen slaves for one horse.

Ludamar has for its northern boundary the great desert of Sahara. From the best inquiries I could make, this vast ocean of sand, which occupies so large a space in northern Africa, may be pronounced almost destitute of inhabitants, except where the scanty vegetation which appears in certain spots affords pasturage for the flocks of a few miserable Arabs, who wander from one well to another. In other places, where the supply of water and pasturage is more abundant, small parties of the Moors have taken up their residence. Here they live, in independent poverty, secure from the tyrannical government of Barbary. But the greater part of the desert, being totally destitute of water, is seldom visited by any human being, unless where the trading caravans trace out their toilsome and dangerous route across it. In some parts of this extensive waste the ground is covered with low stunted shrubs, which serve as landmarks for the caravans, and furnish the camels with a scanty forage. In other parts the disconsolate wanderer, wherever he turns, sees nothing around him but a vast interminable expanse of sand and sky—a gloomy and barren void, where the eye finds no particular object to rest upon, and the mind is filled with painful apprehensions of perishing with thirst.

The few wild animals which inhabit these melancholy regions are the antelope and the ostrich; their swiftness of foot enabling them to reach the distant watering-places. On the skirts of the desert, where water is more plentiful, are found lions, panthers, elephants, and wild bears.

Of domestic animals, the only one that can endure the fatigue of crossing the desert is the camel. By the particular conformation of the stomach he is enabled to carry a supply of water sufficient for ten or twelve days; his broad and yielding foot is well adapted for a sandy country; and, by a singular motion of his upper lip, he picks the smallest leaves from the thorny shrubs of the desert as he passes along. The camel is therefore the only beast of burden employed by the trading caravans which traverse the desert in different directions, from Barbary to Nigritia. As this useful and docile creature has been sufficiently described by systematical writers, it is unnecessary for me to enlarge upon his properties. I shall only add that his flesh, though to my own taste dry and unsavory, is preferred by the Moors to any other; and that the milk of the female is in universal esteem, and is indeed sweet, pleasant, and nutritive.

I have observed that the Moors, in their complexion, resemble the mulattoes of the West Indies; but they have something unpleasant in their aspect which the mulattoes have not. I fancied that I discovered in the features of most of them a disposition towards cruelty and low cunning; and I could never contemplate their physiognomy without feeling sensible uneasiness. From the staring wildness of their eyes a stranger would immediately set them down as a nation of lunatics. The treachery and malevolence of their character are manifest in their plundering excursions against the negro villages. Oftentimes without the smallest provocation, and sometimes under the fairest professions of friendship, they will suddenly seize upon the negroes' cattle, and even on the inhabitants themselves. The negroes very seldom retaliate.

Like the roving Arabs, the Moors frequently remove from one place to another, according to the season of the year or the convenience of pasturage. In the month of February, when the heat of the sun scorches up every sort of vegetation in the desert, they strike their tents and approach the negro country to the south, where they reside until the rains commence, in the month of July. At this time, having purchased corn and other necessaries from the negroes, in exchange for salt, they again depart to the northward, and continue in the desert until the rains are over, and that part of the country becomes burnt up and barren.

This wandering and restless way of life, while it inures them to hardships, strengthens at the same time the bonds of their little society, and creates in them an aversion toward strangers which is almost insurmountable. Cut off from all intercourse with civilized nations, and boasting an advantage over the negroes, by possessing, though in a very limited degree, the knowledge of letters, they are at once the vainest and proudest, and perhaps the most bigoted, ferocious, and intolerant of all the nations on the earth—combining in their character the blind superstition of the negro with the savage cruelty and treachery of the Arab.

CHAPTER XVI—VILLAGES ON THE NIGER—DETERMINES TO GO NO FARTHER EASTWARD

Being, in the manner that has been related, compelled to leave Sego, I was conducted the same evening to a village about seven miles to the eastward, with some of the inhabitants of which my guide was acquainted, and by whom we were well received. He was very friendly and communicative, and spoke highly of the hospitality of his countrymen, but withal told me that if Jenné was the place of my destination, which he seemed to have hitherto doubted, I had undertaken an enterprise of greater danger than probably I was apprised of; for, although the town of Jenné was nominally a part of the king of Bambarra's dominions, it was in fact, he said, a city of the Moors—the leading part of the inhabitants being bushreens, and even the governor himself, though appointed by Mansong, of the same sect. Thus was I in danger of falling a second time into the hands of men who would consider it not only justifiable, but meritorious, to destroy me, and this reflection was aggravated by the circumstance that the danger increased as I advanced in my journey, for I learned that the places beyond Jenné were under the Moorish influence in a still greater degree than Jenné itself, and Timbuctoo, the great object of my search, altogether in possession of that savage and merciless people, who allow no Christian to live there. But I had now advanced too far to think of returning to the westward on such vague and uncertain information, and determined to proceed; and being accompanied by the guide, I departed from the village on the morning of the 24th. About eight o'clock we passed a large town called Kabba, situated in the midst of a beautiful and highly cultivated country, bearing a greater resemblance to the centre of England than to what I should have supposed had been the middle of Africa. The people were everywhere employed in collecting the fruit of shea trees, from which they prepare the vegetable butter mentioned in former parts of this work. These trees grow in great abundance all over this part of Bambarra. They are not planted by the natives, but are found growing naturally in the woods; and in clearing woodland for cultivation every tree is cut down but the shea. The tree itself very much resembles the American oak, and the fruit— from the kernel of which, being first dried in the sun, the butter is prepared by boiling the kernel in water—has somewhat the appearance of a Spanish olive. The kernel is enveloped in a sweet pulp, under a thin green rind; and the butter produced from it, besides the advantage of its keeping the whole year without salt, is whiter, firmer, and, to my palate, of a richer flavor, than the best butter I ever tasted made from cow's milk. The growth and preparation of this commodity seem to be among the first objects of African industry in this and the neighboring states, and it constitutes a main article of their inland commerce.

We passed, in the course of the day, a great many villages inhabited chiefly by fishermen, and in the evening about five o'clock arrived at Sansanding, a very large town, containing, as I was told, from eight to ten thousand inhabitants. This place is much resorted to by the Moors, who bring salt from Berroo, and beads and coral from the Mediterranean, to exchange here for gold dust and cotton cloth. This cloth they sell to great advantage in Berroo, and other Moorish countries, where, on account of the want of rain, no cotton is cultivated.

I desired my guide to conduct me to the house in which we were to lodge by the most private way possible. We accordingly rode along between the town and the river, passing by a creek or harbor, in which I observed twenty large canoes, most of them fully loaded, and covered with mats to prevent the rain from injuring the goods. As we proceeded, three other canoes arrived, two with passengers and one with goods. I was happy to find that all the negro inhabitants took me for a Moor, under which character I should probably have passed unmolested, had not a Moor, who was sitting by the river-side, discovered the mistake, and, setting up a loud exclamation, brought together a number of his countrymen.

When I arrived at the house of Counti Mamadi, the dooty of the town, I was surrounded with hundreds of people speaking a variety of different dialects, all equally unintelligible to me. At length, by the assistance of my guide, who acted as interpreter, I understood that one of the spectators pretended to have seen me at one place, and another at some other place; and a Moorish woman absolutely swore that she had kept my house three years at Gallam, on the river Senegal. It was plain that they mistook me for some other person, and I desired two of the most confident to point toward the place where they had seen me.

They pointed due south; hence I think it probable that they came from Cape Coast, where they might have seen many white men. Their language was different from any I had yet heard. The Moors now assembled in great number, with their usual arrogance, compelling the negroes to stand at a distance. They immediately began to question me concerning my religion, but finding that I was not master of Arabic, they sent for two men, whom they call Ilhuidi (Jews), in hopes that they might be able to converse with me. These Jews, in dress and appearance, very much resemble the Arabs; but though they so far conform to the religion of Mohammed as to recite in public prayers from the Koran, they are but little respected by the negroes; and even the Moors themselves allowed that, though I was a Christian, I was a better man than a Jew. They, however, insisted that, like the Jews, I must conform so far as to repeat the Mohammedan prayers; and when I attempted to waive the subject by telling them that I could not speak Arabic, one of them, a shereef from Tuat, in the Great Desert, started up and swore by the Prophet that if I refused to go to the mosque, he would be one that would assist in carrying me thither; and there is no doubt that this threat would have been immediately executed had not my landlord interposed on my behalf. He told them that I was the king's stranger, and he could not see me ill-treated whilst I was under his protection. He therefore advised them to let me alone for the night, assuring them that in the morning I should be sent about my business. This somewhat appeased their clamor, but they compelled me to ascend a high seat by the door of the mosque, in order that everybody might see me, for the people had assembled in such numbers as to be quite ungovernable, climbing upon the houses, and squeezing each other, like the spectators at an execution. Upon this seat I remained until sunset, when I was conducted into a neat little hut, with a small court before it, the door of which Counti Mamadi shut, to prevent any person from disturbing me. But this precaution could not exclude the Moors. They climbed over the top of the mud wall and came in crowds into the court, "in order," they said, "to see me perform my evening devotions, and eat eggs." The former of these ceremonies I did not think proper to comply with, but I told them I had no objection to eat eggs, provided they would bring me eggs to eat. My landlord immediately brought me seven hens' eggs, and was much surprised to find that I could not eat them raw; for it seems to be a prevalent opinion among the inhabitants of the interior that Europeans subsist almost entirely on this diet. When I had succeeded in persuading my landlord that this opinion was without foundation, and that I would gladly partake of any victuals which he might think proper to send me, he ordered a sheep to be killed, and part of it to be dressed for my supper. About midnight, when the Moors had left me, he paid me a visit, and with much earnestness desired me to write him a saphie. "If a Moor's saphie is good," said this hospitable old man, "a white man's must needs be better." I readily furnished him with one, possessed of all the virtues I could concentrate, for it contained the Lord's Prayer. The pen with which it was written was made of a reed; a little charcoal and gum-water made very tolerable ink, and a thin board answered the purpose of paper.

July 25.—Early in the morning, before the Moors were assembled, I departed from Sansanding, and slept the ensuing night at a small town called Sibili, from whence on the day following I reached Nyara, a large town at some distance from the river, where I halted the 27th, to have my clothes washed, and recruit my horse. The dooty there has a very commodious house, flat-roofed, and two storeys high. He showed me some gunpowder of his own manufacturing; and pointed out, as a great curiosity, a little brown monkey that was tied to a stake by the door, telling me that it came from a far distant country called Kong.

July 28.—I departed from Nyara, and reached Nyamee about noon. This town is inhabited chiefly by Foulahs from the kingdom of Masina. The dooty, I know not why, would not receive me, but civilly sent his son on horseback to conduct me to Modiboo, which he assured me was at no great distance.

We rode nearly in a direct line through the woods, but in general went forward with great circumspection. I observed that my guide frequently stopped and looked under the bushes. On inquiring the reason of this caution he told me that lions were very numerous in that part of the country, and frequently attacked people traveling through the woods. While he was speaking, my horse started, and looking round, I observed a large animal of the camelopard kind standing at a little distance. The neck and fore-legs were very long; the head was furnished with two short black horns, turning backwards; the tail, which reached down to the ham joint, had a tuft of hair at the end. The animal was of a mouse color, and it trotted away from us in a very sluggish manner—moving its head from side to side, to see if we were pursuing it. Shortly after this, as we were crossing a large open plain, where there were a few scattered bushes, my guide, who was a little way before me, wheeled his horse round in a moment, calling out something in the Foulah language which I did not understand. I inquired in Mandingo what he meant; "Wara billi billi!" ("A very large lion!") said he, and made signs for me to ride away. But my horse was too much fatigued; so we rode slowly past the bush from which the animal had given us the alarm. Not seeing anything myself, however, I thought my guide had been mistaken, when the Foulah suddenly put his hand to his mouth, exclaiming, "Soubah an allahi!" ("God preserve us!") and, to my great surprise, I then perceived a large red lion, at a short distance from the bush, with his head couched between his forepaws. I expected he would instantly spring upon me, and instinctively pulled my feet from my stirrups to throw myself on the ground, that my horse might become the victim rather than myself. But it is probable the lion was not hungry; for he quietly suffered us to pass, though we were fairly within his reach. My eyes were so riveted upon this sovereign of the beasts that I found it impossible to remove them until we were at a considerable distance. We now took a circuitous route through some swampy ground, to avoid any more of these disagreeable encounters. At sunset we arrived at Modiboo—a delightful village on the banks of the Niger, commanding a view of the river for many miles both to the east and west. The small green islands (the peaceful retreat of some industrious Foulahs, whose cattle are here secure from the depredations of wild beasts) and the majestic breadth of the river, which is here much larger than at Sego, render the situation one of the most enchanting in the world. Here are caught great plenty of fish, by means of long cotton nets, which the natives make themselves, and use nearly in the same manner as nets are used in Europe. I observed the head of a crocodile lying upon one of the houses, which they told me had been killed by the shepherds in a swamp near the town. These animals are not uncommon in the Niger, but I believe they are not oftentimes found dangerous. They are of little account to the traveler when compared with the amazing swarms of mosquitoes, which rise from the swamps and creeks in such numbers as to harass even the most torpid of the natives; and as my clothes were now almost worn to rags, I was but ill prepared to resist their attacks. I usually passed the night without shutting my eyes, walking backwards and forwards, fanning myself with my hat; their stings raised numerous blisters on my legs and arms, which, together with the want of rest, made me very feverish and uneasy.

July 29.—Early in the morning, my landlord, observing that I was sickly, hurried me away, sending a servant with me as a guide to Kea. But though I was little able to walk, my horse was still less able to carry me; and about six miles to the east of Modiboo, in crossing some rough clayey ground, he fell, and the united strength of the guide and myself could not place him again upon his legs. I sat down for some time beside this worn-out associate of my adventures, but finding him still unable to rise, I took off the saddle and bridle, and placed a quantity of grass before him. I surveyed the poor animal, as he lay panting on the ground, with sympathetic emotion, for I could not suppress the sad apprehension that I should myself, in a short time, lie down and perish in the same manner, of fatigue and hunger. With this foreboding I left my poor horse, and with great reluctance followed my guide on foot along the bank of the river until about noon, when we reached Kea, which I found to be nothing more than a small fishing village. The dooty, a surly old man, who was sitting by the gate, received me very coolly; and when I informed him of my situation, and begged his protection, told me with great indifference that he paid very little attention to fine speeches, and that I should not enter his house. My guide remonstrated in my favor, but to no purpose, for the dooty remained inflexible in his determination. I knew not where to rest my wearied limbs, but was happily relieved by a fishing canoe belonging to Silla, which was at that moment coming down the river. The dooty waved to the fisherman to come near, and desired him to take charge of me as far as Moorzan. The fisherman, after some hesitation, consented to carry me, and I embarked in the canoe in company with the fisherman, his wife, and a boy. The negro who had conducted me from Modiboo now left me. I requested him to look to my horse on his return, and take care of him if he was still alive, which he promised to do.

Departing from Kea, we proceeded about a mile down the river, when the fisherman paddled the canoe to the bank and desired me to jump out. Having tied the canoe to a stake, he stripped off his clothes, and dived for such a length of time that I thought he had actually drowned himself, and was surprised to see his wife behave with so much indifference upon the occasion; but my fears were over when he raised up his head astern of the canoe and called for a rope. With this rope he dived a second time, and then got into the canoe and ordered the boy to assist him in pulling. At length they brought up a large basket, about ten feet in diameter, containing two fine fish, which the fisherman—after returning the basket into the water—immediately carried ashore and hid in the grass. We then went a little farther down and took up another basket, in which was one fish. The fisherman now left us to carry his prizes to some neighboring market, and the woman and boy proceeded with me in the canoe down the river.

About four o'clock we arrived at Moorzan, a fishing town on the northern bank, from whence I was conveyed across the river to Silla, a large town, where I remained until it was quite dark, under a tree, surrounded by hundreds of people.

With a great deal of entreaty the dooty allowed me to come into his baloon to avoid the rain, but the place was very damp, and I had a smart paroxysm of fever during the night. Worn down by sickness, exhausted with hunger and fatigue, half-naked, and without any article of value by which I might procure provisions, clothes, or lodging, I began to reflect seriously on my situation. I was now convinced, by painful experience, that the obstacles to my further progress were insurmountable. The tropical rains were already set in with all their violence—the rice grounds and swamps were everywhere overflowed—and in a few days more, traveling of every kind, unless by water, would be completely obstructed. The kowries which remained of the king of Bambarra's present were not sufficient to enable me to hire a canoe for any great distance, and I had but little hopes of subsisting by charity in a country where the Moors have such influence. But, above all, I perceived that I was advancing more and more within the power of those merciless fanatics, and, from my reception both at Sego and Sansanding, I was apprehensive that, in attempting to reach even Jenné (unless under the protection of some man of consequence amongst them, which I had no means of obtaining), I should sacrifice my life to no purpose, for my discoveries would perish with me. The prospect either way was gloomy. In returning to the Gambia, a journey on foot of many hundred miles presented itself to my contemplation, through regions and countries unknown. Nevertheless, this seemed to be the only alternative, for I saw inevitable destruction in attempting to proceed to the eastward. With this conviction on my mind I hope my readers will acknowledge that I did right in going no farther.

Having thus brought my mind, after much doubt and perplexity, to a determination to return westward, I thought it incumbent on me, before I left Silla, to collect from the Moorish and negro traders all the information I could concerning the farther course of the Niger eastward, and the situation and extent of the kingdoms in its vicinage; and the following few notices I received from such various quarters as induce me to think they are authentic:-

Two short days' journey to the eastward of Silla is the town of Jenné, which is situated on a small island in the river, and is said to contain a greater number of inhabitants than Sego itself, or any other town in Bambarra. At the distance of two days more, the river spreads into a considerable lake, called Dibbie (or the Dark Lake), concerning the extent of which all the information I could obtain was that in crossing it from west to east the canoes lose sight of land one whole day. From this lake the water issues in many different streams, which terminate in two large branches, one whereof flows toward the north-east, and the other to the east; but these branches join at Kabra, which is one day's journey to the southward of Timbuctoo, and is the port or shipping-place of that city. The tract of land which the two streams encircle is called Jinbala, and is inhabited by negroes; and the whole distance by land from Jenné to Timbuctoo is twelve days' journey.

From Kabra, at the distance of eleven days' journey down the stream, the river passes to the southward of Houssa, which is two days' journey distant from the river. Of the farther progress of this great river, and its final exit, all the natives with whom I conversed seemed to be entirely ignorant. Their commercial pursuits seldom induce them to travel farther than the cities of Timbuctoo and Houssa, and as the sole object of those journeys is the acquirement of wealth, they pay little attention to the course of rivers or the geography of countries. It is, however, highly probable that the Niger affords a safe and easy communication between very remote nations. All my informants agreed that many of the negro merchants who arrive at Timbuctoo and Houssa from the eastward speak a different language from that of Bambarra, or any other kingdom with which they are acquainted. But even these merchants, it would seem, are ignorant of the termination of the river, for such of them as can speak Arabic describe the amazing length of its course in very general terms, saying only that they believe it runs to the world's end.

The names of many kingdoms to the eastward of Houssa are familiar to the inhabitants of Bambarra. I was shown quivers and arrows of very curious workmanship, which I was informed came from the kingdom of Kassina.

On the northern bank of the Niger, at a short distance from Silla, is the kingdom of Masina, which is inhabited by Foulahs. They employ themselves there, as in other places, chiefly in pasturage, and pay an annual tribute to the king of Bambarra for the lands which they occupy.

To the north-east of Masina is situated the kingdom of Timbuctoo, the great object of European research—the capital of this kingdom being one of the principal marts for that extensive commerce which the Moors carry on with the negroes. The hopes of acquiring wealth in this pursuit, and zeal for propagating their religion, have filled this extensive city with Moors and Mohammedan converts. The king himself and all the chief officers of state are Moors; and they are said to be more severe and intolerant in their principles than any other of the Moorish tribes in this part of Africa. I was informed by a venerable old negro, that when he first visited Timbuctoo, he took up his lodging at a sort of public inn, the landlord of which, when he conducted him into his hut, spread a mat on the floor, and laid a rope upon it, saying, "If you are a Mussulman, you are my friend—sit down; but if you are a kafir, you are my slave, and with this rope I will lead you to market." The present king of Timbuctoo is named Abu Abrahima. He is reported to possess immense riches. His wives and concubines are said to be clothed in silk, and the chief officers of state live in considerable splendor. The whole expense of his government is defrayed, as I was told, by a tax upon merchandise, which is collected at the gates of the city.

The city of Houssa (the capital of a large kingdom of the same name, situated to the eastward of Timbuctoo), is another great mart for Moorish commerce. I conversed with many merchants who had visited that city, and they all agreed that it is larger—and more populous than Timbuctoo. The trade, police, and government are nearly the same in both; but in Houssa the negroes are in greater proportion to the Moors and have some share in the government.

Concerning the small kingdom of Jinbala I was not able to collect much information. The soil is said to be remarkably fertile, and the whole country so full of creeks and swamps that the Moors have hitherto been baffled in every attempt to subdue it. The inhabitants are negroes, and some of them are said to live in considerable affluence, particularly those near the capital, which is a resting-place for such merchants as transport goods from Timbuctoo to the western parts of Africa.

To the southward of Jinbala is situated the negro kingdom of Gotto, which is said to be of great extent. It was formerly divided into a number of petty states, which were governed by their own chiefs; but their private quarrels invited invasion from the neighboring kingdoms. At length a politic chief of the name of Moossee had address enough to make them unite in hostilities against Bambarra; and on this occasion he was unanimously chosen general—the different chiefs consenting for a time to act under his command. Moossee immediately dispatched a fleet of canoes, loaded with provisions, from the banks of the lake Dibbie up the Niger toward Jenné, and with the whole of his army pushed forward into Bambarra. He arrived on the bank of the Niger opposite to Jenné before the townspeople had the smallest intimation of his approach. His fleet of canoes joined him the same day, and having landed the provisions, he embarked part of his army, and in the night took Jenné by storm. This event so terrified the king of Bambarra that he sent messengers to sue for peace; and in order to obtain it consented to deliver to Moossee a certain number of slaves every year, and return everything that had been taken from the inhabitants of Gotto. Moossee, thus triumphant, returned to Gotto, where he was declared king, and the capital of the country is called by his name.

On the west of Gotto is the kingdom of Baedoo, which was conquered by the present king of Bambarra about seven years ago, and has continued tributary to him ever since.

West of Baedoo is Maniana, the inhabitants of which, according to the best information I was able to collect, are cruel and ferocious—carrying their resentment toward their enemies so far as never to give quarter, and even to indulge themselves with unnatural and disgusting banquets of human flesh.

JOHN H. SPEKE
INTRODUCTION TO AFRICA

John Hanning Speke was a soldier in the British colonial Indian Army. He made three voyages in Africa around the middle of the century. He traveled in Somalia with the famous explorer Richard Burton and then made two trips from Zanzibar to the Great Lakes region. He was one of the first Europeans to reach Lake Victoria Nyanza, and he argued that it was the source of the White Nile. Speke saw African landscapes in decidedly racial terms. He is credited with popularizing the "Hamitic" myth, which argued that all signs of development in precolonial Africa came from outside. The brief excerpt here gives an account of the fauna of Africa that Speke encountered in his travel. The excerpt is from John H. Speke, Journal of the Discovery of the Source of the Nile *(Edinburg: 1863 [1st ed.]).*

Fauna of Africa
by John Hanning Speke
Now, descending to the inferior order of creation, I shall commence with the domestic animals first, to show what the traveler may expect to find for his usual support. Cows, after leaving the low lands near the coast, are found to be plentiful everywhere, and to produce milk in small quantities, from which butter is made. Goats are common all over Africa; but sheep are not so plentiful, nor do they show such good breeding—being generally lanky, with long fat tails. Fowls, much like those in India, are abundant everywhere. A few Muscovy ducks are imported, also pigeons and cats. Dogs, like the Indian pariah, are very plentiful, only much smaller; and a few donkeys are found in certain localities. Now, considering this good supply of meat, whilst all tropical plants will grow just as well in central equatorial Africa as they do in India, it surprises the traveler there should be any famines; yet such is too often the case, and the negro, with these bounties within his reach, is sometimes found eating dogs, cats, rats, porcupines, snakes, lizards, tortoises, locusts, and white ants, or is forced to seek the seeds of wild grasses, or to pluck wild herbs, fruits, and roots; whilst at the proper seasons they hunt the wild elephant, buffalo, giraffe, zebra, pigs, and antelopes; or, going out with their arrows, have battles against the guinea-fowls and small birds.

The frequency with which collections of villages are found all over the countries we are alluding to leaves but very little scope for the runs of wild animals, which are found only in dense jungles, open forests, or prairies generally speaking, where hills can protect them, and near rivers whose marshes produce a thick growth of vegetation to conceal them from their most dreaded enemy—man. The prowling, restless elephant, for instance, though rarely seen, leaves indications of his nocturnal excursions in every wilderness, by wantonly knocking down the forest-trees. The morose rhinoceros, though less numerous, are found in every thick jungle. So is the savage buffalo, especially delighting in dark places, where he can wallow in the mud and slake his thirst without much trouble; and here also we find the wild pig.

The gruff hippopotamus is as widespread as any, being found wherever there is water to float him; whilst the shy giraffe and zebra affect all open forests and plains where the grass is not too long; and antelopes, of great variety in species and habits, are found wherever man will let them alone and they can find water. The lion is, however, rarely heard—much more seldom seen. Hyenas are numerous, and thievishly inclined. Leopards, less common, are the terror of the villagers. Foxes are not numerous, but frighten the black traveler by their ill-omened bark. Hares, about half the size of English ones—there are no rabbits—are widely spread, but not numerous; porcupines the same. Wild cats, and animals of the ferret kind, destroy game. Monkeys of various kinds and squirrels harbor in the trees, but are rarely seen. Tortoises and snakes, in great variety, crawl over the ground, mostly after the rains. Rats and lizards—there are but few mice—are very abundant, and feed both in the fields and on the stores of the men.

The wily ostrich, bustard, and florikan affect all open places. The guineafowl is the most numerous of all game-birds. Partridges come next but do not afford good sport; and quails are rare. Ducks and snipe appear to love Africa less than any other country; and geese and storks are only found where water most abounds. Vultures are uncommon; hawks and crows much abound, as in all other countries; but little birds, of every color and note, are discoverable in great quantities near water and by the villages. Huge snails and small ones, as well as fresh-water shells, are very abundant, though the conchologist would find but little variety to repay his labors; and insects, though innumerable, are best sought for after the rains have set in.

SAMUEL W. BAKER
THE ALBERT NYANZA

Samuel Baker was another British soldier and explorer during the middle of the nineteenth century. He traveled in Ethiopia, Egypt, and the Nile Valley during the 1860s and eventually became an officer in the army of the nominally independent ruler, the Khedive of Egypt. He met Speke in what became the Sudan in 1864 and following Speke's advice, found Lake Albert, on the present-day border of Uganda and the Democratic Republic of the Congo. Baker's description of the lake makes it seem larger than it is, but he accurately describes several of the African states in the region. After the journey from which this excerpt comes, he became involved first in the Anglo-Egyptian conquest of Sudan and then in efforts to reconquer the Mahdist state in Sudan that had driven Anglo-Egyptian forces out under his successor. Baker's account of finding Lake Albert betrays an imperial vision of African landscapes. For him, the landscape is not real until a "civilized European" saw it. His companions and the peoples he met are mere backdrop for his "discoveries." Rare among nineteenth-century explorers, Baker's wife, Florence von Sass Baker, participated in the journey. This ex-

cerpt comes from Sir Samuel W. Baker, The Albert Nyanza, Great Basin of the Nile, and Explorations of the Nile Sources *(London: Macmillan and co. 1866).*

Lake Albert
Samuel W. Baker
The 14th March.—The sun had not risen when I was spurring my ox after the guide, who, having been promised a double handful of beads on arrival at the lake, had caught the enthusiasm of the moment. The day broke beautifully clear, and having crossed a deep valley between the hills, we toiled up the opposite slope. I hurried to the summit. The glory of our prize burst suddenly upon me! There, like a sea of quicksilver, lay far beneath the grand expanse of water,—a boundless sea horizon on the south and southwest, glittering in the noonday sun; and on the west, at fifty or sixty miles' distance, blue mountains rose from the bosom of the lake to a height of about 7,000 feet above its level.

It is impossible to describe the triumph of that moment;—here was the reward for all our labor—for the years of tenacity with which we had toiled through Africa. England had won the sources of the Nile! Long before I reached this spot, I had arranged to give three cheers with all our men in English style in honor of the discovery, but now that I looked down upon the great inland sea lying nestled in the very heart of Africa, and thought how vainly mankind had sought these sources throughout so many ages, and reflected that I had been the humble instrument permitted to unravel this portion of the great mystery when so many greater than I had failed, I felt too serious to vent my feelings in vain cheers for victory, and I sincerely thanked God for having guided and supported us through all dangers to the good end. I was about 1,500 feet above the lake, and I looked down from the steep granite cliff upon those welcome waters—upon that vast reservoir which nourished Egypt and brought fertility where all was wilderness—upon that great source so long hidden from mankind; that source of bounty and of blessings to millions of human beings; and as one of the greatest objects in nature, I determined to honor it with a great name. As an imperishable memorial of one loved and mourned by our gracious Queen and deplored by every Englishman, I called this great lake "the Albert N'yanza." The Victoria and the Albert lakes are the two sources of the Nile.

The zigzag path to descend to the lake was so steep and dangerous that we were forced to leave our oxen with a guide, who was to take them to Magungo and wait for our arrival. We commenced the descent of the steep pass on foot. I led the way, grasping a stout bamboo. My wife in extreme weakness tottered down the pass, supporting herself upon my shoulder, and stopping to rest every twenty paces. After a toilsome descent of about two hours, weak with years of fever, but for the moment strengthened by success, we gained the level plain below the cliff. A walk of about a mile through flat sandy meadows of fine turf interspersed with trees and bush, brought us to the water's edge. The waves were rolling upon a white pebbly beach: I rushed into the lake, and thirsty with heat and fatigue, with a heart full of gratitude, I drank deeply from the Sources of the Nile. Within a quarter of a mile of the lake was a fishing village named Vacovia, in which we now established ourselves. Everything smelt of fish—and everything looked like fishing; not the "gentle art" of England with rod and fly, but harpoons were leaning against the huts, and lines almost as thick as the little finger were hanging up to dry, to which were attached iron hooks of a size that said much for the monsters of the Albert lake. On entering the hut I found a prodigious quantity of tackle; the lines were beautifully made of the fiber of the plantain stem, and were exceedingly elastic, and well adapted to withstand the first rush of a heavy fish; the hooks were very coarse, but well barbed, and varied in size from two to six inches. A number of harpoons and floats for hippopotami were arranged in good order, and the tout ensemble of the hut showed that the owner was a sportsman.

The harpoons for hippopotami were precisely the same pattern as those used by the Hamran Arabs on the Taka frontier of Abyssinia, having a narrow blade of three-quarters of an inch in width, with only one barb. The rope fitted to the harpoon was beautifully made of plantain fiber, and the float was a huge piece of ambatch-wood about fifteen inches in diameter. They speared the hippopotamus from canoes, and these large floats were necessary to be easily distinguished in the rough waters of the lake.

My men were perfectly astounded at the appearance of the lake. The journey had been so long, and "hope deferred" had so completely sickened their hearts, that they had long since disbelieved in the existence of the lake, and they were persuaded that I was leading them to the sea. They now looked at the lake with amazement—two of them had already seen the sea at Alexandria, and they unhesitatingly declared that this was the sea, but that it was not salt.

Vacovia was a miserable place, and the soil was so impregnated with salt that no cultivation was possible. Salt was the natural product of the country; and the population were employed in its manufacture, which constituted the business of the lake shores—being exchanged for supplies from the interior. I went to examine the pits: these were about six feet deep, from which was dug a black sandy mud that was placed in large earthenware jars; these were supported upon frames, and mixed with water, which filtering rapidly through small holes in the bottom, was received in jars beneath: this water was again used with fresh mud until it became a strong brine, when it was boiled and evaporated. The salt was white, but very bitter. I imagine that it has been formed by the decay of aquatic plants that have been washed ashore by the waves; decomposing, they have formed a mud deposit, and much potash is combined with the salt. The flat sandy meadow that extends from the lake for about a mile to the foot of the precipitous cliffs of 1,500 feet, appears to have formed at one period the bottom of the lake—in fact, the flat land of Vacovia looks like a bay, as the mountain cliffs about five miles south and north descend abruptly to the water, and the flat is the bottom of a horseshoe formed by the cliffs. Were the level of the lake fifteen feet higher, this flat would be flooded to the base of the hills.

I procured a couple of kids from the chief of the village for some blue beads, and having received an ox as a present from the headman of Parkani in return for a number of beads and bracelets, I gave my men a grand feast in honor of the discovery; I made them an address, explaining to them how much trouble we should have been saved had my whole party behaved well from the first commencement and trusted to my guidance, as we should have arrived here twelve mouths ago; at the same time I told them, that it was a greater honor to have achieved the task with so small a force as thirteen men, and that as the lake was thus happily reached, and Mrs. Baker was restored to health after so terrible a danger, I should forgive them past offenses and wipe out all that had been noted against them in my journal. This delighted my people, who ejaculated "El hamd el Illah!" (thank God!) and fell to immediately at their beef.

At sunrise on the following morning I took the compass, and accompanied by the chief of the village, my guide Rabonga, and the woman Bacheeta, I went to the borders of the lake to survey the country. It was beautifully clear, and with a powerful telescope I could distinguish two large waterfalls that cleft the sides of the mountains on the opposite shore. Although the outline of the mountains was distinct upon the bright blue sky, and the dark shades upon their sides denoted deep gorges, I could not distinguish other features than the two great falls, which looked like threads of silver on the dark face of the mountains. No base had been visible, even from an elevation of 1,500 feet above the water level, on my first view of the lake, but the chain of lofty mountains on the west appeared to rise suddenly from the water. This appearance must have been due to the great distance, the base being below the horizon, as dense columns of smoke were ascending apparently from the surface of the water: this must have been produced by the burning of prairies at the foot of the mountains. The chief assured me that large canoes had been known to cross over from the other side, but that it required four days and nights of hard rowing to accomplish the voyage, and that many boats had been lost in the attempt. The canoes of Unyoro were not adapted for so dangerous a journey; but the western shore of the lake was comprised in the great kingdom of Malegga, governed by King Kajoro, who possessed large canoes, and traded with Kamrasi from a point opposite to Magungo, where the lake was contracted to the width of one day's voyage. He described Malegga as a very powerful country, and of greater extent than either Unyora or Uganda. . . . South of Malegga was a country named Tori, governed by a king of the same name: beyond that country to the south on the western shore no intelligence could be obtained from any one.

The lake was known to extend as far south as Karagwe; and the old story was repeated, that Rumanika, the king of that country, was in the habit of sending ivory-hunting parties to the lake at Utumbi, and that formerly they had navigated the lake to Magungo. This was a curious confirmation of the report given me by Speke at Gondokoro, who wrote: "Rumanika is constantly in the habit of sending ivory-hunting parties to Utumbi."

The eastern shores of the lake were, from north to south, occupied by Chopi, Unyoro, Uganda, Utumbi, and Karagwe: from the last point, which could not be less than about two degrees south latitude, the lake was reported to turn suddenly to the west and to continue in that direction for an unknown distance. North of Malegga, on the west of the lake, was a small country called M'Caroli; then Koshi, on the west side of the Nile at its exit from the lake; and on the east side of the Nile was the Madi, opposite to Koshi. Both the guide and the chief of Vacovia informed me that we should be taken by canoes to Magungo, to the point at which the Somerset that we had left at Karuma joined the lake; but that we could not ascend it, as it was a succession of cataracts the whole way from Karuma until within a short distance of Magungo. The exit of the Nile from the lake at Koshi was navigable for a considerable distance, and canoes could descend the river as far as the Madi.

They both agreed that the level of the lake was never lower than at present, and that it never rose higher than a mark upon the beach that accounted for an increase of about four feet. The beach was perfectly clean sand, upon which the waves rolled like those of the sea, throwing up weeds precisely as seaweed may be seen upon the English shore. It was a grand sight to look upon this vast reservoir of the mighty Nile, and to watch the heavy swell tumbling upon the beach, while far to the southwest the eye searched as vainly for a bound as though upon the Atlantic. It was with extreme emotion that I enjoyed this glorious scene. My wife, who had followed me so devotedly, stood by my side pale and exhausted—a wreck upon the shores of the great Albert lake that we had so long striven to reach. No European foot had ever trod upon its sand, nor had the eyes of a white man ever scanned its vast expanse of water. We were the first; and this was the key to the great secret that even Julius Caesar yearned to unravel, but in vain. Here was the great basin of the Nile that received EVERY DROP OF WATER, even from the passing shower to the roaring mountain torrent that drained from Central Africa toward the north. This was the great reservoir of the Nile!

The first coup d'oeil from the summit of the cliff 1,500 feet above the level had suggested what a closer examination confirmed. The lake was a vast depression far below the general level of the country, surrounded by precipitous cliffs, and bounded on the west and southwest by great ranges of mountains from five to seven thousand feet above the level of its waters—thus it was the one great reservoir into which everything MUST drain; and from this vast rocky cistern the Nile made its exit, a giant in its birth. It was a grand arrangement of Nature for the birth of so mighty and important a stream as the river Nile. The Victoria N'yanza of Speke formed a reservoir at a high altitude, receiving a drainage from the west by the Kitangule river, and Speke had seen the M'fumbiro mountain at a great distance as a peak among other mountains from which the streams descended, which by uniting formed the main river Kitangule, the principal feeder of the Victoria lake from the west, in about the 2 degrees S. latitude: thus the same chain of mountains that fed the Victoria on the east must have a watershed to the west and north that would flow into the Albert lake. The general drainage of the Nile basin tending from south to north, and the Albert lake extending much farther north than the Victoria, it receives the river from the latter lake, and thus monopolizes the entire headwaters of the Nile. The Albert is the grand reservoir, while the Victoria is the eastern source, the parent streams that form these lakes are from the same origin, and the Kitangule sheds its waters to the Victoria to be received eventually by the Albert, precisely as the highlands of M'fumbiro and the Blue Mountains pour their northern drainage direct into the Albert lake. The entire Nile system, from the first Abyssinian tributary the Atbara in N. latitude 17 deg. 37 min. even to the equator, exhibits a uniform drainage from S.E. to N.W., every tributary flowing in that direction to the main stream of the Nile; this system is persisted in by the Victoria Nile, which having continued a northerly course from its exit from the Victoria lake to Karuma in lat. 2 degrees 16' N. turns suddenly to the west and meets the Albert lake at Magungo; thus, a line drawn from Magungo to the Ripon Falls from the Victoria lake will prove the general slope of the country to be the same as exemplified throughout the entire system of the eastern basin of the Nile, tending from S.E. to N.W.

That many considerable effluents flow into the Albert lake there is no doubt. The two waterfalls seen by telescope upon the western shore descending from the Blue Mountains must be most important streams, or they could not have been distinguished at so great a distance as fifty or sixty miles; the natives assured me that very many streams, varying in size, descended the mountains upon all sides into the general reservoir.

I returned to my hut: the flat turf in the vicinity of the village was strewn with the bones of immense fish, hippopotami, and crocodiles; but the latter reptiles were merely caught in revenge for any outrage committed by them, as their flesh was looked upon with disgust by the natives of Unyoro. They were so numerous and voracious in the lake, that the natives cautioned us not to allow the women to venture into the water even to the knees when filling their water jars.

DAVID LIVINGSTONE
MISSIONARY TRAVELS IN SOUTHERN AFRICA

Perhaps no explorer of the late nineteenth century was more famous than David Livingstone, the Scottish doctor and missionary. He came to Africa in 1840 to work among the Tswana people of what is now northern South Africa and Botswana. He proved more adept at geography than at missionizing, however. He traveled through the Kalahari. While he married the daughter of another prominent (and more successful) missionary in southern Africa, he soon began to think strategically rather than spiritually. He crossed Africa from Luanda to Mozambique. The account comes from his work from his early years in Africa. After these journeys, he commanded the Zambezi expedition where he tried to prove that the Zambezi River could be navigated and would open Central Africa to commerce and Christianity. The mission was an abject failure, with all the European members of the expedition either leaving or dying (including his wife). He then sought to explore eastern and Central Africa with the aim of gaining knowledge necessary to spread Christianity and end the slave trade. Sick, he wound up living off the generosity of the very merchants (who dealt in slaves among other things) he sought to combat. He was "found" by Henry Stanley at Ujiji on Lake Tanganyika in 1871 but died soon after, in 1872, in Africa. This account of the Kalahari of northern South Africa and Botswana shows Livingstone reading the landscape as a missionary seeing the landscape as a field of people to bring to God and in need of both spiritual guidance and civilization. The excerpt is from David Livingstone, Missionary Travels and Researches in South Africa *(London: J. Murray, 1857).*

The Kalahari
David Livingstone
Before narrating the incidents of this journey, I may give some account of the great Kalahari Desert, in order that the reader may understand in some degree the nature of the difficulties we had to encounter.

The space from the Orange River in the south, lat. 29 Degrees, to Lake Ngami in the north, and from about 24 Degrees east long. to near the west coast, has been called a desert simply because it contains no running water, and very little water in wells. It is by no means destitute of vegetation and inhabitants, for it is covered with grass and a great variety of creeping plants; besides which there are large patches of bushes, and even trees. It is remarkably flat, but intersected in different parts by the beds of ancient rivers; and prodigious herds of certain antelopes, which require little or no water, roam over the trackless plains. The inhabitants, Bushmen and Bakalahari, prey on the game and on the countless rodentia and small species of the feline race which subsist on these. In general, the soil is light-colored soft sand, nearly pure silica. The beds of the ancient rivers contain much alluvial soil; and as that is baked hard by the burning sun, rain-water stands in pools in some of them for several months in the year.

The quantity of grass which grows on this remarkable region is astonishing, even to those who are familiar with India. It usually rises in tufts with bare spaces between, or the intervals are occupied by creeping plants, which, having their roots buried far beneath the soil, feel little the effects of the scorching sun. The number of these which have tuberous roots is very great; and their structure is intended to supply nutriment and moisture, when, during the long droughts, they can be obtained nowhere else. Here we have an example of a plant, not generally tuber-bearing, becoming so under circumstances where that appendage is necessary to act as a reservoir for preserving its life; and the same thing occurs in Angola to a species of grape-bearing vine, which is so furnished for the same purpose. The plant to which I at present refer is one of the cucurbitaceae, which bears a small, scarlet-colored, eatable cucumber. Another plant, named Leroshua, is a blessing to the inhabitants of the Desert. We see a small plant with linear leaves, and a stalk not thicker than a crow's quill; on digging down a foot or eighteen inches beneath, we come to a tuber, often as large as the head of a young child; when the rind is removed, we find it to be a mass of cellular tissue, filled with fluid much like that in a young turnip. Owing to the depth beneath the soil at which it is found, it is generally deliciously cool and refreshing. Another kind, named Mokuri, is seen in other parts of the country, where long-continued heat parches the soil. This plant is an herbaceous creeper, and deposits under ground a number of tubers, some as large as a man's head, at spots in a circle a yard or more, horizontally, from the stem. The natives strike the ground on the circumference of the circle with stones, till, by hearing a difference of sound, they know the water-bearing tuber to be beneath. They then dig down a foot or so, and find it.

But the most surprising plant of the Desert is the "Kengwe or Keme" ('Cucumis caffer'), the watermelon. In years when more than the usual quantity of rain falls, vast tracts of the country are literally covered with these melons; this was the case annually when the fall of rain was greater than it is now, and the Bakwains sent trading parties every year to the lake. It happens commonly once every ten or eleven years, and for the last three times its occurrence has coincided with an extraordinarily wet season. Then animals of every sort and name, including man, rejoice in the rich supply. The elephant, true lord of the forest, revels in this fruit, and so do the different species of rhinoceros, although naturally so diverse in their choice of pasture. The various kinds of antelopes feed on them with equal avidity, and lions, hyaenas, jackals, and mice, all seem to know and appreciate the common blessing. These melons are not, however, all of them eatable; some are sweet, and others so bitter that the whole are named by the Boers the "bitter watermelon." The natives select them by striking one melon after another with a hatchet, and applying the tongue to the gashes. They thus readily distinguish between the bitter and sweet. The bitter are deleterious, but the sweet are quite wholesome. This peculiarity of one species of plant bearing both sweet and bitter fruits occurs also in a red, eatable cucumber, often met with in the country. It is about four inches long, and about an inch and a half in diameter. It is of a bright scarlet color when ripe. Many are bitter, others quite sweet. Even melons in a garden may be made bitter by a few bitter kengwe in the vicinity. The bees convey the pollen from one to the other.

The human inhabitants of this tract of country consist of Bushmen and Bakalahari. The former are probably the aborigines of the southern portion of the continent, the latter the remnants of the first emigration of Bechuanas. The Bushmen live in the Desert from choice, the Bakalahari from compulsion, and both possess an intense love of liberty. The Bushmen are exceptions in language, race, habits, and appearance. They are the only real nomads in the country; they never cultivate the soil, nor rear any domestic animal save wretched dogs. They are so intimately acquainted with the habits of the game that they follow them in their migrations, and prey upon them from place to place, and thus prove as complete a check upon their inordinate increase as the other carnivora. The chief subsistence of the Bushmen is the flesh of game, but that is eked out by what the women collect of roots and beans, and fruits of the Desert. Those who inhabit the hot sandy plains of the Desert possess generally thin, wiry forms, capable of great exertion and of severe privations. Many are of low stature, though not dwarfish; the specimens brought to Europe have been selected, like costermongers' dogs, on account of their extreme ugliness; consequently, English ideas of the whole tribe are formed in the same way as if the ugliest specimens of the English were exhibited in Africa as characteristic of the entire British nation. That they are like baboons is in some degree true, just as these and other simiae are in some points frightfully human.

The Bakalahari are traditionally reported to be the oldest of the Bechuana tribes, and they are said to have possessed enormous herds of the large horned cattle mentioned by Bruce, until they were despoiled of them and driven into the Desert by a fresh migration of their own nation. Living ever since on the same plains with the Bushmen, subjected to the same influences of climate, enduring the same thirst, and subsisting on similar food for centuries, they seem to supply a standing proof that locality is not always sufficient of itself to account for difference in races. The Bakalahari retain in undying vigor the Bechuana love for agriculture and domestic animals. They hoe their gardens annually, though often all they can hope for is a supply of melons and pumpkins. And they carefully rear small herds of goats, though I have seen them lift water for them out of small wells with a bit of ostrich egg-shell, or by spoonfuls. They generally attach themselves to influential men in the different Bechuana tribes living adjacent to their desert home, in order to obtain supplies of spears, knives, tobacco, and dogs, in exchange for the skins of the animals they may kill. These are small carnivora of the feline species, including two species of jackal, the dark and the golden; the former, "motlose" ('Megalotis capensis' or 'Cape fennec'), has the warmest fur the country yields; the latter, "pukuye" ('Canis mesomelas' and 'C. aureus'), is very handsome when made into the skin mantle called kaross. Next in value follow the "tsipa" or small ocelot ('Felis nigripes'), the "tuane" or lynx, the wild cat, the spotted cat, and other small animals. Great numbers of 'puti' ('duiker') and 'puruhuru' ('steinbuck') skins are got too, besides those of lions, leopards, panthers, and hyaenas. During the time I was in the Bechuana country, between twenty and thirty thousand skins were made up into karosses; part of them were worn by the inhabitants, and part sold to traders: many, I believe, find their way to China. The Bakwains bought tobacco from the eastern tribes, then purchased skins with it from the Bakalahari, tanned them, and sewed them into karosses, then went south to purchase heifer-calves with them, cows being the highest form of riches known, as I have often noticed from their asking "if Queen Victoria had many cows." The compact they enter into is mutually beneficial, but injustice and wrong are often perpetrated by one tribe of Bechuanas going among the Bakalahari of another tribe, and compelling them to deliver up the skins which they may be keeping for their friends. They are a timid race, and in bodily development often resemble the aborigines of Australia. They have thin legs and arms, and large, protruding abdomens, caused by the coarse, indigestible food they eat. Their children's eyes lack lustre. I never saw them at play. A few Bechuanas may go into a village of Bakalahari, and domineer over the whole with impunity; but when these same adventurers meet the Bushmen, they are fain to change their manners to fawning sycophancy; they know that, if the request for tobacco is refused, these free sons of the Desert may settle the point as to its possession by a poisoned arrow.

The dread of visits from Bechuanas of strange tribes causes the Bakalahari to choose their residences far from water; and they not unfrequently hide their supplies by filling the pits with sand and making a fire over the spot. When they wish to draw water for use, the women come with twenty or thirty of their water-vessels in a bag or net on their backs. These water-vessels consist of ostrich egg-shells, with a hole in the end of each, such as would admit one's finger. The women tie a bunch of grass to one end of a reed about two feet long, and insert it in a hole dug as deep as the arm will reach; then ram down the wet sand firmly round it. Applying the mouth to the free end of the reed, they form a vacuum in the grass beneath, in which the water collects, and in a short time rises into the mouth. An egg-shell is placed on the ground alongside the reed, some inches below the mouth of the sucker. A straw guides the water into the hole of the vessel, as she draws mouthful after mouthful from below. The water is made to pass along the outside, not through the straw. If any one will attempt to squirt water into a bottle placed some distance below his mouth, he will soon perceive the wisdom of the Bushwoman's contrivance for giving the stream direction by means of a straw. The whole stock of water is thus passed through the woman's mouth as a pump, and, when taken home, is carefully buried. I have come into villages where, had we acted a domineering part, and rummaged every hut, we should have found nothing; but by sitting down quietly, and waiting with patience until the villagers were led to form a favorable opinion of us, a woman would bring out a shellful of the precious fluid from I know not where.

The so-called Desert, it may be observed, is by no means a useless tract of country. Besides supporting multitudes of both small and large animals, it sends something to the market of the world, and has proved a refuge to many a fugitive tribe—to the Bakalahari first, and to the other Bechuanas in turn—as their lands were overrun by the tribe of true Caffres, called Matebele. The Bakwains, the Bangwaketze, and the Bamangwato all fled thither; and the Matebele marauders, who came from the well-watered east, perished by hundreds in their attempts to follow them. One of the Bangwaketze chiefs, more wily than the rest, sent false guides to lead them on a track where, for hundreds of miles, not a drop of water could be found, and they perished in consequence. Many Bakwains perished too. Their old men, who could have told us ancient stories, perished in these flights. An intelligent Mokwain related to me how the Bushmen effectually balked a party of his tribe which lighted on their village in a state of burning thirst. Believing, as he said, that nothing human could subsist without water, they demanded some, but were coolly told by these Bushmen that they had none, and never drank any. Expecting to find them out, they resolved to watch them night and day. They persevered for some days, thinking that at last the water must come forth; but, notwithstanding their watchfulness, kept alive by most tormenting thirst, the Bakwains were compelled to exclaim, "Yak! yak! these are not men; let us go." Probably the Bushmen had been subsisting on a store hidden under ground, which had eluded the vigilance of their visitors.

JOSEPH CONRAD, *THE HEART OF DARKNESS*

Joseph Conard, the Polish émigré/British author, wrote The Heart of Darkness, *one of the most compelling, and some would say pernicious, accounts of Africa during the early colonial era. Using the metaphor taken from the now cliché term for Africa as the "Dark Continent," Conrad uses the landscape of the Congo River as the setting for his exploration of human corruption. At one level, the book is a critique of colonialism, especially of Belgian control over the "Congo Free State." The excerpt here recounts the narrator's journey up the Congo River to the trading station of Krutz, the master trader, who has "gone native." Conrad uses the "jungle" as the symbol for the darkness of colonialism and human souls. The book, though overtly critical of colonial exploitation, also portrays Africa and Africans as unremittingly savage. Conrad's veracity comes from his one voyage up the Congo, after which he quit the company for which he worked in disgust. The excerpt is from Joseph Conrad,* The Heart of Darkness *(London: 1902).*

Up the Congo River
Joseph Conrad

"In a few days the Eldorado Expedition went into the patient wilderness, that closed upon it as the sea closes over a diver. Long afterwards the news came that all the donkeys were dead. I know nothing as to the fate of the less valuable animals. They, no doubt, like the rest of us, found what they deserved. I did not inquire. I was then rather excited at the prospect of meeting Kurtz very soon. When I say very soon I mean it comparatively. It was just two months from the day we left the creek when we came to the bank below Kurtz's station.

"Going up that river was like traveling back to the earliest beginnings of the world, when vegetation rioted on the earth and the big trees were kings. An empty stream, a great silence, an impenetrable forest. The air was warm, thick, heavy, sluggish. There was no joy in the brilliance of sunshine. The long stretches of the waterway ran on, deserted, into the gloom of overshadowed distances. On silvery sand-banks hippos and alligators sunned themselves side by side. The broadening waters flowed through a mob of wooded islands; you lost your way on that river as you would in a desert, and butted all day long against shoals, trying to find the channel, till you thought yourself bewitched and cut off for ever from everything you had known once—somewhere—far away—in another existence perhaps. There were moments when one's past came back to one, as it will sometimes when you have not a moment to spare for yourself; but it came in the shape of an unrestful and noisy dream, remembered with wonder amongst the overwhelming realities of this strange world of plants, and water, and silence. And this stillness of life did not in the least resemble a peace. It was the stillness of an implacable force brooding over an inscrutable intention. It looked at you with a vengeful aspect. I got used to it afterwards; I did not see it any more; I had no time. I had to keep guessing at the channel; I had to discern, mostly by inspiration, the signs of hidden banks; I watched for sunken stones; I was learning to clap my teeth smartly before my heart flew out, when I shaved by a fluke some infernal sly old snag that would have ripped the life out of the tin-pot steamboat and drowned all the pilgrims; I had to keep a lookout for the signs of dead wood we could cut up in the night for next day's steaming. When you have to attend to things of that sort, to the mere incidents of the surface, the reality—the reality, I tell you—fades. The inner truth is hidden—luckily, luckily. But I felt it all the same; I felt often its mysterious stillness watching me at my monkey tricks, just as it watches you fellows performing on your respective tight-ropes for—what is it? half-a-crown a tumble-"

"Try to be civil, Marlow," growled a voice, and I knew there was at least one listener awake besides myself.

"I beg your pardon. I forgot the heartache which makes up the rest of the price. And indeed what does the price matter, if the trick be well done? You do your tricks very well. And I didn't do badly either, since I managed not to sink that steamboat on my first trip. It's a wonder to me yet. Imagine a blindfolded man set to drive a van over a bad road. I sweated and shivered over that business considerably, I can tell you. After all, for a seaman, to scrape the bottom of the thing that's supposed to float all the time under his care is the unpardonable sin. No one may know of it, but you never forget the thump—eh? A blow on the very heart. You remember it, you dream of it, you wake up at night and think of it— years after—and go hot and cold all over. I don't pretend to say that steamboat floated all the time. More than once she had to wade for a bit, with twenty cannibals splashing around and pushing. We had enlisted some of these chaps on the way for a crew. Fine fellows—cannibals—in their place. They were men one could work with, and I am grateful to them. And, after all, they did not eat each other before my face: they had brought along a provision of hippo-meat which went rotten, and made the mystery of the wilderness stink in my nostrils. Phoo! I can sniff it now. I had the manager on board and three or four pilgrims with their staves—all complete. Sometimes we came upon a station close by the bank, clinging to the skirts of the unknown, and the white men rushing out of a tumble-down hovel, with great gestures of joy and surprise and welcome, seemed very strange—had the appearance of being held there captive by a spell. The word ivory would ring in the air for a while—and on we went again into the silence, along empty reaches, round the still bends, between the high walls of our winding way, reverberating in hollow claps the ponderous beat of the stern-wheel. Trees, trees, millions of trees, massive, immense, running up high; and at their foot, hugging the bank against the stream, crept the little begrimed steam-boat, like a sluggish beetle crawling on the floor of a lofty portico. It made you feel very small, very lost, and yet it was not altogether depressing, that feeling. After all, if you were small, the grimy beetle crawled on—which was just what you wanted it to do. Where the pilgrims imagined it crawled to I don't know. To some place where they expected to get something. I bet! For me it crawled to-wards Kurtz—exclusively; but when the steam-pipes started leaking we crawled very slow. The reaches opened before us and closed behind, as if the forest had stepped leisurely across the water to bar the way for our return. We penetrated deeper and deeper into the heart of darkness. It was very quiet there. At night sometimes the roll of drums behind the curtain of trees would run up the river and remain sustained faintly, as if hovering in the air high over our heads, till the first break of day. Whether it meant war, peace, or prayer we could not tell. The dawns were heralded by the descent of a chill stillness; the wood-cutters slept, their fires burned low; the snapping of a twig would make you start. We were wanderers on a prehistoric earth, on an earth that wore the aspect of an un-known planet. We could have fancied ourselves the first of men taking posses-sion of an accursed inheritance, to be subdued at the cost of profound anguish and of excessive toil. But suddenly, as we struggled round a bend, there would be a glimpse of rush walls, of peaked grass-roofs, a burst of yells, a whirl of black limbs, a mass of hands clapping, of feet stamping, of bodies swaying, of eyes rolling, under the droop of heavy and motionless foliage. The steamer toiled along slowly on the edge of a black and incomprehensible frenzy. The prehistoric man was cursing us, praying to us, welcoming us—who could tell? We were cut off from the comprehension of our surroundings; we glided past like phantoms, wondering and secretly appalled, as sane men would be before an enthusiastic outbreak in a madhouse. We could not understand because we were too far and could not remember because we were traveling in the night of first ages, of those ages that are gone, leaving hardly a sign—and no memories.

"The earth seemed unearthly. We are accustomed to look upon the shackled form of a conquered monster, but there—there you could look at a thing mon-strous and free. It was unearthly, and the men were—No, they were not inhuman. Well, you know, that was the worst of it—this suspicion of their not being inhuman. It would come slowly to one. They howled and leaped, and spun, and made horrid faces; but what thrilled you was just the thought of their humanity—like yours—the thought of your remote kinship with this wild and passionate uproar. Ugly. Yes, it was ugly enough; but if you were man enough you would admit to yourself that there was in you just the faintest trace of a response to the terrible frankness of that noise, a dim suspicion of there being a meaning in it which you—you so remote from the night of first ages—could comprehend. And why not? The mind of man is capable of anything—because everything is in it, all the past as well as all the future. What was there after all? Joy, fear, sorrow, devotion, valour, rage—who can tell?—but truth—truth stripped of its cloak of time. Let the fool gape and shudder—the man knows, and can look on without a wink. But he must at least be as much of a man as these on the shore. He must meet that truth with his own true stuff—with his own inborn strength. Principles won't do. Acquisitions, clothes, pretty rags—rags that would fly off at the first good shake. No; you want a deliberate belief. An appeal to me in this fiendish row—is there? Very well; I hear; I admit, but I have a voice, too, and for good or evil mine is the speech that cannot be silenced. Of course, a fool, what with sheer fright and fine sentiments, is always safe. Who's that grunting? You wonder I didn't go ashore for a howl and a dance? Well, no—I didn't. Fine sentiments, you say? Fine sentiments, be hanged! I had no time. I had to mess about with white-lead and strips of woolen blanket helping to put bandages on those leaky steam-pipes—I tell you. I had to watch the steering, and circumvent those snags, and get the tin-pot along by hook or by crook. There was surface-truth enough in these things to save a wiser man. And between whiles I had to look after the savage who was fireman. He was an improved specimen; he could fire up a vertical boiler. He was there below me, and, upon my word, to look at him was as edifying as seeing a dog in a parody of breeches and a feather hat, walking on his hind-legs. A few months of training had done for that really fine chap. He squinted at the steam-gauge and at the water-gauge with an evident effort of intrepidity—and he had filed teeth, too, the poor devil, and the wool of his pate shaved into queer patterns, and three ornamental scars on each of his cheeks. He ought to have been clapping his hands and stamping his feet on the bank, instead of which he was hard at work, a thrall to strange witchcraft, full of improving knowledge. He was useful because he had been instructed; and what he knew was this—that should the water in that transparent thing disappear, the evil spirit inside the boiler would get angry through the greatness of his thirst, and take a terrible vengeance. So he sweated and fired up and watched the glass fearfully (with an impromptu charm, made of rags, tied to his arm, and a piece of polished bone, as big as a watch, stuck flatways through his lower lip), while the wooded banks slipped past us slowly, the short noise was left behind, the interminable miles of silence—and we crept on, towards Kurtz.

But the snags were thick, the water was treacherous and shallow, the boiler seemed indeed to have a sulky devil in it, and thus neither that fireman nor I had any time to peer into our creepy thoughts."

CONVENTION ON INTERNATIONAL TRADE IN ENDANGERED SPECIES OF WILD FAUNA AND FLORA

The Convention on International Trade in Endangered Species of Wild Fauna and Flora (CITES for short) was created in 1973 as a major international effort to promote wildlife conservation. Eventually, most nations have become parties to the pact, which regulates trade in animal products such as meat, hides, and ivory. The movement to create the organization grew out of the global conservation movement that began in the early twentieth century. Its proponents argued that the greatest danger to wildlife in Africa lay in hunting and in the trade of animal products. In Africa, the most important early organization was the British Society for the Preservation of the Wild Fauna of the Empire, which pressured the British government to create game reserves and national parks in its African colonies. Other colonial empires followed suit. With independence for most of Africa during the 1960s, many feared that lack of colonial control would quickly lead to decimation of wildlife. African governments responded that demand outside of Africa, especially for elephant ivory and rhino horn, drove the destruction of game. CITES was an attempt to regulate demand for wild animal products. Eventually, the organization agreed to outlaw completely trade in ivory and rhino horn. The following excerpts are from the original convention and from a regulation passed that would allow the partial marketing of ivory from areas with stable elephant populations. The easing of the ban on ivory occurred as some African nations argued that the total ban deprived them of resources garnered through wise management. The issue remains divisive, with changes to the regulations coming regularly. Most authorities now agree that hunting, legally or illegally, presents less of a threat to wildlife populations than habitat loss. Hence, emphasis has shifted to community-based conservation programs. The excerpts here come from the CITES Web site, www.cites.org.

Convention on International Trade in Endangered Species of Wild Fauna and Flora

Signed at Washington, D.C., on 3 March 1973

Amended at Bonn, on 22 June 1979

The Contracting States,

Recognizing that wild fauna and flora in their many beautiful and varied forms are an irreplaceable part of the natural systems of the earth which must be protected for this and the generations to come; Conscious of the ever-growing value of wild fauna and flora from aesthetic, scientific, cultural, recreational and economic points of view;

Recognizing that peoples and States are and should be the best protectors of their own wild fauna and flora;

Recognizing, in addition, that international co-operation is essential for the protection of certain species of wild fauna and flora against over-exploitation through international trade; Convinced of the urgency of taking appropriate measures to this end; Have agreed as follows:

Article I

Definitions

For the purpose of the present Convention, unless the context otherwise requires:

(a) "Species" means any species, subspecies, or geographically separate population thereof;

(b) "Specimen" means:

(i) any animal or plant, whether alive or dead;

(ii) in the case of an animal: for species included in Appendices I and II, any readily recognizable part or derivative thereof; and for species included in Appendix III, any readily recognizable part or derivative thereof specified in Appendix III in relation to the species; and

(iii) in the case of a plant: for species included in Appendix I, any readily recognizable part or derivative thereof; and for species included in Appendices II and III, any readily recognizable part or derivative thereof specified in Appendices II and III in relation to the species;

(c) "Trade" means export, re-export, import and introduction from the sea;

(d) "Re-export" means export of any specimen that has previously been imported;

(e) "Introduction from the sea" means transportation into a State of specimens of any species which were taken in the marine environment not under the jurisdiction of any State;

(f) "Scientific Authority" means a national scientific authority designated in accordance with Article IX;

(g) "Management Authority" means a national management authority designated in accordance with Article IX;

(h) "Party" means a State for which the present Convention has entered into force.

Article II

Fundamental Principles

1. Appendix I shall include all species threatened with extinction which are

or may be affected by trade. Trade in specimens of these species must be subject to particularly strict regulation in order not to endanger further their survival and must only be authorized in exceptional circumstances.

2. Appendix II shall include:

(a) all species which although not necessarily now threatened with extinction may become so unless trade in specimens of such species is subject to strict regulation in order to avoid utilization incompatible with their survival; and

(b) other species which must be subject to regulation in order that trade in specimens of certain species referred to in sub-paragraph (a) of this paragraph may be brought under effective control.

3. Appendix III shall include all species which any Party identifies as being subject to regulation within its jurisdiction for the purpose of preventing or restricting exploitation, and as needing the co-operation of other Parties in the control of trade.

4. The Parties shall not allow trade in specimens of species included in Appendices I, II and III except in accordance with the provisions of the present Convention.

Article III

Regulation of Trade in Specimens of Species Included in Appendix I

1. All trade in specimens of species included in Appendix I shall be in accordance with the provisions of this Article.

2. The export of any specimen of a species included in Appendix I shall require the prior grant and presentation of an export permit. An export permit shall only be granted when the following conditions have been met:

(a) a Scientific Authority of the State of export has advised that such export will not be detrimental to the survival of that species;

(b) a Management Authority of the State of export is satisfied that the specimen was not obtained in contravention of the laws of that State for the protection of fauna and flora;

(c) a Management Authority of the State of export is satisfied that any living specimen will be so prepared and shipped as to minimize the risk of injury, damage to health or cruel treatment; and

(d) a Management Authority of the State of export is satisfied that an import permit has been granted for the specimen.

3. The import of any specimen of a species included in Appendix I shall require the prior grant and presentation of an import permit and either an export permit or a re-export certificate. An import permit shall only be granted when the following conditions have been met:

(a) a Scientific Authority of the State of import has advised that the import will be for purposes which are not detrimental to the survival of the species involved;

(b) a Scientific Authority of the State of import is satisfied that the proposed recipient of a living specimen is suitably equipped to house and care for it; and

(c) a Management Authority of the State of import is satisfied that the specimen is not to be used for primarily commercial purposes.

4. The re-export of any specimen of a species included in Appendix I shall require the prior grant and presentation of a re-export certificate. A re-export certificate shall only be granted when the following conditions have been met:

(a) a Management Authority of the State of re-export is satisfied that the specimen was imported into that State in accordance with the provisions of the present Convention;

(b) a Management Authority of the State of re-export is satisfied that any living specimen will be so prepared and shipped as to minimize the risk of injury, damage to health or cruel treatment; and

(c) a Management Authority of the State of re-export is satisfied that an import permit has been granted for any living specimen.

5. The introduction from the sea of any specimen of a species included in Appendix I shall require the prior grant of a certificate from a Management Authority of the State of introduction. A certificate shall only be granted when the following conditions have been met:

(a) a Scientific Authority of the State of introduction advises that the introduction will not be detrimental to the survival of the species involved;

(b) a Management Authority of the State of introduction is satisfied that the proposed recipient of a living specimen is suitably equipped to house and care for it; and

(c) a Management Authority of the State of introduction is satisfied that the specimen is not to be used for primarily commercial purposes.

Article IV

Regulation of Trade in Specimens of Species Included in Appendix II

1. All trade in specimens of species included in Appendix II shall be in accordance with the provisions of this Article.

2. The export of any specimen of a species included in Appendix II shall require the prior grant and presentation of an export permit. An export permit shall only be granted when the following conditions have been met:

(a) a Scientific Authority of the State of export has advised that such export will not be detrimental to the survival of that species;

(b) a Management Authority of the State of export is satisfied that the specimen was not obtained in contravention of the laws of that State for the protection of fauna and flora; and

(c) a Management Authority of the State of export is satisfied that any living specimen will be so prepared and shipped as to minimize the risk of injury, damage to health or cruel treatment.

3. A Scientific Authority in each Party shall monitor both the export permits granted by that State for specimens of species included in Appendix II and the actual exports of such specimens. Whenever a Scientific Authority determines that the export of specimens of any such species should be limited in order to maintain that species throughout its range at a level consistent with its role in the ecosystems in which it occurs and well above the level at which that species might become eligible for inclusion in Appendix I, the Scientific Authority shall advise the appropriate Management Authority of suitable measures to be taken to limit the grant of export permits for specimens of that species.

4. The import of any specimen of a species included in Appendix II shall require the prior presentation of either an export permit or a re-export certificate.

5. The re-export of any specimen of a species included in Appendix II shall require the prior grant and presentation of a re-export certificate. A re-export certificate shall only be granted when the following conditions have been met:

(a) a Management Authority of the State of re-export is satisfied that the specimen was imported into that State in accordance with the provisions of the present Convention; and

(b) a Management Authority of the State of re-export is satisfied that any living specimen will be so prepared and shipped as to minimize the risk of injury, damage to health or cruel treatment.

6. The introduction from the sea of any specimen of a species included in Appendix II shall require the prior grant of a certificate from a Management Authority of the State of introduction. A certificate shall only be granted when the following conditions have been met:

(a) a Scientific Authority of the State of introduction advises that the introduction will not be detrimental to the survival of the species involved; and

(b) a Management Authority of the State of introduction is satisfied that any living specimen will be so handled as to minimize the risk of injury, damage to health or cruel treatment.

7. Certificates referred to in paragraph 6 of this Article may be granted on the advice of a Scientific Authority, in consultation with other national scientific authorities or, when appropriate, international scientific authorities, in respect of periods not exceeding one year for total numbers of specimens to be introduced in such periods.

Article V

Regulation of Trade in Specimens of Species Included in Appendix III

1. All trade in specimens of species included in Appendix III shall be in accordance with the provisions of this Article.

2. The export of any specimen of a species included in Appendix III from

any State which has included that species in Appendix III shall require the prior grant and presentation of an export permit. An export permit shall only be granted when the following conditions have been met:

(a) a Management Authority of the State of export is satisfied that the specimen was not obtained in contravention of the laws of that State for the protection of fauna and flora; and

(b) a Management Authority of the State of export is satisfied that any living specimen will be so prepared and shipped as to minimize the risk of injury, damage to health or cruel treatment.

3. The import of any specimen of a species included in Appendix III shall require, except in circumstances to which paragraph 4 of this Article applies, the prior presentation of a certificate of origin and, where the import is from a State which has included that species in Appendix III, an export permit.

4. In the case of re-export, a certificate granted by the Management Authority of the State of re-export that the specimen was processed in that State or is being re-exported shall be accepted by the State of import as evidence that the provisions of the present Convention have been complied with in respect of the specimen concerned.

Convention on the International Trade in Endangered Species

Tenth meeting of the Conference of the Parties

Harare (Zimbabwe), 9 to 20 June 1997

10.1 Conditions for the resumption of trade in African elephant ivory from populations transferred to Appendix II at the 10th meeting of the Conference of the Parties

Part A

Trade in raw ivory shall not resume unless:

a) deficiencies identified by the CITES Panel of Experts (established pursuant to Resolution Conf. 7.9, replaced by Resolution Conf. 10.9) in enforcement and control measures have been remedied;

b) the fulfillment of the conditions in this Decision has been verified by the CITES Secretariat in consultation with the African regional representatives on the Standing Committee, their alternates and other experts as appropriate;

c) the Standing Committee has agreed that all of the conditions in this Decision have been met;

d) the reservations entered by the range States1 with regard to the transfer of the African elephant to Appendix I were withdrawn by these range States prior to the entry into force of the transfer to Appendix II;

e) the relevant range States1 support and commit themselves to international cooperation in law enforcement through such mechanisms as the Lusaka Agreement;

f) the relevant range States1 have strengthened and/or established mechanisms to reinvest trade revenues into elephant conservation;

g) the Standing Committee has agreed to a mechanism to halt trade and immediately retransfer to Appendix I populations that have been transferred to Appendix II2, in the event of non-compliance with the conditions in this Decision or of the escalation of illegal hunting of elephants and/or trade in elephant products owing to the resumption of legal trade;

h) all other precautionary undertakings by the relevant range States in the supporting statements to the proposals adopted at the 10th meeting of the Conference of the Parties have been complied with; and

i) the relevant range States1, the CITES Secretariat, TRAFFIC International and any other approved party agree to:

i) an international system for reporting and monitoring legal and illegal international trade, through an international database in the CITES Secretariat and TRAFFIC International; and

ii) an international system for reporting and monitoring illegal trade and illegal hunting within or between elephant range States, through an international database in the CITES Secretariat, with support from

TRAFFIC International and institutions such as the IUCN/SSC African Elephant Specialist Group and the Lusaka Agreement.

Part B

a) If all of the conditions in this Decision are met, the Standing Committee shall make available its evaluation of legal and illegal trade and legal offtake pursuant to the implementation of Resolution Conf. 10.10 as soon as possible after the experimental trade has taken place.

b) The Standing Committee shall identify, in cooperation with the range States, any negative impacts of this conditional resumption of trade and determine and propose corrective measures.

10.2 Conditions for the disposal of ivory stocks and generating resources for conservation in African elephant range States.

a) The African elephant range States recognize:

i) the threats that stockpiles pose to sustainable legal trade;

ii) that stockpiles are a vital economic resource for them;

iii) that various funding commitments were made by donor countries and agencies to offset the loss of assets in the interest of unifying these States regarding the inclusion of African elephant populations in Appendix I;

iv) the significance of channelling such assets from ivory into improving conservation and community-based conservation and development programmes;

v) the failure of donors to fund elephant conservation action plans drawn up by the range States at the urging of donor countries and conservation organizations; and

vi) that, at its ninth meeting, the Conference of the Parties directed the Standing Committee to review the issue of stockpiles and to report back at the 10th meeting.

b) Accordingly, the African elephant range States agree that all revenues from any purchase of stockpiles by donor countries and organizations will be deposited in and managed through conservation trust funds, and that:

i) such funds shall be managed by Boards of Trustees (such as representatives of governments, donors, the CITES Secretariat, etc.) set up, as appropriate, in each range State, which would direct the proceeds into enhanced conservation, monitoring, capacity building and local community-based programmes; and

ii) these funds must have a positive rather than harmful influence on elephant conservation.

c) It is understood that this decision provides for a one-off purchase for non-commercial purposes of government stocks declared by African elephant range States to the CITES Secretariat within the 90-day period before the transfer to Appendix II of certain populations of the African elephant takes effect. The ivory stocks declared should be marked in accordance with the ivory marking system approved by the Conference of the Parties in Resolution Conf. 10.10. In addition, the source of ivory stocks should be given. The stocks of ivory should be consolidated in a pre-determined number of locations. An independent audit of any declared stocks shall be undertaken under the auspices of TRAFFIC International, in co-operation with the CITES Secretariat.

d) The African elephant range States that have not yet been able to register their ivory stocks and develop adequate controls over ivory stocks require priority assistance from donor countries to establish a level of conservation management conducive to the long-term survival of the African elephant.

e) The African elephant range States therefore urge that this matter be acted upon urgently since any delays will result in illegal trade and the premature opening of ivory trade in nonproponent range States.

f) This mechanism only applies to those range States wishing to dispose of ivory stocks and agreeing to and participating in:

i) an international system for reporting and monitoring legal and illegal international trade, through an international database in the CITES Secretariat and TRAFFIC International; and

ii) an international system for reporting and monitoring illegal trade and illegal hunting within or between elephant range States, through an international database in the CITES Secretariat, with support from TRAFFIC International and institutions such as the IUCN/SSC African Elephant Specialist Group and the Lusaka Agreement.

Notes from the Secretariat:

1 This is understood to mean the States whose populations of African elephant have been transferred to Appendix II (as in paragraph h).

2 This decision is in conflict with the text of the Convention. The mechanism for the transfer of species (including populations) from Appendix II to Appendix I is specified in Article XV of the Convention. Any such transfer can be done only if it is proposed by a Party and is agreed by the Conference of the Parties, either at a regular meeting or by the postal procedure, and will enter into force only 90 days after the proposal is adopted by the Conference. An appropriate action for the Standing Committee would be to request a Party (such as the Depositary Government) to submit the required proposal.

Thirteenth meeting of the Conference of the Parties

Bangkok (Thailand), 2 to 14 October 2004

Annex 2

Action plan for the control of trade in African elephant ivory.

1. All African elephant range States1 should urgently:

a) prohibit the unregulated domestic sale of ivory (raw, semi-worked or worked). Legislation should include a provision which places the onus of proof of lawful possession upon any person found in possession of ivory in circumstances from which it can reasonably be inferred that such possession was for the purpose of unauthorized transfer, sale, offer for sale, exchange or export or any person transporting ivory for such purposes;

b) issue instructions to all law enforcement and border control agencies to enforce existing or new legislation rigorously; and

c) engage in public awareness campaigns publicizing existing or new prohibitions on ivory sales.

2. Parties should, by 31 March 2005, report to the Secretariat on progress made. Such reports should include details of seizures, copies of new legislation, copies of administrative instructions or orders to enforcement agencies and details of awareness campaigns. The Secretariat should report on Parties' progress at the 53rd meeting of the Standing Committee.

3. In the interim, the Secretariat should work with the relevant countries in Africa to provide any technical assistance that may be necessary to aid the implementation of this action plan.

4. The Secretariat should also engage in efforts to publicize the present action plan and the subsequent halting of domestic ivory sales in individual African countries through contacting relevant organizations such as airlines and IATA. It should also, via ICPO-Interpol and the World Customs Organization, communicate with the heads of police and Customs authorities in Africa, advis-

ing them of this initiative. Furthermore the Secretariat should request all Parties worldwide to publicize the action plan, particularly to discourage persons who are travelling to Africa from purchasing raw, semi-worked or worked ivory and to encourage border control authorities to be alert to illegal imports of ivory and to make every effort to intercept illicit movements of ivory.

5. All elephant range States are recommended to cooperate with existing research projects studying the identification of ivory, especially by supplying relevant samples for DNA and other forensic science profiling.

6. The Secretariat should seek the assistance of Governments, international organizations and non-governmental organizations in supporting the work to eradicate illegal exports of ivory from the African continent and the unregulated domestic markets that contribute to illicit trade.

7. At the 13th meeting of the Conference of the Parties, the Secretariat should seek the agreement of the Parties that it would, from 1 June 2005, ensure that work is undertaken, including *in situ* verification missions where appropriate, to assess, on a country-by-country basis, progress made with the implementation of the action plan. Priority should be given to those Parties that are identified during research by the Secretariat and through other appropriate sources of information to have active and unregulated internal markets for ivory. Priority should be given to Cameroon, the Democratic Republic of the Congo, Djibouti, Nigeria and any other country identified through ETIS.

8. In cases where Parties or non-Parties are found not to implement the action plan, or where ivory is found to be illegally sold, the Secretariat should issue a Notification to the Parties advising them that the Conference of the Parties recommends that Parties should not engage in commercial trade in specimens of CITES-listed species with the country in question.

9. The Secretariat should continue to monitor all domestic ivory markets outside Africa to ensure that internal controls are adequate and comply with the relevant provisions of Resolution Conf. 10.10 (Rev. CoP12) on Trade in elephant specimens. Priority should be given to China, Japan and Thailand, with particular attention being paid to any Party that has notified the Secretariat that it wishes to authorize imports of ivory for commercial purposes.

10. The Secretariat should report upon the implementation of the action plan at each meeting of the Standing Committee.

CONSERVATION IN INDEPENDENT AFRICA
TANZANIA'S WILDLIFE POLICY

The East African nation of Tanzania has one of the largest proportions of its territory under some kind of conservation protection in the world. This process be-

gan under German colonial rule and expanded during the British era between 1920 and 1961. At independence, many in the country hoped to open up land for human settlement and ease restriction on the use of protected areas for grazing cattle and other activities. The then prime minister and later longtime president of the country, Julius K. Nyerere, undertook discussions with both conservationists and political leaders in the country and in 1961 issued the Arusha manifesto (named after the town in northern Tanzania that serves as the jumping-off point for the Serengeti) calling for continued conservation efforts. The manifesto, quoted in the following document, remains the cornerstone of conservation policy in Tanzania and is an important statement of general African support for conservation. Since independence, Tanzania's official policy has been fraught with conflict over access to resources. This restatement of its general policy, issued in 1998, seeks to balance conservation with human needs. These excerpts are from the Ministry of Natural Resources and Tourism, The Wildlife Policy of Tanzania (Dar es Salaam, March 1998) and is available online at http://www.tzonline.org/pdf/wildlifepolicy.pdf

The Wildlife Policy of Tanzania
Mission and Vision
The Ministry of Natural Resources and Tourism is charged with formulating a wildlife policy, overseeing its administration and co-ordinating the development of the wildlife sector in Tanzania. The vision of the wildlife sector for the next twenty (20) years conforms with the Development Vision 2025 for Tanzania on environmental sustainability and socio-economic transformation. The vision for the wildlife sector is to:

- promote conservation of biological diversity,
- administer, regulate and develop wildlife resources,
- involve all stakeholders in wildlife conservation and sustainable utilisation, as well as in fair and equitable sharing of benefits,
- promote sustainable utilisation of wildlife resources,
- raise the contribution of the wildlife sector in country's Gross Domestic Product (GDP) from about 2% to 5%,
- contribute to poverty alleviation and improve the quality of life of the people of Tanzania, and,
- promote exchange of relevant information and expertise nationally, regionally and internationally,

    . . .

1.0 HISTORICAL BACKGROUND
The wildlife conservation in Tanzania dates back in 1891 when laws controlling hunting were first enacted by the German rule. These laws regulated

the off-take, the hunting methods and the trade in wildlife, with some endangered species being fully protected. The first Game Reserves were established in 1905 by the Germans in the area which now forms the Selous Game Reserve. Game Reserves were chosen mainly for their concentrations of big game rather than their biological diversity.

In 1921 the British Government established the Game Department followed by the gazettement of the first Game Reserve, the Selous Game Reserve in 1922. The roles of the Game Department were to administer the Game Reserves, enforce the hunting regulations and protect people and crops from raiding animals. Later on, the then Ngorongoro Crater closed and Serengeti Game Reserves were established in 1928 and 1929 respectively.

Tanganyika was always famous for its variety of big game, wildlife numbers and diversity of landscapes and in those early days, attracted a steady stream of wealthy hunters. The tourist hunting industry dates back to 1946 when Game Controlled Areas (GCAs) were established and divided into hunting blocks, where professional hunters and their clients could hunt trophy animals.

The present framework of Wildlife Protected Areas (WPAs) in Tanzania comprising of National Parks, Game Reserves and Game Controlled Areas was started after World War II. In 1951 the Serengeti National Park which incorporated the Ngorongoro Crater was gazetted followed by several National Parks (NPs) and Game Reserves (GRs).

In 1961 there were three (3) National Parks and nine (9) Game Reserves and the Ngorongoro Conservation Area. After independence it was the policy of the Government to continue with the extension of the Game Reserves and National Parks, and many new parks and reserves were gazetted.

At independence, Tanzania showed her commitment to wildlife conservation when the then President of Tanganyika released a statement, the famous "Arusha Manifesto" as quoted hereunder:

The survival of our wildlife is a matter of grave concern to all of us in Africa. These wild creatures amid the wild places they inhabit are not only important as a source of wonder and inspiration but are an integral part of our natural resources and of our future livelihood and well being.

In accepting the trusteeship of our wildlife we solemnly declare that we will do everything in our power to make sure that our children's grand-children will be able to enjoy this rich and precious inheritance.

The conservation of wildlife and wild places calls for specialist knowledge, trained manpower, and money, and we look to other nations to co-operate with us in this important task the success or failure of which not only affects the continent of Africa but the rest of the world as well.

Mwalimu J. K. Nyerere, 1961

The "Arusha Manifesto" has been used to guide wildlife conservation in Tanzania to date.

Despite her longstanding history of wildlife conservation, Tanzania never had any comprehensive wildlife policy. Wildlife was all along being protected and utilised by use of guidelines, regulations and laws implemented by the department of wildlife and other institutions entrusted with the responsibility of conserving the same. At independence in 1961, Tanzania's human population was relatively low (only 8 million) making land use conflicts uncalled for, especially under conditions of inadequate technological and scientific development. Parts of land could easily be set aside for the protection of wildlife without seriously inconveniencing local people. Today the Tanzanian human population is about 30 million, with advancement of science and technology, both of which make land scarce and necessitate land use plans and an elaborate wildlife conservation policy.

SASAKAWA/GLOBAL 2000
INCREASING AGRICULTURAL TECHNOLOGY IN AFRICA

Sasakawa/Global 2000 is an organization dedicated to increasing food output in Africa through the application of higher-technology agricultural input. It grew out of the Sahelian famines of the 1970s, when Ryoichi Sasakawa, a prominent business leader in Japan, contacted Nobel Laureate Norman Borlaug about the possibility of adopting Green Revolution techniques used to increase productivity in Asian agriculture to Africa. Backed by Sasakawa's Nippon Foundation and with the support of former U.S. president Jimmy Carter, Borlaug has led an effort to bring new production techniques to African smallholders. The project has focused primarily on maize, the most important food crop in much of Africa, and has targeted smallholders. Since 1985, projects in several African countries have led to increased productivity. However, the project has focused on the use of fertilizer and hybrid seed, and as a result, it has come under criticism for fostering dependence on markets for the expansion of output. Ghana is usually cited as one of the most successful participants in the program. The following excerpts are from the Sasakawa Africa Association Web site at the addresses listed below.

Sasakawa/Global 2000

http://www.saa-tokyo.org/english/sg2000/index.html

Project Components

Although SG 2000 projects are located in diverse countries, their central approach to the problems of small farmers is similar. Before launching a new project,

SG 2000 determines that there is a pool of technology appropriate for the country that could have a significant impact, that the citizens are poor, that the country is food insecure, and that the government is committed to agricultural development.

On that basis, SG 2000 and the government draw up a memorandum of understanding that lays out the responsibilities of both parties. SG 2000 insists on working through government agencies—rather than setting up a parallel organization outside the government. That helps ensure that the national extension agency benefits from the lessons learned and will retain them when the SG 2000 projects draw to a close.

Lean staffing allows SG 2000 to concentrate its funds on operations rather than expatriate salaries. SG 2000 projects have at most one or two expatriate advisors per country. The bulk of the work of SG 2000 projects is carried out by national extension workers under the supervision of senior extension officials.

Initially, SG 2000 projects focus on training extension workers and farmers in better production technology for the principal food crops. They shortly take on such critical issues as improved on-farm storage, input availability, and credit services. After several years, when extension training and staffing are stronger, SG 2000 shifts from direct involvement in production demonstration programmes and expands work in grain storage, draught power, agro-processing, and seed production. Such programmes are in operation in Ghana and Tanzania.

Sasakawa/Global 2000 in Ghana

http://www.saa-tokyo.org/english/country/ghana.html

The first project of SG 2000 began in 1986 in Ghana. The project worked with Ghana's Ministry of Food and Agriculture, through the Agricultural Extension Services Department.

Technology demonstration is the heart of the project, and the principal tool is a large (approximately 0.5 ha) production test plot (PTP). The PTP is grown by a participating farmer using a package of recommended production practices. The PTP serves as a classroom where both farmers and extension agents are introduced to new technology.

Alongside the PTP, the farmer is asked to grow a second plot using his or her conventional farming practices to help in making comparisons. The crop growth and yield in the PTP is almost always overwhelmingly superior, and year after year the project attracted large numbers of additional farmers who wanted to learn to use the improved technology.

The profitability and simplicity of the recommended technological packages are the key to the success of the SG 2000 strategy. The participating farmer gets recommend inputs on credit for up to three seasons. After that, participants are considered to have graduated from the demonstration programme, and they turn to conventional channels to acquire the inputs they have learned to use.

In the first year, 40 farmers grew maize demonstration plots. By 1988, the project had attracted nearly 15,000 small-scale farmers, and the next year there were 80,000 participants. Because further expansion would strain the project's capacity for effective supervision, and because the objective of widely demonstrating the payoff from improved technology appeared to have been achieved, the number of plots were reduced to 17,000 in 1990. The project then shifted considerable emphasis to improving on-farm postharvest technology and grain storage, and also became involved in promoting the development of a private seed industry.

The Ghana project spearheaded a credit scheme involving formal banking institutions. The Agricultural Development Bank (ADB) issues credit for inputs to the farmers' group that have fully repaid the previous season's loans. The scheme ensures that the farmers are able to get input credit after they have "graduated" from SG 2000 project.

Assisting in the development and diffusion of quality protein maize (QPM) varieties is one of the most significant achievements of SG 2000 Ghana. The protein of QPM varieties is better than that of normal maize varieties, an important benefit for diets in countries like Ghana where maize is the staple food. Ghana's farmers plant over 100,000 hectares to QPM varieties and hybrids released by the national Crop Research Institute. The varieties have been spreading to many other African countries, as well.

Conservation tillage, or no-tillage farming, has been fostered by the Ghana project. By using an herbicide to kill vegetation before crop planting period, farmers can sow their crop without plowing. The dead vegetation forms a mulch that retains moisture and suppresses weeds. Through conservation tillage, farmers save the cost of ploughing and land preparation, they are better able to plant when soil moisture conditions are right, and their land is less vulnerable to erosion from heavy rainfall.

In 1995, the SG 2000 Ghana project entered into a new mode of operation called Phase II. Under the new mode, the project no longer has an expatriate country director; instead the project is guided by a national project coordinator. SG 2000 Ghana operated with the District Assemblies and the Rural Banks which is close to the farming communities.

CHRONOLOGY

DATE	EVENT	COMMENTS
5 million years ago	Emergence of hominid line	Evolutionary divergence between hominids and great apes, appearance of australopithecines east of the Great Rift Valley
2.7 to 2.5 million years ago	Severe dry period	Forests retreat throughout Africa and savannahs expand
2.6 million years ago	Appearance of *Homo habilis*	First toolmakers, Olduvai culture appears
1.8 million years ago	Appearance of *Homo ergaster*	Larger hominids appear, development of "hand axes" as signature tool
1.5 million years ago	Appearance of *Homo erectus*	Migration of first hominids out of Africa
1.5 million to 600,000 years ago	Spread of hominid populations across much of the Old World	Development of several different species in Europe, Asia, and Africa
1.5 million years ago to 130,000 years ago	Alternating periods of glaciation and warming	Isolation of African flora and fauna south of the Sahara with only occasional contact during warmer periods
130,000 years ago	Emergence of *Homo sapiens* in Africa	Development of more complex tool kit
100,000 years ago	First spread of *Homo sapiens* out of Africa	Temporary population moves to Southwest Asia
75,000 years ago	Shift to drier, colder climate associated with glaciation	Die-off of *Homo sapiens* outside of Africa and reduction of *Homo sapiens* population in Africa

DATE	EVENT	COMMENTS
60,000 years ago	Spread of *Homo sapiens* out of Africa	Development of different tool kits indicative of development of different lifeways based on environmental conditions
40,000 to 20,000 years ago	Pluvial period	Sahara contracts, forests and lakes expand, last of other hominid populations disappear
18,000 to 11,000 B.C.E.	Last glacial maximum	Sahara expands, forests shrink, savannahs expand, many lakes dry out
11,000 to 8000 B.C.E.	Early Holocene	Pluvial phase with some variability, forests expand, Sahara contracts, appearance of foraging populations deep in the Sahara
8000 to 6000 B.C.E.	Holocene Optimum	Pluvial phase with extreme wetness throughout African continent
8000 to 6000 B.C.E.	Appearance of first cattle-herding populations	Sites in southwestern Egypt show distinct sign of domesticated cattle
6000 to 2000 B.C.E.	Aquatic civilization	Permanent settlements along lakes and rivers indicate populations extensively exploiting plants and marine food sources
6000 B.C.E.	Expansion of Middle Eastern agriculture into Nile Valley	First fully agricultural society in Africa, based on the Middle Eastern crops of wheat and barley, also spread of sheep and goats into Africa
4,000 B.C.E.	Cushitic agropastoralism	Combines livestock keeping with either cultivation or extensive exploitation of grain crops in northeast Africa
4000 B.C.E.	Sudanic agropastoralism	Combines livestock keeping with either cultivation or extensive exploitation of grain crops in central Sudan

DATE	EVENT	COMMENTS
3500 B.C.E.	West African planting agriculture	Based on cultivation of yams and oil palms in forest-savannah border areas
3000 B.C.E.	Transition to drier conditions	Beginning of transition to current conditions, Sahara dries, lakes shrink, population moves out of Sahara to south and to the Nile Valley
3000 B.C.E.	Development of food production in Ethiopian Highlands	First agriculture in the area based on ensete, a bananalike plant
2500 B.C.E.	Final desiccation of the Sahara	Population almost completely empties in the desert
2500 B.C.E. to 0 C.E.	Spread of agricultural societies	Gradual filling in of agropastoral societies across Sudan and into eastern Africa, spread of West African planting societies across northern forest-savannah border region to the Great Lakes
1000 B.C.E	Invention of iron working	Possible independent invention of iron working somewhere between Lake Chad and the Great Lakes, iron working appears in Northern Nigeria and in the Great Lakes region by 2,500 years ago
1000 B.C.E. to 0 C.E.	Intensification of food production	Spread of grain production and stock keeping across western and eastern Africa and of agriculture into the Central African rainforest
1000 B.C.E.	Development of plow agriculture in Ethiopian Highlands	Creation of first terraces to support animal-drawn plows in region
800 B.C.E.	Urbanization in Middle Niger Valley	Development of towns in the Middle Niger, introduction of rice cultivation

DATE	EVENT	COMMENTS
After 500 B.C.E.	Spread of camel into Sahara	Herds of animals able to survive better in the desert increases population and contact across desert
500 B.C.E. to 500 C.E.	Rapid expansion of agricultural societies southward	Movement of agriculture, stock keeping and iron working south through both the Central African forest and the plains of eastern and southern Africa
300 B.C.E. to 300 C.E.	Extreme dry and cool phase	Sahara expands, contact across the desert limited, population retreats south
0 C.E.	Domesticated stock reach southern Africa	Sheep and cattle spread in advance of agricultural societies
0 to 500 C.E.	Introduction of bananas in eastern Africa	Plantain-type bananas spread into Central and West African forest, allowing for expansion of population in the region
0 C.E.	Movement of Malagasy-speaking peoples to East African coast and island of Madagascar	Evidence of ocean-borne trade across the Indian Ocean, arrival of chickens and sugar cane in Africa
0 to 500 C.E.	Spread of agriculture and iron working into southern Africa	Coincides with the arrival of Bantu languages
300 to 1000 C.E.	Moist phase, especially in the Sahel	Expansion of savannahs into the Sahara Desert
300 to 800 C.E.	Reopening of trans-Saharan trade	Moister climate allows development of camel-borne trade network across the Sahara
500 C.E.	Deforestation in Great Lakes Region	Loss of forest cover in areas around the Great Lakes caused by expansion of agricultural and pastoral production and iron working
700 C.E.	Muslim conquest of North Africa	Leads to intensification of trade both across the Sahara and along the East African coast

DATE	EVENT	COMMENTS
900 C.E.	Beginnings of large-scale urbanization in southern Africa	Towns develop that eventually lead to the rise of the Zimbabwe civilization based on an extensive pastoral and agricultural economy and the trade of gold with the Swahili coast
1200 C.E.	Development of extensive banana cultivation in East African Highlands	Intensive banana cultivation leads to rapid expansion of population in highlands around the Great Lakes and on the Eastern Arc Mountains
1400 C.E.	Collapse of population in western Sahel	Arid phase moves frontier of Sahel southward, arrival of bubonic plague
1400 C.E.	Collapse of Great Zimbabwe	Capital at Great Zimbabwe abandoned and state fractured into several smaller ones, perhaps caused by a subsistence crisis due to overexploitation of pastoral and agricultural land
1450 C.E.	First direct contact between Europeans and sub-Saharan Africa	Portuguese voyages along the west coast of Africa begin
1500 C.E.	Beginning of Columbian Exchange	Portuguese dominate trade with Africa, buying gold, tropical produce, and slaves with horses, metal goods, and captives brought from other regions of Africa. Portuguese also occupy Swahili city-states in East Africa, causing decline in trade
1500 to 1700 C.E.	Spread of New World crops in Africa	Maize, cassava, and peanuts spread unevenly throughout parts of Africa from contact with Europeans bringing them from the Americas. May have helped the spread of population in the forest regions of West Africa

DATE	EVENT	COMMENTS
1600 C.E.	Development of plantation system of sugar production	Developed first in the Mediterranean, it spread to Atlantic Islands and then to the New World. Its rapid expansion after 1600 created a demand for labor that Europeans filled by purchasing slaves in Africa
1600 to 1850 C.E.	Atlantic slave trade	Around 13 million Africans taken from the west and west-central regions of the continent. Loss of population is highest in West Central Africa
1600 C.E.	Expansion of Masai pastoralist in East Africa	Abandonment of plains around East African highlands as Masai expand their more pastoral way of life into the plains. Irrigated agricultural works at places like Engaruka abandoned
1652 C.E.	Establishment of Dutch colony at Cape Town	Disease and conquest destroy population of Africans in the Cape region
1800 C.E.	Beginnings of commodity production by Africans for export	Production of agricultural commodities begins in several regions to meet demand generated by European industrial growth for commodities. Important exports include oil from oil palms and peanuts in West Africa
1800 to 1850 C.E.	Massive population shifts in southern Africa	Disruptions caused by expansion of white farmers and the rise of the Zulu state in South Africa
1850 C.E.	Expansion of plantation production in East Africa	Increase in slave raiding from Mozambique to Ethiopia and as far west as eastern Congo leads to population loss and disruption

DATE	EVENT	COMMENTS
1850 to 1950 C.E.	Expansion of ivory trade	Introduction of more power-ful and accurate firearms leads to massive increase in elephant hunting. Elephant populations throughout the continent decline sharply
1870s C.E.	Scramble for Africa	European powers begin con-quest of large portions of Africa. Conquest leads to two decades of warfare, much more rapid spread of disease, and severe popula-tion loss in Central, eastern, and southern Africa
1870s C.E.	Rubber boom	Surge in demand for rubber leads to destruction of forests in West and espe-cially Central Africa
1870 to 1890 C.E.	Discovery of precious minerals in South Africa	Discovery of diamonds in the 1860s and gold in the 1880s leads to influx of Euro-pean immigrants. Africans are confined to reserves and forced into migrant labor on plantations and white-owned farms. Expansion of farming almost totally destroys South African wildlife populations
1880s to 1900s C.E.	Establishment of colonial states	Despite differences among colonial powers, colonial states generally tried to claim control over natural resources such as forests and minerals. Colonial states tried to promote sustainable exploitation of such re-sources and expanded agri-cultural production

DATE	EVENT	COMMENTS
1890s C.E.	Rinderpest pandemic	Spread of rinderpest brought by Italians during their attempt to conquer Ethiopia decimates cattle and wildlife herds throughout the continent. Leads to the expansion of bush as pasture, and fields are abandoned. Tsetse fly habitat expands, and sleeping sickness becomes a major killer disease in much of Africa where it previously had been under control
1900s C.E.	Establishment of settler colonies	Migration of European settlers to eastern and southern Africa drives Africans into reserves in Kenya, Zimbabwe, Zambia, German East Africa, southwest Africa, and the Belgian Congo
1900s C.E.	Beginning of "Safaris"	Publicized in part by ex-U.S. president Theodore Roosevelt's hunting trip to East Africa, hunting safaris by wealthy tourists become popular, especially in East Africa. This development leads to support for wildlife conservation
1905 C.E.	Genocide of the Herero	German response to a revolt in Southwest Africa is to almost exterminate the Herero people
1910s C.E.	Drought in West Africa	Worst drought in history in the Sahel and West African Sudan leads to famine
1917 to 1920 C.E.	Famine in East Africa	East Africa Campaign during World War I leads to massive famine in large parts of East Africa
1930s C.E.	Extensive concern over soil erosion	Extensive efforts by colonial states to promote soil conservation

DATE	EVENT	COMMENTS
1930s C.E.	Establishment of game reserves, forest reserves, and national parks	Concern over decline of wildlife and forests leads to efforts to create preserves in many colonies. First national park is established in the Belgian Congo in the 1920s. National parks increase in numbers through the 1930s, 1940s, and 1950s
1940s C.E.	Sustained population growth	African populations begin to increase at dramatic rates after 1930
1945 C.E.	Second colonial occupation	Post–World War II efforts by colonial states to promote development lead to large increases in investment in agriculture, much of it into failed, large-scale mechanized farming projects
1950s C.E.	African resistance to colonial conservation programs	Protests against conservation schemes that often brought little benefit to African producers help fuel anticolonial nationalism
1960s C.E.	Independence for much of Africa	After 1960, most African colonies become independent. They in general continue colonial era conservation policies
1970s C.E.	Drought in the Sahel	Decade-long drought in the Sahel covering an area stretching from West Africa to Ethiopia leads to widespread concern about desertification
1970s C.E.	Decline in game populations	Loss of habitat to agriculture and poaching, especially of elephants and rhinoceroses, leads to bans on the sale of ivory and rhino horn and renewed efforts at conservation

DATE	EVENT	COMMENTS
2004 C.E.	African environmentalist wins Nobel Peace Prize	Wangari Maathai of Kenya, a biologist who campaigned to preserve forests and green space in urban areas, wins the Nobel Peace Prize

GLOSSARY

African rice   *Oryza glaberrima.* This plant was domesticated from a relative of the plant from which Asian rices developed. It is also called red rice (red is the color of the hull, not of the rice itself). It was domesticated in the Inner Delta of the Niger River around 3500 B.C.E. It spread to the western half of the West African rainforest as a stable crop and to the east as a supplemental crop. African farmers developed a wide variety of cultivars including floating rice along the Niger. Asian rice came to East Africa sometime after the beginning of the current era and gradually spread westward. The Portuguese apparently first brought Asian rice to the region of Angola, and from there it spread into the Central African rainforest. Asian and African rice are cultivated in the same ways including paddy cultivation and dry land cultivation. Asian rice usually yields better (except when rainfall is erratic) and handles easier. Hence, by a very early date even in West Africa, Asian rice began to replace African rice. Today, African rice is grown as a specialty crop in some areas. Africans from the West African rainforest brought knowledge of rice cultivation to the Americas. The establishment of rice production in colonial South Carolina came directly as a result of Africans cultivating the crop for their own use under slavery and then was adopted by plantations as a major export crop.

Afro-Asiatic languages   Also called Afrasan. A family of languages spoken in the Sahel, the Sahara, North Africa, Ethiopia, East Africa, and Southwest Asia. Its subfamilies include Berber (Sahara and North Africa), Chadic (Sahel), Cushitic (Ethiopia and East Africa), Ancient Egyptian, and Semetic (Asia and North Africa). Most scholars believe that the languages spread from Africa, perhaps originating in the Sahara or Ethiopia, although a few argue for an Asian origin. Speakers of these languages were among the first to practice animal husbandry and agriculture in Africa.

Ahaggar Mountains   A mountain range in the central Sahara, it is located in southern Algeria. Although part of the desert, it contains oases where Tuareg people live and from which camel-borne caravans traveled connecting sub-Saharan Africa with the Mediterranean.

Akan   Name of a language family spoken in the West African rainforests. The languages include Twi, Fante, and Asante.

Aquatic civilization   The name given to an archaeological civilization that appeared about 9000 B.C.E. and spread all across the southern Sahara. During this period, most of Africa went through a much wetter phase, and lakes expanded and rivers ran permanently throughout much of what is now desert. A new way of life developed among river dwellers, perhaps first in the area of the middle Nile that relied on the waters for their sources of food. People not only fished but also collected the grains and seeds of grasses that grew along the water's edge year-round. People of the Aquatic civilization lived in permanent settlements. The lifeway spread rapidly across the now grown streams and lakes of the south Sahara. An arid phase between 6500 and 5500 B.C.E. dried up many of the lakes and rivers, forcing the abandonment of many areas. When more humid conditions returned, pastoral and agricultural people expanded into the southern Sahara bringing with them new methods of subsistence in the area.

Atlas Mountains   A chain of mountains running from Morocco to Algeria. Part of the chain stretches south into the Sahara Desert.

Australopithecines   The name given to a family of hominids that lived in Africa from about 3.9 million years ago. Four species are currently recognized as belonging to the genus: *A. afarensis*, *A. aficanus*, *A. amanensis*, and *A. garhi*. Although the exact relationships are still unclear, as some point out the first members of the genus *Homo* evolved from Australopithecines. Other species formerly though members of the genus are now considered part of a separate genus that also evolved from Australopithecines, the Paranthropos, including *P. aethiopicus*, *P. robustus*, and *P. boisei*.

Bananas and plantains   *Musu* sp. Bananas and plantains are closely related trees that produce significant amounts of food in Africa. First domesticated in Southeast Asia, they spread to Africa sometime shortly after the beginning of the current era, perhaps brought by Malagasy-speaking traders and immigrants. Bananas and plantains require warm and moist environments, but they provide exceptionally high yields in the right conditions and, with the use of organic matter as fertilizer, allow for almost indefinite utilization of the same land. Plantains spread westward through the Central African rainforest. In the highlands of East Africa, farmers developed a spectacularly successful system of farming bananas that allowed for extremely dense populations on the mountains of East Africa and in the highlands around the Great Lakes.

Bantu   Name given to a family of languages spoken all over Central, eastern, and southern Africa. Bantu languages began to expand perhaps as early as

3000 B.C.E. from a heartland near the Nigeria-Cameron border. The earliest Bantu speakers practiced a style of agriculture based on yams and oil palms, and their settlements hugged the forest-savannah border along the northern frontier of the Central African rainforest. While some Bantu-speaking communities gradually moved south through the rainforest, those at the extreme east of the rainforest in the Great Lakes region gradually acquired cattle, iron working, and grain agriculture. Sometime just before the beginning of the current era, Bantu languages began to expand rapidly southwards in East Africa and this branch reached southern Africa, bringing with them agriculture and iron working.

Benguela current A cold stream of ocean water that flows north from the Antarctic along the west coast of Africa. Where it meets the Indian Ocean Aqulhas current off the Cape of Good Hope, it creates the stormy conditions for which the Cape is famous. It also creates the conditions that prevent much moisture from rising along the coast of Namibia and thus the Kalahari.

Bight of Benin A bay extending along the west coast of Africa comprising the coastlines of Benin and western Nigeria. It is part of the Gulf of Guinea. It was also known as the Slave Coast during the era of the slave trade.

Bight of Bonny Formerly know as the Bight of Biafra, it extends from the Niger Delta to Gabon. It is part of the Gulf of Guinea.

Camel Camels first arrived in Africa sometime before the beginning of the current era. Camels can conserve water and hence are well suited to arid environments. Their husbandry gradually spread in the Sahara and they became the basis of the lifeway of the nomadic peoples of the desert by about the fourth century of the current era. They revolutionized transport in the desert. Communities herded camels from Somalia to the Atlantic.

Cameroon Highlands Volcanic mountains in the modern Republic of Cameroon. Mount Cameroon is the highest in West Africa. In 1986 a release of poisonous, volcanic gas from Lake Nios killed over 2,000 people.

Cape of Good Hope The extreme southern tip of Africa. It was named the Cape of Storms by the earliest Portuguese navigators to reach it because the meeting of a warm and a cold ocean current creates quite stormy weather off the coast. However, the same conditions combine to create a Mediterranean-type climate in a tiny area around what is now Cape Town with dry, warm summers and wet, cool winters. This temperate climate helps explain why the area became the first place in sub-Saharan Africa to see the establishment of a sizable European-settler population beginning in 1652 when the Dutch East India Company established a base there.

Cassava *Manihot esculenta*, it is also commonly called manioc. It is a New World crop first domesticated in Brazil. The Portuguese brought it to Africa sometime in the 16th century. Cassava produces a tuber than can be cooked like potatoes or dried and ground into flour. Its leaves are also edible. It survives drought well by going into hibernation and tolerates poor soils. It is not very nutritious by itself, and "bitter" varieties must be treated to remove poisons. It has expanded dramatically in both forest and savannah environments to become one of the most important food crops in Africa. It has replaced African yams in forested areas and grains in savannahs. It is often regarded as a famine-prevention crop, giving a yield even in the driest years, and as a crop to be grown on land with reduced fertility.

Cattle   Cattle were domesticated at least three times in world history, once being in the Sahara about 8000 B.C.E. In sub-Saharan Africa, cattle keeping preceded agriculture unlike in Southwest Asia and India, the other areas where cattle were domesticated. Cattle spread throughout the Sahara and Sudan in the next thousand or so years, but further expansion south took much longer. Diseases such as trypanosomiasis, carried by tsetse flies and harbored in wild game, prevented their expansion until bushland had begun to be cleared by humans using fire. They gradually spread south, reaching South Africa around the beginning of the current era. Cattle remain a critical part of subsistence strategies in Africa and have come to occupy a central place in many African societies. In some areas specialized pastoralist communities, such as Fulani in West Africa and Masai in East Africa, practice pastoralism and trade with agricultural communities. In many other areas, cattle keeping is integrated with agriculture and small-stock raising, and is a major element of personal wealth for individuals. In some states, such as Rwanda and perhaps Great Zimbabwe, the ruling class controlled cattle and used the wealth they represented to maintain political power. African cattle keepers had quite sophisticated knowledge about their work, but until the twentieth century they tended to focus on subsistence, which meant keeping as many head alive as possible, rather than the quality of the stock. During the twentieth century, these practices came under attack as inefficient and destructive of rangelands. Colonial and independent government efforts to promote "rational" grazing often led to conflict over stock numbers and efforts to reduce them through de-stocking campaigns.

Chad, Lake   A lake on the borders of Chad, Nigeria, Niger, and Cameroon. It is very shallow and has declined in size dramatically in the last forty years due to drought and increased use of its waters for irrigation. During pluvial periods, such as the Holocene Optimum of about 9,000 years ago, it was much larger and is sometimes referred to as "Mega-Chad."

Coffee   A tree crop probably domesticated in the Ethiopian highlands. It spread into Asia, where it became a source of stimulants for many centuries. Outside of Ethiopia, it was little cultivated in Africa until the twentieth century. Europeans encouraged the cultivation of coffee in both East and West Africa by African farmers and by settlers on estates. It went on to become one of the most important export crops for many African states. Grown on estates, it often requires the clearing of forest. African farmers, though, have developed systems of cultivation that mix coffee with other crops, especially bananas in East Africa and the Great Lakes, that are quite efficient and ecologically sound.

Colonial era   Generally, the period from the end of the nineteenth century to the 1960s in Africa. It was only then that advances in medical knowledge and in military technology gave European nations the power to conquer and occupy Africa. The "Scramble for Africa" developed as a race to control natural resources and African peoples. Conquest itself was often fiercely resisted. Colonial states in some cases alienated land from African communities for European settlement, but in others sought to rule over African populations. The fifty-three colonies of Africa eventually became the independent nations of the continent with very little change in borders as independence came.

Community based conservation   A name for conservation programs developed starting in the 1980s that seek to involve local communities in both the management and any benefits from conservation programs. The approach developed as a reaction to the resistance to top-down, often militarized conservation. Two of the most well-known examples are the Ngorongoro Conservation Area in Tanzania and the CAMPFIRE (Communal Areas Management Programme for Indigenous Resources) in Zimbabwe. Such programs have shown some success in bringing benefits to local communities, but they have not necessarily found a permanent way to address the often competing and contradictory claims of conservation for protection of wildlife and habitat and those of local communities for access to and use of resources.

Congo River   Formerly known as the Zaire River. It is the second longest river in Africa (to the Nile). It is second in the world in both flow and size of watershed (to the Amazon). It flows west from the East African Rift Valley to the Atlantic Ocean through the Central African rainforest. Its major tributaries include the Lualaba, the Kasai, the Ubangi, and the Sangha Rivers. It is navigable from Kisangani in the east to the Malebo Pool in the west. It and its tributaries were long used as transport routes with large canoes carrying produce and fish along its route. The Kingdom of the Kongo, south of the mouth of the river, gave its name to the river. The main course of the river became the center of King Leopold of Belgium's personal empire in the Congo Free State in the late nineteenth century. The Belgian government took over the colony in 1905 after a scandal over forced labor. The northern tributaries became part of French Equatorial Africa.

Convention on International Trade in Endangered Species of Wild Fauna and Flora (CITES)   It is a convention signed by the nations that are members of the World Conservation Union in 1963. It regulates the conservation of wild species and has moved to ban trade in endangered animals and plants. It most famously banned the trade in ivory from Africa in the 1980s after poaching placed many elephant populations under extreme threat.

Dahomey Gap   A gap in the West African rainforest covering roughly the area of modern Benin made up of savannah and forest-savannah mosaic. It is caused by ocean currents bringing cooler water to the shore that reduces regular rainfall in the region. Its existence allowed for more livestock in the area and the use of horse cavalry for at least short periods.

Debt for conservation   A process where international debt held by a developing country is purchased by governments and nongovernmental organizations and then repayment in local currency is used to fund conservation activities. It has been widely used to fund conservation efforts in several African countries including Madagascar, the Central African Republic, and others. In many African countries, Nongovernmental Organizations (NGOs) such as the World Wide Fund for Nature (WWF), the Nature Conservatory, and the Frankfurt Zoo have reached agreements with national governments to assist in managing controlled areas and natural parks, often at the insistence of donor governments. Such programs have led to the expansion of protected areas in Africa but have also often led to conflicts with local communities who are forced to move or not allowed access to resources they had previously used.

Deforestation   A major concern in rainforest and forest-savannah mosaic areas of Africa for many years. Clearing of forests for agriculture and exploitation of timber for export and fuel first became an issue at the beginning of the twentieth century when colonial officials began to claim that much of Africa faced the threat of permanent degradation due to forest-cover loss. Efforts at conservation often generally involved limited African access to forest resources while reserving the right to profit from timber to colonially sanctioned concessions. These efforts often provoked resistance from African populations denied access to resources. In many cases, at independence, new African governments opened some areas to exploitation, but most quickly reverted to strict controls on paper of forest resources. In many parts of West Africa, forest loss has continued while the political instability of Central Africa has meant that much of the Central African rainforest has been less extensively cut. Recent research has indicated that forest-cover loss may be less extensive in the colonial era than previously thought.

Desiccation-desertification   An argument that overgrazing, loss of fallow, and overuse of water resources, particularly in the Sahel on the southern border of the Sahara, was leading to the irreversible expansion of the desert. French colonial officials and scientists first made the argument in the aftermath of the Sahelian drought of the last half of the 1910s. It again became a major issue in the 1970s' drought in the Sahel. The increase in rainfall experienced since the mid-1980s has changed the terms of the drought as vegetation has recovered in areas previously thought to have undergone permanent desiccation in the Sahel. Concerns remain about degradation.

Donkey   One of the animals domesticated in the Saharan region of Africa. Domesticated donkeys spread to the rest of the Old World.

Drakensberg Mountains   A mountain range in South Africa, Lesotho, and Swaziland running roughly parallel to the Indian Ocean coast. Below the Drakensberg Mountains lie tropical lowlands and behind them the more temperate highveld of South Africa.

East Coast fever   Also called Theileriosis, is a disease in cattle cased by a tick-borne parasite, *Theileria parva*. It is found from southern Sudan down East Africa as far south as South Africa. It is carried by buffalo. It causes high mortality in newly introduced cattle and was a major factor in slowing the spread of domestic livestock and determining the range of livestock.

Eastern Arc Mountains   A series of mountain ranges stretching from the Taita Hills in southern Kenya through the Pare, Usambara, Nguru, Sagara, Uluguru, and Poroto in southern Tanzania. Formed over two million years ago, they have received regular rainfall from the monsoons and hence have a much more regular climate than the surrounding plains in both the short run, when the plains suffer regular drought, and in the long run as arid and humid phases have less effect on them. Mount Kilimanjaro, Kenya, and Meru and other volcanic mountains along the Rift Valley are not geologically part of the Eastern Arc but have similar climates. The volcanic peaks are much younger. All have become sites of species endemism. They have also over thousands of years become sites of relatively high human-population density because of their regular rainfall and often fertile soils. They face extreme pressure on their forest cover from logging and expanding populations.

Ecotourism   The practice by relatively wealthy tourists of visiting game reserves and national parks in order to see or photograph wildlife and scenic landscapes. It developed in Africa in the 1960s as sport hunting fell out of favor in the industrialized west and governments began to limit hunting in preserved areas. It has become the major form of tourism in East Africa and is very important in southern Africa. Despite the ideal of helping conserve nature by not shooting it, the development of infrastructure to support tourism such as roads and lodges has led to concerns about its effects on the parks and reserves. Popular parks like Masai Mara in Kenya and Kruger National Park in South Africa receive tens of thousands of visitors each year. Sport hunting remains legal, though, on some private lands in southern Africa and in some national parks and reserves in places like Tanzania.

Enset   *Ensete ventricosum.* A banana-like plant found in eastern Africa. Peoples in the Ethiopian highlands began to cultivate the plant perhaps as early as around 6000 B.C.E. The bulb and inner stalk are edible. After becoming the basis of one of the oldest agricultural traditions, it is still grown today as a food crop in the highlands.

Equatorial rainforest   One of the largest rainforests in the world spreads from Cameroon south and east to cover most of the Congo River Basin. Despite receiving regular rainfall that allows it to be a rainforest, it is one of the drier rainforests in the world. The size of this forest has varied over time, shrinking to several isolated refugee regions during dry glacial ages, the last of which occurred about 20,000 years ago, and expanding during more humid times such as the Holocene Optimum about 9,000 years ago. Although foragers have lived in the forests for at least tens of thousands of years and agriculture spread to the equatorial rainforest about four thousand years ago, its general human population has remained relatively small, less in density for example than the West African rainforest. Hence, its degree of tree cover is today much higher than in West Africa. Many parts now, though, are being deforested to open agricultural land and for logging.

Ethiopian Highlands   A high plateau region covering about two thirds of modern Ethiopia. It is split into two by the Great Rift Valley and surrounded by extremely arid plains. The area was home to some of the earliest human ancestors. The highlands were a very early center of the development of agriculture and several crops were domesticated there, including perhaps teff, finger millet, and coffee. The highlands were occupied by speakers of Afro-Asiatic languages. Farmers over thousands of years have developed an intensive agricultural system based on terracing the slopes of the highlands and the use of oxen plows. Over the last several decades, the region has been subject to repeated droughts and food shortages that on several occasions have become major famines such as in 1974 and 1985. Many fear the land worked so intensively for several thousand years has lost the ability to support the growing population. A drought in the late 1990s did not cause a full-blown famine only because of sizable outside assistance.

Finger millet   *Eleusine coracana.* It was probably domesticated in northeastern Africa or the Ethiopian Highlands sometime after 6000 B.C.E. It is a very nutritious grain, but requires care in handling. It became, by historic times, a major crop in Uganda and southern Sudan. It has also been widely grown for beer making. It is not very drought tolerant but grows well in well watered areas, especially in highland regions of eastern and Central Africa.

Fish River   The river runs from the Drakensberg Mountains down to the Indian Ocean Coast of South Africa. It is sometimes called the Great Fish River to distinguish it from another Fish River in Namibia. It formed the rough boundary in southern Africa between the arid Karoo to the west and the better-watered lands to the east. Hence, it became the frontier between the agricultural Xhosa-speaking peoples and the pastoral and forager Khoi-San populations. After Europeans conquered the Cape of Good Hope and spread their settlements into the Karoo, the Fish River became the frontier between European settlement and independent African societies for over a century. Only in the mid-nineteenth century did the British establish rough control over the peoples to the east of the Fish River.

Fonio   *Digitaria exilis* (white fonio) and *Digitaria iburua* (black fonio). A small grain cultivated in regions of west Africa, especially northern Nigeria. Domesticated in the southern Sahara, some believe it to be the first crop domesticated in Africa. It tolerates poor soils well and has become a crop cultivated on worn-out plots. It produces a very small grain that makes handling difficult. It has declined sharply in importance over the last several centuries.

Foragers   Foraging, or hunting and gathering, was the original human lifeway and persisted in parts of Africa into the twentieth century. The spread of agriculture and pastoralism, which allowed for higher population densities, relegated foraging to extremely marginal environments or to specialist communities living in close connection with agricultural and pastoral ones. Some of the most important historic foraging communities in Africa include the Batwa/Ba'aka peoples of the equatorial rainforest. These communities now have close relations with agricultural communities. They are thought in part to be descended from the original inhabitants of the rainforest, at least in terms of lifeway, and are often called "Pygmies." However, all known communities in the region speak versions of the same Bantu languages as their agriculturalist neighbors, so they have lived in close contact with Bantu speakers for over 3,000 years and there has also been intermarriage as well as cultural exchange. The San of southern Africa also are thought to be descendants of original inhabitants, and versions of their languages were spoken as far north as East Africa until a few thousand years ago as well as by communities that practiced pastoralism with sheep and cattle, the Khoi. They lived in the arid reaches of southern Africa, although some had close relations with the agricultural Tswana speakers of what is now Botswana and northern South Africa. The term Dorobo is used in eastern Africa for foraging communities that often speak different Cushitic, Nilotic, or Bantu languages. Some of the languages seem descended from original inhabitants, others related to surrounding peoples. In all of these cases, though, foraging should be seen as a subsistence strategy pursued by some groups within the context of relations with other groups practicing agriculture and pastoralism. They are often "dependent" on agricultural peoples and have specialized functions in the production of hunting products or even as iron workers.

Forestry   Forestry in the modern sense began during the early colonial era in Africa. Many African peoples had cultural practices that served to protect forests, especially in sensitive areas such as hillsides. These included beliefs that ancestors or spirits lived in the forest, or that ancestors required forests for their propagation. These practices often resulted in "sacred groves" where people could not clear fields or even cut wood. Likewise, in some areas, complex systems of land tenure where for example heads of extended clans controlled distribution of land to clan members and newcomers often were geared toward maintaining a degree of forest cover and promoting proper cycles of long fallowing. Colonial regimes that European nations developed in the late nineteenth and early twentieth centuries sought to impose what they considered scientific management to at least parts of African forests. They sought to promote rational exploitation and to prevent haphazard clearing. In short, they sought to take control of forests from African communities. In many areas, this meant strict limits on the ways Africans could use forests. Large areas in some colonies were declared forest reserves, closed to agriculture. Africans in many areas could cut timber without a permit or even collect firewood. Valuable species were reserved for licensed loggers who paid fees to the colonial government. Colonial governments also experimented with exotic species. Black wattle and eucalyptus both became widespread and planted by African farmers, due to their efforts. Africans often resisted forestry policy and often ignored rules concerning forest use. They often saw it as a theft of their resources by the colonial government. Colonial governments also promoted tree planting by Africans and sought to encourage agroforestry in some areas. At independence, many African governments eased forestry rules and opened some reserves. However, both because forests represented resources and because of conservation concerns, after a few years, government forestry departments became often more restrictive than colonial ones had been. The weakness of several African states and the need for foreign currency has often served to undermine conservation efforts since independence. Likewise, expanding populations have cleared large areas of forest, especially in West Africa, for agriculture.

Fossey, Dian An American animal behavior specialist who became famous for her work in Rwanda with mountain gorillas. She promoted the preservation of their habitat and fought against poaching. This campaign led to conflict with some people in Rwanda, and she was murdered in 1985.

Goats Domesticated in Southwest Asia, goats came to Africa perhaps 5,000 years ago. They quickly spread, often in conjunction with sheep and cattle. Goats had resistance to several livestock diseases and could live on resources that could not support cattle. They have become a major part of African subsistence strategies across the continent, providing both meat and milk.

Goodall, Jane An American anthropologist who is famous for her work on chimpanzee behavior in western Tanzania. Through tireless effort she has created an international effort to protect chimpanzee habitat in Gombe National Park and Mahale Mountains National Park. Efforts there have included community conservation and development programs with local communities.

Great Lakes of Eastern Africa These lakes are perhaps more properly called Rift Valley lakes since they all lie with the two arms of the Rift Valley. The three largest are Lake Victoria Nyanza, Lake Tanganyika, and Lake Malawi (formerly Nyassa) and are generally called the Great Lakes. Many other lakes lie within the Rift Valley from Ethiopia down to Lake Malawi. Many of those in the northern stretch feed the Nile River. Several in the western Rift in Kenya are alkaline including Lakes Magadi, Natron, and Eyasi. Lake Victoria Nyanza is the largest and feeds the White Nile as do Lakes Albert and Edwards. Just to the south of these Lake Kivu feeds Lake Tanganyika, which in turn helps feed the Congo River system. Lake Tanganyika is the oldest and deepest of the lakes. Lake Malawi feeds the Zambezi River through Lake Malombo. The lakes are home to a large number of endemic species of fish, including over 800 chichids. They support a large bird life and have provided a livelihood for people for thousands of years. Some of the mostly densely populated parts of Africa lie around the Great Lakes especially in the well watered area to the west of Lake Victoria Nyanza in Rwanda, Burundi, Uganda, and northwestern Tanzania.

Groundnut Scheme The name of a massive project in British-ruled Tanganyika after World War II to produced groundnuts (peanuts) for oil using mechanized farming techniques on land alienated from local populations. The scheme was a gigantic failure. The land picked in central Tanganyika was too arid and the soil hardened to the point that it broke the metal implements used.

Grzimek, Bernard Together with his son Michael, the German director of the Frankfurt Zoo did much to promote the ideal of wildlife conservation in independent Africa. He first came to Africa in the 1950s to collect animals for his zoo. He wrote several books on the weakness of conservation efforts, including *Serengeti Shall Not Die*, which also was the name of an Academy Award–winning documentary made by Bernard and Michael. Michael died in a plane crash in the Serengeti during the making of the film. He did much to support the extension of conservation, especially in Tanzania. He helped found the Serengeti Research Institute and Mweka Wildlife College, which have become important training grounds for researchers and managers across the continent. Part of his legacy is the Frankfurt Zoological Society, which provides major support for conservation efforts at Serengeti and across the globe.

Guinea fowl   *Numida meleagris.* A bird related to other game birds such as turkeys, domesticated in Africa. It was the only fowl domesticated in Africa.

Holocene Optimum   The term refers to a humid phase in the Earth's climate that lasted from about 12,000 to about 9,000 years ago. During the time, the monsoon rains reached deep into what is now the Sahara Desert and the desert shrank to a much smaller area than it is now. Lakes filled in many areas and channels took water from the desert to the Niger and Senegal Rivers in the west and the Nile River in the east. Farther south, forests expanded, both in the rainforest regions and down the slopes of highland areas, and lakes and rivers expanded greatly. As the Earth's climate cooled, the phase gradually ended. Conditions reached about those of current times around 4,500 years ago. Some scholars suggest that the cycle of dry periods and renewed moist periods helped hasten the spread of both pastoralism and agriculture in the Sahara and southwards.

Homo egaster   Name given to a population of hominids in Africa that emerged about 1.9 million years ago. This species is associated with the development of a more complex stone tool kit, called the Achulean industry in Africa. This population is contemporary with the *H. erectus* populations that spread into Asia. It is thought to be directly ancestral to modern humans who emerged from them in Africa.

Homo erectus   Name given to hominids who emerged from the *H. egaster* populations of Africa and spread across the Old World to Asia. Fossils have been found especially in Indonesia and China. Most populations of *H. erectus* disappeared after about 400,000 years ago, replaced in many areas by successor populations that themselves were replaced by *H. sapiens* populations arriving from Africa. Recent discoveries seem to indicate that some relic populations survived in isolated areas in Southeast Asia until as late as 50,000 years ago.

Homo habilis   Name given to a species whose fossils have been found at Olduvai Gorge in Tanzania and associated with the earliest and simplest tools.

The species was long thought to be the direct ancestor of all later hominids; however, recent contemporary finds in other parts of Africa show species with distinct differences from both the Olduvai finds and other australopithecines, leading to the argument that this species, called *H. rudolfensis*, is directly ancestral to *H. egaster* and that *H. habilis* is an evolutionary dead end.

Homo sapiens   Modern humans. The most recent evidence indicates that modern humans emerged from African hominid populations about 200,000 years ago and slowly and gradually spread first across the continent and then across the globe. The oldest, clearly dated fossils of *H. sapiens* date to about 130,000 years ago and have been found in Africa. Appearance of *H. Sapiens* is also associated with the appearance of Middle Stone Age artifacts, which have been dated to about 180,000 years ago in Africa. This technology used more complex and composite tools.

Horn of Africa   The peninsula lies in the northeastern corner of Africa that juts into the Arabian Sea. The broader region includes the countries of Somalia, Ethiopia, Eritrea, and Djibouti.

Inner Delta of the Niger River   An area just to the west of the great bend of the Niger River where it turns south after flowing into the Sahara, the Inner Delta is an area where the river breaks into several different channels. The Inner Delta becomes inundated each year during the rainy season. It became an early center of human settlement and of agricultural diffusion. African rice was probably domesticated in the area.

Iron working   Iron working, and metal working generally, was a critical technological development in human history. The earliest known iron working occurred in what is now Turkey about 1500 B.C.E. and spread from there throughout the Mediterranean, Asia, and Europe. Archaeologists long believed iron working in Africa spread from the Mediterranean through perhaps two routes, one up the Nile from Egypt to Nubia and perhaps another across the Sahara from Carthage to West Africa. Recent discoveries of iron working in Nigeria and especially in the Great Lakes region of East Africa have called this scenario into question. In Nigeria, finds at the site of Nok show an advanced iron working culture dating to about 800 B.C.E., which seems much too early for diffusion from North Africa to have occurred. In the Great Lakes region, some cites have tentatively been dated to 1000 B.C.E., much earlier than the earliest cites in Nubia where iron working definitively spread from Egypt by about 500 B.C.E.. The dates for the Great Lakes sites are much more insecure than those from Nok. In addition, early bronzeworking sites have been found in the Sahara dating to before 1000 B.C.E., which would support independent invention in Africa. Elsewhere in Africa, iron working spread in advance of working other metals. Iron working gave peoples the ability to make more effective tools, especially hoes and axes. The rapid spread of iron working across after Africa after about 500 B.C.E. attests to its importance and iron working is associated with the spread of agriculture and the Bantu languages in eastern and southern Africa.

Ivory   The tusks of elephants, although the term is also sometimes used for hippopotamus teeth. In Africa, elephant ivory had long been worked for decorative purposes and became a major export to Egypt and the Mediterranean during the classical age. Ivory continued to be a major export both into the trans-Sahara trade and into the Indian Ocean trade. Elephant hunting became a specialized activity at a very early date as a result. The opening of the Atlantic trade brought new demands for ivory from European traders, and elephant populations seem to have declined sharply in West Africa as a result. The nineteenth century was the heyday of the eastern and southern African ivory trades. Improved firearms massed in larger numbers made elephant hunting more methodical, and thousands were slaughtered to feed caravans from the East African coast and merchants operating from South Africa and Mozambique. Elephant populations were decimated, although significant herds remained in eastern and southern Africa. Gradually, colonial regimes placed controls on hunting, although loss of habitat prevented elephants from recovering to former levels in the twentieth century. In the last of half of the twentieth century, new markets for ivory opened in Asia particularly, and poaching became a major concern. In 1989, the Convention on International Trade in Endangered Species of Wild Fauna and Flora (CITES) banned the international trade in ivory completely. Since then, elephant populations have recovered in eastern and southern Africa to the point were southern African nations feel they have too many elephants and have sought permission to trade ivory legally, collected or culled. Poaching and illegal trade remain problems, given the insecure status of much of Africa and the continuing demand, especially in Asia.

Jenné Jeno   The name means "ancient Jenne." Historic Jenné (also known as Djenné) is a city on the Niger River in the Inner Delta in the modern nation of Mali. During the period between about 800 C.E. and the nineteenth century it was a major trading town, with gold coming from the south and fish and rice from nearby and then trans-shipped via the river to desert-side towns like Timbuktu for consumption and trade. It also became a center of Islamic scholarship under the empires of Mali and Songhai. Jenné Jeno was the original site of the town that dated to about 200 B.C.E. It is the first known urban area in West Africa and developed as a result of interaction between peoples moving south during the periodic episodes of dry climate that moved the boundary of the Sahara southward.

Kalahari Desert   The Kalahari is a large, semi-arid and arid area covering most of Botswana and parts of South Africa, Zimbabwe, and Namibia. It is not a true desert in many areas, receiving around 250 mm of irregular rain. It however has no permanent surface water. San (or Bushmen) foragers have lived in the area for at least 20,000 years. In addition to several reserves and national parks, it is also home to extensive mineral exploitation and exploration.

Karoo   The large, semi-arid area in South Africa stretching from the Cape of Good Hope east to about the Fish River. Before the arrival of the Dutch at Cape Town, it had been populated by herders and foragers, the Khoi and San or Hottentots and Bushmen as the Dutch called them. Gradually white settlers (Boers, farmers in Dutch) killed, drove out, or absorbed that population and expanded into the area, developing an extensive pastoral lifestyle. The area was taken over by the British during the Napoleonic wars and made part of the Cape Colony. In the 1830s, many Boers left the colony and created independent republics in lands to the north that they eventually conquered from the local African populations.

Khoi   The people who lived in the Cape region of South Africa before European conquest and practiced a pastoral lifestyle based on herding cattle and sheep. Closely related linguistically to San forager groups in the area, their languages had been spoken all over the southern half of Africa in the distant past, before being replaced by Bantu languages as agriculture and iron working spread. Living in an arid and isolated environment before European contact, they died of disease, were killed by Europeans, or were absorbed into colonial society as part of a subordinate group labeled "Coloureds."

Kilimanjaro   Tallest mountain in Africa, it is located in northern Tanzania near the border with Kenya. It is an inactive volcano with two peaks, Mawenzi and Kibo. Its highest point rises 5,895 meters off the plain. It is also the world's largest free-standing mountain. Its lower elevations are densely populated, and its soils and climate make it very productive.

Kintampo   The name of an archaeological culture found around the Volta River that appeared about 3500 B.C.E. It marked the extension of agriculture into the forest regions of West Africa.

Kruger National Park   One of the largest and most successful game reserves in Africa, Kruger National Park is located on the northern border of South Africa bordering Mozambique and Zimbabwe. Since 2000 it has been part of the Great Limpopo Transfrontier Park along with Limpopo National Park of Mozambique and Zimbabwe's Gonarezhou National Park. The area of the park is described as "lowveld" with a hot climate. Population had remained relatively light in the area before the last half of the nineteenth century because of malaria and tsetse fly. In the last half of the nineteenth century, the discovery of gold in the area resulted in an influx of whites into the area and a sharp decline of the animal population. The rinderpest epidemic in 1896 further reduced animals as well as forcing much of the white and African population out of the area. The independent Transvaal (the Afrikaner Republic also called the South African Republic) set aside part of the area as a game reserve in 1896 to allow wildlife populations to recover for hunting. The area was expanded after the Anglo-Boer war, and all human residents were forced to leave. Gradually, animal populations increased, and the park became a favorite destination for white holiday seekers in South Africa. In 1926, the area was declared a national park and renamed after the founder of the Transvaal. By the late twentieth century, the animal population of the park had increased beyond the capacity of the land to support it. The creation of the Peace Park and the removal of the boundary fence of the park adjoining private hunting preserves in South Africa have raised hope that the animals will be able to move into new land, and that the legal hunting on private land will keep the population of the park at a more sustainable level.

Leakey, Mary, Louis, and Richard   A mother, father (Mary and Louis), and son (Richard) who were paleontologists as well as being active in public affairs in Kenya and East Africa generally. Mary and Louis's work at sites like Olduvai Gorge in Tanzania during the 1950s helped begin the process of establishing current knowledge about human evolution and the centrality of Africa in the process. Richard Leakey continued his parents work, leading the expedition that discovered the "Turkana boy," a nearly complete skeleton of a 12-year-old boy dating to about 1.6 million years ago and now classified as an example of species *H. egaster*, the hominid species directly ancestral to modern humans. Louis and Richard have both played important roles in Kenyan public affairs, Louis serving the colonial government and becoming an advocate for conservation. Richard followed his father and became head of the Kenya Wildlife Service, where he advocated strict conservation measures and a complete ban of ivory sales. He moved in and out of several positions in the Kenyan government, often accused of insensitivity, and survived a plane crash that cost him both of his legs.

Limpopo River   A river that rises in the Kalahari and then flows eastward to the Indian Ocean. For part of its length it serves as the boundary between the Republic of South Africa and Zimbabwe and Botswana. It then flows through Mozambique and enters the Indian Ocean at the port of Xai Xai. Since 2000 it flows through the Great Limpopo Transfrontier Park created by South Africa, Zimbabwe, and Mozambique.

Lucy   Name given to the almost complete skeleton of a female *Australopithecus afarensis* dating from 3.2 million years ago discovered in 1974 by David

Johanson and Tom Gray in Hadar, Ethiopia. The fossils proved that australopithecines walked upright and the species *A. afarensis* known from fossils found in Tanzania and Ethiopia is directly ancestral to later hominids, including *H. sapiens.*

Maathai, Wangari A Kenya professor of biological sciences and conservation activist who became the first African woman to win the Nobel Peace Prize in 2004 for her campaigns to promote conservation and green space in Kenya. Dr. Maathai was educated in the United States and Kenya and because a professor at the University of Nairobi. She founded the Green Belt movement in Kenya to promote tree planting and preserve the remaining green space in that city. Her activism brought her into conflict with the government, and she was arrested and harassed. She also campaigned for democracy, human rights, and women's rights. In 2002 she was elected to the Kenyan Parliament as the opposition took control of the government and became an assistant minister for the environment, natural resources, and wildlife.

Madagascar Madagascar is the fourth largest island in the world. It is a remnant of the super continent (along with the Seychelles Islands) of Gondwana, where it joined what became Africa to India. As a result, its flora and fauna are highly endemic. Humans arrived on the island fairly recently, perhaps as late as 2,000 years ago. Within a brief period, many endemic species, including seventeen species of lemur, an ostrich-like bird, and a pygmy hippopotamus became extinct. Malagasy, the language of many of the people of Madagascar, is related to the languages of Southeast Asia and demonstrates that some of the immigrants to the island came via sea. Others came from Africa in the period after the establishment of regular sea contact across the Indian Ocean.

Maghreb The northwest of the African continent currently occupied by the countries of Morocco, Algeria, Tunisia, and Libya. The original inhabitants were Berber speaking and the coastal areas came under the control of the Phoenicians and then Romans. Muslim conquest brought it into the Islamic world. After about 1100 C.E., a gradual process of Arabization occurred with Berber languages remaining only in more isolated areas.

Maize American corn; is a New World Crop that came to Africa in the aftermath of the Columbian exchange. It gradually spread from the Atlantic and Indian Ocean coasts and across the Sahara to become the most widely grown grain crop in Africa. Maize generally yields well with enough moisture and thrives in many parts of Africa, especially cleared forests. However, it is not as drought resistance as several African grains. It expanded greatly in the twentieth century as the result of the development of markets in the continent. It was favored by white settlers in southern and eastern Africa, and adopted by African farmers all over the continent. Local cultivars have in many areas been driven out by commercially derived white maize varieties.

Malaria   Malaria is one of the oldest diseases known. It appears to have emerged along with humans in Africa. It is caused by organisms of the plasmodium genus of protazoal (mainly *P. falciparum*, *P. vivax*, *P. ovale*, and *P. malariae*). With rare exceptions, it infects only humans. It is deadly, especially to those first exposed, and even today causes around 2 million deaths a year in Africa. *P. falciparum* causes the most deadly form of malaria, especially in Africa. Mosquitoes of the Anopholine family are the only ones that carry the disease. The disease spread out of Africa into other areas of the tropical Old World with humans, but did not reach the New World until after the beginning of the Columbian exchange. Malaria inhibited human settlement into the wetter areas of Africa. Population densities in moist areas remained lower than in the savannahs until quite recently. In fact, the expansion of agriculture may have encouraged the spread of malaria by creating more opportunities for mosquitoes to breed. Human populations in Africa developed several biological defenses against malaria. Vivax malaria, a common form of the disease in Asia, is not present in Africa because of a genetic mutation in many African populations that prevents its survival. People constantly exposed, as is often the case in the tropical regions of Africa, to malaria develop temporary immunity to it. Children and those whose systems are otherwise weakened through other diseases or malnutrition remain at risk. Malaria became the main reason both the outsiders could not long occupy tropical Africa and that Africans came to suffer the Atlantic slave trade. The spread of malaria into the tropical New World helped decimate indigenous populations and at the same time lowered European survival rates. Africans had a slightly higher survival rate and came to be favored as laborers in the developing New World plantations in the tropics. Knowledge of the use of the bark of the chichon tree (*Cinchona ledgeriana*, native to South America) as a curative dates to perhaps the sixteenth century. Its active ingredient, quinine, was isolated in the 1820s, and the gradual spread of its use as a preventative treatment allowed higher survival rates for Europeans by the late nineteenth century. By about 1900, both the cause and the vector of malaria had been identified. Aggressive use of insecticides gradually drove malaria out of several parts of the world. However, concerns over the effects of the most effective pesticide, DDT, on other animals, led to its widespread abandonment. In the period since 1960, malaria has become even more important as a killer as populations have expanded in malaria areas, despite the development of several fairly effective treatments. Malaria parasites have also evolved resistance to some of the most common treatments.

Mande   Mande is a term used for a large family of languages in West Africa. The Mande have spread in a diaspora across the West African Sudan and Sahel and even into the forest regions. Some of them created the great empires of Ghana and Mali and others developed trade networks that linked West Africa with the Saharan trade.

Miombo woodlands   Miombo is the term used for open tropical woodlands that occur across eastern, Central and southern Africa. They stretch in several different subcategories from Tanzania in the east to Angola in the west and from the Central African Rain forest to South Africa. They provided the natural habitat for tsetse and hence human action in clearing the land had to occur before agriculture and pastoralism could thrive. The great rinderpest epidemic of the 1890s reduced human and cattle populations and allowed bush to re-emerge, facilitating the spread of tsetse. Miombo woodlands are also home to the largest populations of large game in eastern and southern Africa and some of them have become parts of national parks and game reserves.

Monsoon   The term refers to the regular annual shift in the predominant direction of the winds around the equator that determines the weather of most tropical areas. Monsoons are caused by the shift in the Earth's axis over the course of the year. The hemisphere tilted toward the Sun becomes warmer and warm, dry air pushes off land masses and into the oceans. In the hemisphere tilted away, the opposite occurs, and moist air flows from the oceans to the landmasses. Geomorphologic features can break up these patterns. Around Africa, two monsoon systems operate, both connected by the movement of the Intertropical Convergence Zone (ITCZ). The shifting of the ITCZ draws moisture north into West Africa and its return south takes moisture to East Africa. This gives much of the African continent a bimodal wet season/dry season climate. The shift also causes wind patterns to shift, particularly along the East African coast. In the past sailing ships could move from East Africa to Asia on the monsoon and return in the same year by sailing with the shifting winds.

Nbata Playa   Nbata Playa is a basin now in the Sahara desert in southwestern Egypt. In the period between about 9000 B.C.E. and perhaps as let as 2800 B.C.E. when the Saharan climate was wetter, it was occupied by a series of settlements that took advantage of the water that collects in the playa. The earliest signs of animal husbandry and agricultural in sub-Saharan Africa have been found here, dating to perhaps the origins of the settlement. These finds support the contention of several scholars that cattle were domesti-

cated separately in Africa and that African food production systems developed independently from those of the Mediterranean.

Ngorongoro   An extinct volcano crater in northern Tanzania, it is one of the
most striking landscapes in the world. Wildlife inside the large crater live
basically isolated from their compatriots outside the crater in the Serengeti.
The crater has also long been home to humans, most recently Masai pastoralists who occupied it for several centuries with their herds. After several
failed attempts under German and British colonialism, the British set up and
the independent government maintained the Ngorongoro Conservation
Area (NCA) separated from the Serengeti National Park. This move allowed
Masai groups to continue living in the area as long as they followed rules developed by conservation officials. The process has not necessarily been an
easy one, with local residents chafing as restrictions on agriculture and conservation officials claiming that population growth among the residents and
their demands to be allowed to use land more intensively threaten wildlife.
However, the great benefit of local communities in the area who benefit
from at least some of the tourism income has meant that the animal populations of the NCA did not face the same pressure from poaching as happened
in the Serengeti National Park. In fact, the only confirmed population of
rhinoceroses in northern Tanzania lives inside the crater, protected by the
Masai, while most of the rest are feared hunted to extinction.

Niger River   The Niger River rises near the Atlantic and then flows eastward
for 4,000 km, making a large bend into the Sahara Desert before being joined
by the Benue River flowing from the Cameroon Highlands and entering the
Gulf of Guinea in what is today Nigeria. The Inner Delta was home to an
early urban civilization, and the river has long served as a major artery of
trade in West Africa. The Niger Delta, in what is now Nigeria, lies in the
rainforest zone. Today it is one of the most densely populated areas of
Africa. It is also home to large reserves of petroleum and the scene of conflict over environmental degradation.

Nile River   The Nile is the longest river in Africa and home to one of the oldest
urban civilizations in human history. The Nile has two principle origins.
The White Nile flows from the Great Lakes of eastern Africa. The Blue Nile
rises in the Ethiopian Highlands. They meet at the city of Khartoum in the
modern nation of Sudan. The level of the Nile rises and falls each year based
on the rains in Ethiopia. Before the construction of the High Dam at Aswan,
this feature had led to regular flooding that formed the basis of human survival along the river. The river, passing through the Sahara Desert, had also
served as the original gateway for humans to leave Africa. Humans developed very early on a sedentary lifestyle along the Nile that has become

known as the Aquatic Tradition and spread across Africa. The introduction of wheat and barley from Southwest Asia to the lower Nile allowed for the development of very productive agriculture based on cultivation as the flood receded. Although the Nile remained a corridor of contact, cataracts prevented easy river communication past the First Cataract. Nubia, based on the upper Nile, remained a separate society from Egypt, sometimes conquered by it, sometimes independent, at least once providing Egypt's ruling dynasty. Control over the waters of the Nile has remained a critical issue. In the twentieth century, the British imposed an understanding between all the colonies and nations through which the Nile flowed that they would not interfere with the flow of the river to Egypt. Since independence, several countries, including Ethiopia, have mooted plans to use the waters of the Nile for irrigation and potentially greatly reduce the flow to Egypt.

Nok   The name of a village in central Nigeria that is the site of a large prehistoric settlement where some of the earliest evidence for iron working occurs in sub-Saharan Africa. Nok-culture artifacts, including terra cotta sculptures, have been found widely across Nigeria. The culture practiced agriculture in the forest-savannah mosaic, and iron working in the region may date back to 800 B.C.E. Some scholars speculate that it was invented in the area as a by-product of pottery making. Others believe it diffused into the area either from North Africa or from an as yet undiscovered site of invention somewhere within Africa.

Office du Niger   The name given to a large project in French West Africa to promote "modern" mechanized farming among farmers along the Niger River in what is now Mali and Niger. Although local farmers did adopt some of the innovations pushed by the project, it proved a costly failure as many of the elements of the project were designed by Europeans based on European experience in agriculture and proved unsuited to the semi-arid environment.

Oil palm   *Elaeis guineensis.* The African oil palm is indigenous to the forests of West Africa between Sierra Leone and Angola. Its fruit can be crushed to produce edible oil. It was domesticated in West Africa perhaps as early as 5000 B.C.E. African farmers spread it throughout the regions of the continent that could support it. Palm oil became a major export crop from West Africa in the nineteenth century as demand for oil increased in rapidly industrializing Europe. Major producing areas include what is now southern Nigeria and Ghana. Oil palm plantations in Southeast Asia now dominate the world market for palm oil.

Olduvai Gorge   It is a thirty-mile-long ravine in northern Tanzania that is part of the Great Rift Valley. Because the ravine allows for a clear view of the stratigraphy, and because during humid phases in the Earth's history it was the site of a large lake, the prehominid remains of many different species have been preserved and dated there. It has become one of the most important sites in the world for understanding human evolution. It was first excavated in the 1950s by Louis and Mary Leakey.

Pastoralism   This term refers to the keeping of livestock, especially cattle, by specialized groups. It has been widely practiced in Africa; however, all pastoral groups have maintained close relations with surrounding agricultural peoples. Trade between them provides farmers with animals and their products, and pastoralists with grains and other food. Important pastoral groups in Africa have included the Masai in East Africa, the Samburu and Turkana of northern Kenya, and the Fulani of West Africa. Many African peoples have also combined animal husbandry with agriculture.

Pearl millet   *Pennisetum glaucum.* Also called bulrush millet. Domesticated sometime around 5000 B.C.E. probably in the southern half of what is now the Sahara Desert. Before the twentieth century, it was perhaps the most widely cultivated grain in Africa. Like other African grains, it also spread at an early date to Asia. Originated from a grass common in the semi-arid region of the Sahel, it is perhaps the most heat and drought resistant of major grains. It grows easily and contains excellent nutrition. However, it does not store particularly well. Its yields generally fall below those of maize or even sorghum with adequate moisture, but it's tolerance for drought means that it provides higher yields during dry years or in dry areas. It is now the third-most grown grain in Africa. It still remains a major crop in the Sahel of West Africa and in drier areas of eastern and southern Africa. Maize has replaced it in large parts of eastern and southern Africa.

Poaching   Poaching has developed as a problem in wildlife conservation in the twentieth century. Many African communities relied on hunting for part of their subsistence before the twentieth century. Over the course of the century colonial governments sought to impose control over hunting by both Africans and European settlers in the name of protecting wildlife. As soon as the rules were imposed, poaching began. For Africans it was often a matter of continuing the practices they had often followed which were now illegal. Hunting provided meat, skins, other items, and protection for livestock and crops in the field. European settlers in eastern and southern Africa also continued to hunt to protect their crops and herds, for sport, and for profit. After World War II, controls on hunting in eastern and southern Africa tightened in the name of protecting wildlife. Sport hunting declined in importance (although it has never totally disappeared) and has been replaced by "photo-safaris" as the main way outsiders now see game in Africa. Africans continued to hunt for food and began in the 1960s to hunt elephants and rhinoceroses for their tusks and horns which were then sold, usually illegally. Stricter international controls on wildlife products have perhaps reduced the demand for them, but hunting for food or to protect crops and herds remains a critical issue, especially as wildlife populations stabilized in the last decades of the twentieth century. Conservationists and park authorities consider poaching driven by human population growth around protected areas one of the gravest threats to conservation in Africa.

Quelea birds   *Quelea quelea.* Also called weaver birds because of their nests. A grain-eating bird that moves in dense flocks. It is a major cause of loss of grains throughout sub-Saharan Africa but especially in eastern and southern Africa. It has expanded its habitat because of the spread of cultivation.

Rift Valley   A great fault in the earth that resulted from the splitting of the Arabian plate from the African plate about 35 million years ago. It begins in Syria and includes the Jordan River Valley and the Red Sea. In Africa it begins by dividing the Ethiopian Highlands in half. South of Ethiopia it splits into the Western and Eastern Rift Valleys. The Western Rift includes the Great Lakes of Africa, the largest being Lakes Victoria Nyanza, Tanganyika (the deepest lake in the world), and Malawi. The Eastern Rift Valley is broad and flat and stretches through Kenya and Tanzania. The valley ends in southern Africa. It is surrounded by highlands and volcanoes.

Rinderpest   Rinderpest is a cattle disease that had long been endemic in Europe and Asia but had never reached Africa because of the barrier of the Sahara. It reached sub-Saharan Africa for the first time during the Italian invasion of Eritrea in the late 1880s. The disease, which is spread through contact with spittle, often on grass or food sources, proved especially virulent in Africa. It spread rapidly down East Africa to South Africa and across the Sudan to West Africa. It caused extremely high mortality among cattle, perhaps killing between 80 and 90 percent of all cattle, and killed large proportions of some species of wild game such as buffalo and wildebeest. In its aftermath, African peoples often faced famine and had to abandon large areas where they had used fire to maintain grazing lands for the cattle. In much of this area, bush quickly regrew and trypanosomiasis carrying tsetse colonized the areas. Some of these depopulated areas became the basis for game reserves and others were taken by colonial governments for white settlement. Rinderpest has remained endemic in much of Africa ever since, moving back and forth between game and domestic stock. The recent introduction of an effective rinderpest vaccine has led to a dramatic decline in incidence and has particularly encouraged the growth of wildlife populations in some areas.

Safari hunting   Safari hunting emerged as the first form of tourism to Africa in the early twentieth century. Wealthy tourists would pay to be taken on hunting trips in eastern and southern Africa. Theodore Roosevelt's safari after he retired from the presidency of the United States. did much to popularize the concept. Hunting declined in importance as conservation concerns rose after World War II, and "photo-safaris" and "ecotourism" replaced much hunting. But big-game hunting in Africa remains legal in certain parts of South Africa, Zimbabwe, and Tanzania.

Sahara Desert   The Sahara is the world's largest desert. It occupies almost a quarter of the continent's space. It has long served as both a barrier and filter to contact between the peoples north and south of it. The Sahara is bounded by the Atlas mountains in the north, the Mediterranean, the Nile Valley, the Ethiopian Highlands, the Sudan, the Niger River, and the Atlantic Ocean. It is divided into several regions: western Sahara, the central Ahaggar Mountains, the Tibesti Massif (a region of desert mountains and high plateaus), and the Libyan Desert (a most arid region). The southern boundary is called the Sahel (shore in Arabic) and is an area of irregular rainfall. The desert part of the Sahara has varied greatly over the long term and even moves its boundaries over the course of decades as climatic changes occur. During colder epochs in the Earth's history, the Sahara expands as less rainfall reaches the area. In warmer, moister times, such as the beginning of the Holocene from about 12,000 years ago to about 5,000 years ago, it shrinks and much of the southern Sahara has a climate that resembles that of East Africa. The climate of the northern part of the Sahara seems to have varied less. During the latest moist phase, the Sahara was home to the first food producers in Africa. Cattle seem to have been domesticated in the desert by about 11,000 years ago. Agriculture developed in the area soon thereafter. As the Sahara dried over the next few millennia, people and their technologies spread south. Contact across the Sahara remained sketchy until the introduction of the camel around the beginning of the current era about 2,000 years ago. Gradually then the Berber speaking peoples of the desert began to herd camels in the oases in the highlands and to develop trade routes across the desert. Gold, ivory, leather, salt, and slaves moved across the desert and led to the spread of Islam into sub-Saharan Africa. Much of the desert, like Egypt, became Arabized in the centuries after Muslim conquest.

Sahel   A broad stretch of plains running from the Atlantic to the Ethiopian Highlands just south of the Sahara. The word means shore in Arabic. The Sahel is defined by its semi-arid climate. The monsoons push moist air up from the Atlantic each year, but the extent of the rains by the time they reach the Sahel vary greatly from year to year. Hence, the area only intermittently supports agriculture, and people who live there have more often survived on herding livestock. People would move with their herds north following the rains, and then move back south into the agricultural lands of the Sudan as the rains ended and the grass disappeared. Given the annual variability of the rains, droughts were very common and the area has served as a particular flash point for longer changes in climate. Events such as the Little Ice Age caused noticeable drying of climate across the globe, and the Sahel accordingly moved south. Wetter periods, such as the nineteenth century and the last decade of the twentieth century have seen the Sahel move north. Historically several major droughts have led to widespread famine in the region. The two largest of the most recent droughts came during the 1910s and between 1974 and the mid-1980s. The latest drought led to fears that overgrazing and overpopulation were causing the desiccation of the Sahel into permanent desert. The wetter years after 1985 have led many observers to suggest that climate variation played a larger role in the perceived desiccation and that degradation was a more localized phenomenon.

San   Name given the historic foraging populations of southern Africa. In fact they are divided into many different groups, speaking different but related languages. Their languages are thought descended from the first languages spoken in the region as well as across much of eastern Africa.

Selous, Frederick C.   A son of the chairman of the London Stock Exchange, he became one of the most famous of all white hunters in Africa. He traveled to South Africa at the age of nineteen in 1870 and began hunting elephant in what was then the Ndebele Kingdom. He wrote several books about his exploits and became a leading promoter of white settlement in southern Africa. He worked for Cecil Rhodes' British South Africa Company and fought in the wars of conquest in what became Southern Rhodesia (now Zimbabwe). He retired to England but continued to be an active hunter and promoter of game reserves. He hunted across the globe and joined Theodore Roosevelt on his famous safari. He returned to Africa again during World War I while in his 60s to serve in the East Africa Campaign against the Germans. He was killed in action in southern German East Africa. The Selous National Park (which contains his grave) was named in his honor by the British and is the largest national park in the world.

Serengeti   The Serengeti is the term used for a large ecosystem in northern Tanzania and southern Kenya dominated by the annual migration of wildebeest across it. Today much of the ecosystem is part of several different protected areas including the Serengeti National Park (which contains the Serengeti plains proper) and the Ngorongoro Conservation Area in Tanzania and the Masai Mara Game Reserve in Kenya. Other areas surrounding these protected areas are "game controlled" areas that allow mixed use. The ecosystem is defined by the wildebeest migration, when around 1,000,000 wildebeest accompanied by smaller numbers of zebra and gazelles move followed by predators such as lions, cheetahs, hyenas, jackals, and wild hunting dogs. The migration is one of the last great annual animal migrations left in Africa (there is a smaller one in the southern Sudan). In order to create the protected areas, colonial and independent governments have had to move people out of the areas. The areas had been home to fairly sizable communities of pastoralists until the great rinderpest epidemic of the 1890s drove them into agricultural lands. The Germans and then the British took advantage of the temporary lack of population to gradually create the protected areas. Tsetse flies recolonized the region, and gradually animal populations recovered from the losses caused by rinderpest. Conservation efforts redoubled in the last half of the twentieth century, resulting in one of the largest areas under protection in the world. However, growing populations and climate change have caused concern over the future of the system. Indeed, its popularity has led to increasing visits by tourists and the development of a tourist infrastructure that pose their own threats to the environment.

Sheep   Sheep came to Africa as a domesticated animal from Southwest Asia. They spread rapidly, being able to survive in drier climates than cattle. They have spread over much of the continent and are particularly important in southern Africa.

Shifting cultivation   This term is used for the type of long-fallow agriculture common in much of Africa, particularly before the end of the twentieth century. It is sometimes called swidden agriculture, or more pejoratively, slash and burn. Farmers clear new fields, farm them for a few years, often using a rotation of crops, and then abandon them to bush fallow for many years until their fertility has restored. This form of agriculture is especially well suited to systems where land was available and the returns to intensification low, as was the case in much of Africa's savannahs. In highland areas with fertile soils, farmers practiced and continue to practice more permanent cultivation. As population has increased, the amount of time allowed for fallow has had to decrease in many areas, and farmers have been forced to develop means of intensification. This process does not always lead to improved living standards as the cost of intensification in the form of fertilizer, improved seed, and/or increased labor inputs often exceeds the value of the increased output.

Soil conservation   During the colonial era, especially from the 1930s, soil conservation became a major concern of colonial governments facing agricultural crises and growing populations. Concerns about soil erosion spread from the United States to South Africa and other parts of the British Empire in the aftermath of the Dust Bowl experience. Governmental concern focused both on the practices of settler farmers in places like South Africa and Kenya and on the "unscientific" farming methods of African farmers. Colonial governments invested heavily in support for settler agriculture. In contrast, their solutions to the perceived problems of African agricultural practices often required large increases in labor by farmers on a variety of hastily designed projects or, even more objectionable to African farmers, reductions in livestock numbers. The programs continued throughout the colonial era and often met great opposition from African farmers. In many cases farmers adopted parts of the proposed solutions, but planners often insisted on total compliance with new rules concerning farming techniques. Development programs in the postcolonial era have continued to often suffer from such a top-down approach.

Sorghum   *Sorghum bicolor.* Sorghum is perhaps the most widely cultivated grain crop in Africa and the most widely cultivated African grain crop outside of Africa. People domesticated sorghum perhaps four different times. The earliest was *sorghum bicolor* in the southern Sahara. Either separate domestications or local breeding gave rise to at least four different "races" of the grain in Africa including dura in Ethiopia, guinea in West Africa, Kafir in East Africa, and caudatum in the northern parts of central Africa. The plant puts in very deep roots that allow it to survive both drought and water-logging. Its yields are generally lower than that of maize under optimum conditions, and it requires care in processing. Sorghum's domestication has come under debate. Some scholars point to the thickness and toughness of its stalk to argue that it took metal tools to truly domesticate it since stone ones could not cut the heads containing the grain off the plants. They suggest that exploitation should be called "intensive gathering" rather than domestication until sometime before 1000 B.C.E. Others argue that both archaeological and linguistic evidence suggest that people cultivated sorghum as far back as 8000 B.C.E. even if they used different methods of harvest such as shaking or beating the heads into baskets to cause the grain to fall off. Sorghum remains an important crop throughout much of Africa, often grown in drier areas.

Springbok migration   The annual migration of large herds of springbok in southern Africa is an example of the way that these events have been disrupted by human activity. Over the course of the nineteenth century, as population density in white-ruled South Africa increased and settlers acquired modern firearms, each year as the springbok would move south, more and more would be killed for food, for sport, or to protect crops. The last migration occurred in 1896, when hunting coupled with rinderpest destroyed the great herd. A few springbok survived in isolated groups or on private farms.

Strangled peasantry   This term was developed by historians to describe the process in South Africa during the twentieth century of impoverishment of rural black populations by the white-dominated state and society. It argues that South African policy sought consciously to reduce the ability of rural Africans to produce both food and cash crops by limiting their access to land and markets. Eventually, Africans who made up the vast majority of the population of South Africa were supposed to be confined to about 13 percent of the land of South Africa unless they had a job in the white areas. This forced them to become labor migrants in the growing mines and factories of white South Africa. Its legacy is still apparent in the overcrowded and agriculturally unproductive, former rural homelands of South Africa.

Sudan   The Sudan is the name given to the broad savannahs and plains that stretch between the Sahel and the forests in West and Central Africa. The modern nation of the Sudan takes its name from its location in this broader region. The region is characterized by a bimodal wet/dry season climate. It generally was the site of early agricultural and political development because its environment hosted fewer diseases and its open nature allowed for easy communication.

Sudd   A large swamp in southern Sudan where the Nile breaks into several channels. It expands greatly during the wet season. Historically, it served as a barrier to easy contact between the northern Nile Valley of Egypt and Nubia and sub-Saharan Africa.

Teff   *Eragrostis tef.* A grain crop grown in large amounts only in Ethiopia (although its cultivation has expanded in the United States because of the popularity of Ethiopian food). It is used to make *njera,* a flat, spongy and slightly sour bread. The grain grows well in drier climates and stores well. Its small grain makes it difficult to handle, though. It was first domesticated sometime after 5000 B.C.E. in Ethiopia and has until recently never become a major crop outside of the highlands.

Timbuktu   Timbuktu is a city on the Niger River in the modern nation of Mali. It lies in the area where the Niger travels through the Sahara. It has long been an important trade center for the trans-Saharan trade that brought salt and goods from the Mediterranean to trade for gold, ivory, leather, and slaves. It also became an important seat of Islamic scholarship in West Africa.

Trypanosomiasis   A disease in humans caused by infection with one of two parasites: *Trypanosoma brucei rhodesiense,* which causes East African sleeping sickness; and *Trypanosoma brucei gambiense,* which causes West African sleeping sickness. In animals, three other trypanosoma cause the disease: *Trypanosoma congolense, T. vivax,* or *T. brucei brucei.* The parasites are all carried by species of tsetse fly: *Glossina morsitans, G. palpalis,* and *G. fusca.* The parasites sometimes seem also to be transmitted mechanically by blood-to-blood contact from other biting flies. All of the versions of trypanosomiasis can cause high rates of mortality in domestic animals and humans. They are carried by wild ruminants and also cause disease in some other animals such as lions. The disease is a very old one, and it long prevented dense human settlement in areas where tsetse lived. The presence of areas of bush suitable for tsetse prevented the spread of livestock into many areas. In the long run, people used bush-clearing techniques, especially burning, to open up tsetse infested areas for agriculture and pasture. These efforts would have been subject to short-run variation. A prime example comes in many areas of eastern and southern Africa in the aftermath of the rinderpest epidemic of the 1890s. Large areas that had previously been occupied or used for grazing had to be abandoned because of the loss of cattle. Tsetse recolonized them, and sleeping sickness among humans spread.

Upemba Depression   The Upemba Depression is a lowland area along the upper Lualaba River in the Katanga region of what is now the Democratic Republic of the Congo (DCR). The area had regularly inundated lands that would support regular agriculture and fishing as well as nearby deposits of salt, iron, and copper. Starting about 600 C.E., political centralization took place in the region leading to the formation of the Luba civilization. A period of agricultural intensification allowed an increase in population. Copper from what would become known as the Copper Belt along the border between present-day Zambia and the DRC became the main marker of wealth. The influence of the Luba civilization spread across the southern savannahs and helped spark the rise of many other states in the centuries afterwards, including the Lunda kingdom to the west on the Kasai River and the Malawi kingdom around Lake Malawi.

Virunga National Park   Virunga is the oldest national park in Africa. It was created by the Belgian colonial government of the Belgian Congo (now Democratic Republic of the Congo [DCR], formerly Zaire) in 1928 as Albert National Park. It is located in northeastern DRC along the Rwenzori mountains and straddles the continental divide between the Nile and Congo watersheds. It contains the Virungu volcano and is home to one of the few populations of mountain gorilla. It has come under increasing pressure from population growth and the disruptions caused by warfare in the region since 1994.

Wilton Culture   A stone tool using archaeological culture found across eastern and southern Africa, also called the East African Microlithic tradition. Sites date back in some areas to about 16,000 years ago in eastern Africa and spread southwards to cross the Limpopo about 8,000 years ago. The tool kit included weighted digging sticks for uprooting edible tubers as well as projectile points and blades. It is thought to have been created by Khoisan speakers.

Yams   African yams were domesticated in West Africa sometime around 6000 B.C.E. This provided the basis for subsistence in the forests and forest-savannah border regions for many thousands of years. Eventually, Asian and New World crops had in large part replaced yams as a staple in West and Central Africa, but they remain an important minor crop.

Zambezi River   The Zambezi rises in Angola and flows eastward, forming the boundary between Zimbabwe and Zambia then into Mozambique where it empties into the Indian Ocean. It also serves as the outlet for Lake Malawi. It flows over several waterfalls, the largest of which is Victoria Falls. It floods a large plain in western Zambia.

Zimbabwe, Great   The term refers to the civilization that developed in what is now the modern nation of Zimbabwe and northern South Africa sometime after about 700 C.E. The culture was most notable for the construction of large stone buildings in a series of settlements straddling the South African/Zimbabwe border. The largest site, Great Zimbabwe, was occupied between about 900 C.E. and 1500 C.E. when it was the center of a large state. Gold produced in the region became a major export, carried by caravan to the Swahili city of Kilwa. The civilization was created by the ancestors of the Shona people who live in Zimbabwe today. Some scholars have suggested that overpopulation around Great Zimbabwe caused an environmental crisis and forced its abandonment. Several successor states arose in the area formerly dominated by Great Zimbabwe, including Mono Mutape and the Rozvi empire.

BIBLIOGRAPHY

Abungu, G. H. O. 1996. "Agriculture and Settlement Formation Along the East African Coast." In *The Growth of Farming Communities in Africa from the Equator Southwards,* ed. J. Sutton, 248–256. Nairobi: British Institute in Eastern Africa.

Adams, J. S., and T. O. McShane. 1996. *The Myth of Wild Africa: Conservation Without Illusions.* Berkeley: University of California Press.

Adams, W. M., and M. J. Mortimore. 1997. "Agricultural Intensification and Flexibility in the Nigerian Sahel." *The Geographic Journal, 163*(2): 150–160.

Akyeampong, E. K. 2001. *Between the Sea and the Lagoon: An Eco-Social History of the Anlo of Southeastern Ghana C. 1850 to Recent Times.* Oxford: James Currey.

Alexander, J., J. McGregor, and T. Ranger. 2000. *Violence and Memory: One Hundred Years in the 'Dark Forests' of Matabeleland.* Portsmouth, NH: Heinemann.

Ambler, C. 1988. *Kenyan Communities in the Age of Imperialism: The Central Region in the Late Nineteenth Century.* New Haven: Yale University Press.

Ambrose, S. H. 1984. "The Introduction of Pastoral Adaptations to the Highlands of East Africa." In *From Hunters to Farmers: The Causes and Consequences of Food Production in Africa,* ed. J. D. Clark and S. A. Brandt, 212–239. Berkeley: University of California Press.

———. 1997. "African Neolithic." In *Encyclopedia of Precolonial Africa: Archaeology, History, Languages, Cultures, and Environments,* ed. J. O. Vogel, 381–386. Walnut Creek: Alta Mira Press.

Andah, B. W. 1993. "Identifying Early Farming Traditions of West Africa." In *The Archaeology of Africa: Food, Metals and Towns,* ed. T. Shaw, P. Sinclair, B. Andah, and A. Okpoko, 240–254. London: Routledge.

Andah, B. W., and R. Grove, eds. 1987. *Conservation in Africa: Peoples, Policies and Practice.* Cambridge: Cambridge University Press.

Anderson, D. M. 1984. "Depression, Dust Bowl, Demography, and Drought: The Colonial State and Soil Conservation in East Africa during the 1930s." *African Affairs, 83:* 321–343.

———. 2002. *Eroding the Commons: The Politics of Ecology in Baringo, Kenya, 1890s–1962.* Oxford: James Currey.

Anderson, D. M., and R. Grove, 1987. *Conservation in Africa: Peoples, Policies and Practice.* Cambridge: Cambridge University Press.

Anquandah, J. 1993. "The Kintampo Comples: A Case Study of Early Sedentism and Food Production in Sub-Sahelian West Africa." In *The Archaeology of Africa: Food, Metals and Towns*, ed. T. Shaw, P. Sinclair, B. Andah, and A. Okpoko, 255–260. London: Routledge.

Arcese, P., J. Hando, and K. Campbell. 1995. "Historical and Present-Day Anti-Poaching Efforts in Serengeti." In *Serengeti II: Dynamics, Management, and Conservation of an Ecosystem*, ed. A. R. E. Sinclair and P. Arcese, 506–533. Chicago: University of Chicago Press.

Asombang, R. N. 1999. "Sacred Centers and Urbanization in West Central Africa." In *Beyond Chiefdoms: Pathways to Complexity in Africa*, ed. S. K. McIntosh, 80–87. Cambridge: Cambridge University Press.

Århem, K. 1985. *Pastoral Man in the Garden of Eden: The Masai of the Ngorongoro Conservation Area, Tanzania*. Uppsala: Uppsala Research Reports in Cultural Anthropology.

Barakat, H., and A. G. el-Din Fahmy. 1999. "Wild Grasses as 'Neolithic' Food Resources in the Eastern Sahara: A Review of the Evidence from Egypt." In *The Exploitation of Plant Resources in Ancient Africa*, ed. M. van der Veen, 33–46. New York: Kluwer Academic.

Barham, L. S. 1997. "Stoneworking Technology: Its Evolution." In *Encyclopedia of Precolonial Africa: Archaeology, History, Languages, Cultures, and Environments*, ed. J. O. Vogel, 109–115. Walnut Creek: Alta Mira Press.

Barich, B. 1997. "Saharan Neolithic." In *Encyclopedia of Precolonial Africa: Archaeology, History, Languages, Cultures, and Environments*, ed. J. O. Vogel, 389–394. Walnut Creek: Alta Mira Press.

Barthelme, J. W. 1984. "Early Evidence for Animal Domestication in Eastern Africa." In *From Hunters to Farmers: The Causes and Consequences of Food Production in Africa*, ed. J. D. Clark and S. A. Brandt, 200–205. Berkeley: University of California Press.

Beach, D. N. 1980. *The Shona and Zimbabwe 900–1850: An Outline of Shona History*. New York: Africana Publishing.

Beinart, W. 1984. "Soil Erosion, Conservationism and Ideas About Development: A Southern African Exploration, 1900–1966." *Journal of Southern African Studies*, 11(1): 52–85.

———. 1989. "Introduction: The Politics of Colonial Conservation." *Journal of Southern African Studies:* 15(2), 143–162.

———. 1996. "Soil Erosion, Animals and Pasture over the Longer Term: Environmental Destruction in Southern Africa." In *The Lie of the Land: Challenging Received Wisdom on the African Environment*, ed. M. Leach and R. Mearns, 54–72. Portsmouth, NH: Heinemann.

———. 2000. "African History and Environmental History." *African Affairs*, 99: 269–302.

Beinart, W., and J. McGregor, eds. 2003. *Social History and African Environments.* Oxford: James Currey.

Beinart, W., P. Delius, and S. Trapido. 1986. *Putting a Plough to the Ground: Accumulation and Dispossession in Rural South Africa, 1850–1930.* Johannesburg: Raven Press.

Beinart, W., and P. Coates. 1995. *Environment and History: The Taming of Nature in the Usa and South Africa.* New York: Routledge.

Beinart, W., and J. McGregor. 2003. "Introduction." In *Social History and African Environments,* ed. W. Beinart and J. McGregor, 1–24. Oxford: James Currey.

Binns, T., ed. 1995. *People and Environment in Africa.* Chichester, NY: Wiley.

Blench, R. 1993. "Ethnographic and Linguistic Evidence for the Prehistory of African Ruminant Livestock, Horses and Ponies." In *The Archaeology of Africa: Food, Metals and Towns,* ed. T. Shaw, P. Sinclair, B. Andah, and A. Okpoko, 71–103. London: Routledge.

——. 1996. "Linguistic Evidence for Cultivated Plants in the Bantu Borderland." In *The Growth of Farming Communities in Africa from the Equator Southwards,* ed. J. Sutton, 83–102. Nairobi: British Institute in Eastern Africa.

Blumenschine, R. J. 1997. "Early Hominid Foraging Strategies." In *Encyclopedia of Precolonial Africa: Archaeology, History, Languages, Cultures, and Environments,* ed. J. O. Vogel, 293–297. Walnut Creek: Alta Mira Press.

Boardman, S. 1999. "The Agricultural Foundation of the Aksumite Empire, Ethiopia." In *The Exploitation of Plant Resources in Ancient Africa,* ed. M. van der Veen, 137–147. New York: Kluwer Academic.

Bradley, D. J. 1991. "Malaria." In *Disease and Mortality in Sub-Saharan Africa,* ed. R. G. Feachem, 190–202. Oxford: Oxford University Press.

Brandt, S. A. 1984. "New Perspectives on the Origins of Food Production in Ethiopia." In *From Hunters to Farmers: The Causes and Consequences of Food Prodution in Africa,* ed. J. D. Clark and S. A. Brandt, 173–190. Berkeley: University of California Press.

Breunig, P., K. Neumann, and W. Van Neer. 1996. "New Research on the Holocele Settlement and Environment of the Chad Basin in Nigeria." *African Archeaology Review, 132:* 111–145.

Brockington, D. 2002. *Fortress Conservation: The Preservation of the Mkomazi Game Reserve, Tanzania.* Bloomington: Indiana University Press.

Broecker, W. S. 1990. "Chaotic Climate." *Scientific American, 284:* 926–927.

Broecker, W. S., S. Sutherland, and T. H. Peng. 1999. "A Possible 20th Century Slowdown of Southern Ocean Deep Water Formation." *Scientific America, 286:* 1132–1135.

Bromage, T. G., and F. Schrenk. 1999. *African Biogeography, Climate Change, and Human Evolution.* Oxford: Oxford University Press.

——. 1999. "Searching for an Interdisciplinary Convergence in Paleoanthropology." In *African Biogeography, Climate Change, and Human Evolution,* ed. T. G. Bromage and F. Schrenk, 3–9. Oxford: Oxford University Press.

Brooks, G. E. 1975. "Peanuts and Colonialism: Consequences of the Commercialization of Peanuts in West Africa, 1830–70." *Journal of African History*, 161: 29–54.

———. 1993. *Landlords and Strangers: Ecology, Society, and Trade in Western Africa, 1000–1630.* Boulder: Westview Press.

Broten, M. D., and M. Said. 1995. "Population Trends of Ungulates in and Around Kenya's Masai Mara Reserve." In *Serengeti II: Dynamics, Management, and Conservation of an Ecosystem*, ed. A. R. E. Sinclair and P. Arcese, 169–193. Chicago: University of Chicago Press.

Bryson, R. A., and R. U. Bryson. 1997. "Macrophysical Climatic Modeling of Africa's Late Quaternary Climates: Site Specific, High Resolution Applications for Archaeology." *African Archaeology Review*, 14(3): 143–159.

Bundy, C. 1979. *The Rise and Fall of the South African Peasantry.* Berkeley: University of California Press.

Bunn, D. 2003. "An Unnatural State: Tourism, Water and Wildlife Photography in the Early Kruger National Park." In *Social History and African Environments*, ed. W. Beinart and J. McGregor, 199–218. Oxford: James Currey.

Campbell, K., and H. Hofer. 1995. "People and Wildlife: Spatial Dynamics and Zones of Interaction." In *Serengeti II: Dynamics, Management, and Conservation of an Ecosystem*, ed. A. R. E. Sinclair and P. Arcese, 533–570. Chicago: University of Chicago Press.

Campbell, K., and M. Borner. 1995. "Population Trends and Distribution of Serengeti Herbivores: Implications for Management." In *Serengeti II: Dynamics, Management, and Conservation of an Ecosystem*, ed. A. R. E. Sinclair and P. Arcese, 117–145. Chicago: University of Chicago Press.

Carney, J. A. 2001. *Black Rice: The African Origins of Rice Cultivation in the Americas.* Cambridge: Harvard University Press.

Carruthers, J. 1995. *Kruger National Park: A Social and Political History.* Pietermaritzburg: University of Natal Press.

———. 1997. "Lessons from South Africa: War and Wildlife Protection in the Southern Sudan, 1917–1921." *Environment and History*, 3(3): 299–322.

———. 2003. "Past and Future Landscape Ideology: The Kalahari Gemsbok National Park." In *Social History and African Environments*, ed. W. Beinart and J. McGregor, 255–266. Oxford: James Currey.

Carswell, G. 2003. "Soil Conservation Policies in Colonial Kigezi, Uganda: Successful Implementation and an Absence of Resistance." In *Social History and African Environments*, ed. W. Beinart and J. McGregor, 131–154. Oxford: James Currey.

Cartwright, C. R. 1999. "Reconstructing the Woody Resources of the Medieval Kingdom of Alwa, Sudan." In *The Exploitation of Plant Resources in Ancient Africa*, ed. M. van der Veen , 241–259. New York: Kluwer Academic.

Cavalli-Sforza, L. L., P. Menozzi, and A. Piazza. 1994. *The History and Geography of Human Genes.* Princeton: Princeton University Press.

Chatty, D., and M. Colcshester, eds. 2002. *Conservation and Mobile Indigenous Peoples: Displacement, Forced Settlement, and Sustainable Development.* New York: Berghahn Books.

Clark, J. D. 1984. "Prehistoric Cultural Continuity and Economic Change in the Central Sudan in the Early Holocene." In *From Hunters to Farmers: The Causes and Consequences of Food Prodution in Africa,* ed. J. D. Clark and S. A. Brandt, 113–126. Berkeley: University of California Press.

Clark, J., and S. A. Brandt. 1984. "An Introduction to an Introduction." In *From Hunters to Farmers: The Causes and Consequences of Food Prodution in Africa,* ed. J. D. Clark and S. A. Brandt, 1–4. Berkeley: University of California Press.

Clutton-Brock, J. 1993. "The Spread of Domestic Animals in Africa." In *The Archaeology of Africa: Food, Metals and Towns,* ed. T. Shaw, P. Sinclair, and B. Andah, and A. Okpoko, 61–70. London: Routledge.

———. 1997. "Animal Domestication in Africa." In *Encyclopedia of Precolonial Africa: Archaeology, History, Languages, Cultures, and Environments,* ed. J. O. Vogel, 418–424. Walnut Creek: Alta Mira Press.

Coast, E. 2002. "Masai Socioeconommic Conditions: A Cross-Border Comparison." *Human Ecology, 30*(1): 79–105.

Cobbing, J. 1988. "The Mfecane as Alibi: Thoughts on Dithakong and Mbolompo." *Journal of African History, 29*(3): 487–519.

Connah, G. 2001. *African Civilizations: Precolonial Cities and States in Tropical Africa: An Archaeological Perspective, Second Edition.* Cambridge: Cambridge University Press.

Conte, C. A. 2004. *Highland Sanctuary: Environmental History in Tanzania's Usambara Mountains.* Athens, OH: Ohio University Press.

Coppens, Y. 1999. "Introduction." In *African Biogeography, Climate Change, and Human Evolution,* ed. T. G. Bromage and F. Schrenk, 13–18. Oxford: Oxford University Press.

Cordell, D. D., and J. W. Gregory. 1994. *African Population and Capitalism: Historical Perspectives, Second Edition.* Madison: University of Wisconsin Press.

Cornelissen, E. 1997. "Central African Transitional Cultures." In *Encyclopedia of Precolonial Africa: Archaeology, History, Languages, Cultures, and Environments,* ed. J. O. Vogel, 312–320. Walnut Creek: Alta Mira Press.

Cowie, J. B. 1998. *Climate and Human Change : Disaster or Opportunity?* New York: Parthenon Publishing Group.

Crosby, A. W. 1986. *Ecological Imperialism: The Biological Expansion of Europe, 900–1900.* Cambridge: Cambridge University Press.

Curtin, P. D. 1968. "Epidemiology and the Slave Trade." *Political Science Quarterly, 83*(2): 190–216.

Curtin, P. D. 1998. *Disease and Empire: The Health of European Troops in the Conquest of Africa.* Cambridge: Cambridge University Press.

Dalfes, H. N., G. Kukla, and H. Weiss. 1996. *Third Millennium BC Climate Change and Old World Collapse.* Berlin: Springer.

D'Andrea, C., D. Lyons, M. Haile, and A. Butler. 1999. "Ethnoarchaeological Approaches to the Study of Prehistoric Agriculture in the Highlands of Ethiopia." In *The Exploitation of Plant Resources in Ancient Africa,* ed. M. van der Veen, 101–122. New York: Kluwer Academic.

Davis, M. 2001. *Late Victorian Holocausts: El Niño Famines and the Making of the Third World.* New York: Verso.

Dawson, M. H. 1979. "Smallpox in Kenya, 1880–1920." *Social Science and Medicine, 13B:* 245.

———. 1987. "Health, Nutrition, and Population in Central Kenya, 1890–1945." In *African Population and Capitalism: Historical Perspectives,* ed. D. Cordell and J. Gregory, 201–217. Boulder: Westview Press.

de Langhe, E., R. Swenne, and D. Vuylsteke. 1996. "Plantain in the Early Bantu World." In *The Growth of Farming Communities in Africa from the Equator Southwards,* ed. J. Sutton, 161–167. Nairobi: British Insitute in Eastern Africa.

de Maret, P. 1996. "Pits, Pots and the Far-West Streams." In *The Growth of Farming Communities in Africa from the Equator Southwards,* ed. J. Sutton, 318–323. Nairobi: British Institute in Eastern Africa.

———. 1999. "The Power of Symbols and the Symbols of Power Through Time: Probing the Luba Past." In *Beyond Chiefdoms: Pathways to Complexity in Africa,* ed. S. K. McIntosh, 151–165. Cambridge: Cambridge University Press.

de Maret, P., and G. Thiry. 1996. "How Old Is the Iron Age in Central Africa?" In *The Culture and Technology of Africa Iron Production,* ed. P. R. Schmidt, 29–39. Gainesville: University Press of Florida.

de Menocal, P. B. 2001. "Cultrual Responses to Climate Change During the Late Holocene." *Science, 292:* 667–673.

Denbow, J. 1999. "Material Culture and the Dialectics of Identity in the Kalahari: AD 700–1700." In *Beyond Chiefdoms: Pathways to Complexity in Africa,* ed. S. K. McIntosh, 110–123. Cambridge: Cambridge University Press.

Denton, G. H. 1999. "Cenozoic Climate Change." In *African Biogeography, Climate Change, and Human Evolution,* ed. T. G. Bromage and F. Schrenk, 94–114. Oxford: Oxford University Press.

de Waal, A. 1987. "The Perception of Poverty and Famines." *International Journal of Moral and Social Studies, 2:* 251–262.

———. 1988. "Famine Early Warning Systems and the Use of Socio-Economic Data." *Disasters, 12:* 81–91.

———. 1989. "Famine Mortality: A Case Study of Darfur, Sudan, 1984–1985." *Population Studies, 43:* 5–24.

———. 1989. *Famine That Kills: Darfur, Sudan, 1984–1985.* Oxford: Clarendon Press.

———. 1997. *Famine Crimes: Politics and the Disaster Relief Industry in Africa.* Bloomington: Indiana University Press.

Diamond, J. M. 1997. *Guns, Germs, and Steel: The Fates of Human Societies.* New York: Norton.

Diaz, H. F., and V. Markgraf, eds. 1992. *El Niño: Historical and Paleoclimatic Aspects of the Southern Oscillation.* Cambridge: Cambridge University Press.

Dobson, A. 1995. "The Ecology and Epidemiology of Rinderpest Virus in Serengeti and Ngorongoro Conservation Area." In *Serengeti II: Dynamics, Management, and Conservation of an Ecosystem*, ed. A. R. E. Sinclair and P. Arcese, 485–505. Chicago: University of Chicago Press.

Dowson, T. A. 1997. "Central African Rock Art." In *Encyclopedia of Precolonial Africa: Archaeology, History, Languages, Cultures, and Environments*, ed. J. O. Vogel, 368–373. Walnut Creek: Alta Mira Press.

———. 1997. "Southern African Rock Art." In *Encyclopedia of Precolonial Africa: Archaeology, History, Languages, Cultures, and Environments*, ed. J. O. Vogel, 373–380. Walnut Creek: Alta Mira Press.

Driver, T. S., and G. Chapman. 1996. *Time-Scales and Environmental Change.* London: Routledge.

Dublin, H. T. 1991. "Dynamics of the Serengeti-Mara Woodlands: An Historical Perspective." *Forest and Conservation History*, 35(4): 168–78.

———. 1995. "Vegetation Dynamics in the Serengeti-Mara Ecosystem: The Role of Elephants, Fire, and Other Factors." In *Serengeti II: Dynamics, Management, and Conservation of an Ecosystem*, ed. A. R. E. Sinclair and P. Arcese, 49–70. Chicago: University of Chicago Press.

Dunbar, R. B. 2000. "Climate Variability During the Holocene: An Update." In *The Way the Wind Blows: Climate, History, and Human Action*, ed. R. J. McIntosh, J. A. Tainter, and S. K. McIntosh, 45–88. New York: Columbia University Press.

Eggert, M. K. H. 1993. "Central Africa and the Archaeology of the Equatorial Rainforest: Reflections on Some Major Topics." In *The Archaeology of Africa: Food, Metals and Towns*, ed. T. Shaw, P. Sinclair, and B. Andah, and A. Okpoko, 289–329. London: Routledge.

———. 1996. "Pots, Farming and Analogy: Early Ceramics in the Equatorial Rainforest." In *The Growth of Farming Communities in Africa from the Equator Southwards*, ed, J. Sutton, 332–338. Nairobi: British Institute in Eastern Africa.

Ehret, C. 1984. "Historical/Linguistic Evidence for Early African Food Production." In *From Hunters to Farmers: The Causes and Consequences of Food Prodution in Africa*, ed. J. D. Clark and S. A. Brandt, 26–36. Berkeley: University of California Press.

———. 1998. *An African Classical Age: Eastern and Southern Africa in World History, 1000 B.C. to A.D. 400.* Charlottesville: University Press of Virginia.

———. 2001. "Bantu Expansions: Re-Envisioning a Central Problem in of Early African

History." *International Journal of African Historical Studies*, 341: 5–41.

———. 2002. *The Civilizations of Africa: A History to 1800*. Charlottesville: University Press of Virginia.

Elphick, R. 1977. *Kraal and Castle: Khoikhoi and Founding of White South Africa*. New Haven: Yale University Press.

Fagan, B. 1997. "Archaeology in Africa: Its Influence." In *Encyclopedia of Precolonial Africa: Archaeology, History, Languages, Cultures, and Environments*, ed. J. O. Vogel, 51–54. Walnut Creek: Alta Mira Press.

———. 1999. *Floods, Famines and Emperors: El Niño and the Fate of Civilizations*. New York: Basic Books.

———. 2000. *The Little Ice Age: How Climate Made History 1300–1850*. New York: Basic Books.

Fairhead, J., and M. Leach. 1996. *Misreading the African Landscape: Society and Ecology in a Forest–Savanna Mosaic*. Cambridge: Cambridge University Press.

———. 1996. "Rethinking the Forest-Savanna Mosaic: Colonial Science and Its Relics in West Africa." In *The Lie of the Land: Challenging Received Wisdom on the African Environment*, ed. M. Leach and R. Mearns, 105–121. Portsmouth, NH: Heinemann.

———. 1998. *Reframing Deforestation: Global Analyses and Local Realities: Studies in West Africa*. London: Routledge.

Feachem, R. G., and D. T. Jamison. 1991. *Disease and Mortality in Sub-Saharan Africa*. Oxford: Oxford University Press.

Fein, J., and P. Stephens, eds. 1987. *Monsoons*. New York: Wiley.

Foley, R. 1999. "Evolutionary Geography of Pliocene African Hominids." In *African Biogeography, Climate Change, and Human Evolution*, ed. T. G. Bromage and F. Schrenk, 328–348. Oxford: Oxford University Press.

Ford, J. 1971. *The Role of the Trypanosomiases in African Ecology: A Study of the Tsetse Fly Problem*. Oxford: Clarendon Press.

Galvin, K. A., J. Ellis, R. B. Boone, Ann L. Magennis, N. M. Smith, S. J. Lynn, et al. 2002. "Compatibility of Pastoralism and Conservation? A Test Case Using Integrated Assessment in the Ngorongoro Conservation Area, Tanzania." In *Conservation and Mobile Indigenous Peoples: Displacement, Forced Settlement and Sustainable Development*, ed. D. Chatty and M. Colchester, 36–60. New York: Berghahn Books.

Garlake, P. S. 1973. *Great Zimbabwe*. London: Thames and Hudson.

Giblin, J. 1986. "Famine and Social Change During the Transition to Colonial Rule in Northeastern Tanzania 1880–1896." *African Economic History*, 15: 85–105.

———. 1990. "Trypanosomiasis Control in African History: An Evaded Issue?" *Journal of African History*, 31(1): 59–80.

———. 1993. *The Politics of Environmental Control in Northeastern Tanzania, 1840–1940*. Philadelphia: University of Pennsylvania Press.

Giblin, J. L. 1990. "East Coast Fever in Socio-Historical Context: A Case Study from

Tanzania." *International Journal of African Historical Studies, 23*(3): 401–421.

Gifford-Gonzalez, D. 1998. "Early Pastoralists in East Africa: Ecological and Social Dimensions." *Journal of Anthropological Archaeology, 17*(2): 166–200.

———. 2000. "Animal Disease Challenges to the Emergence of Pastoralism in Sub-Saharan Africa." *African Archaeological Review, 17*(3): 95–139.

Giles-Vernick, T. 2000. "*Doli:* Translating an African Environmental History of Loss in the Sangha River Basin of Equatorial Africa." *Journal of African History,* 41(3): 373–394.

———. 2002. *Cutting the Vines of the Past: Environmental Histories of the Central African Rain Forest.* Charlottesville: University Press of Virginia.

Glacken, C. 1967. *Traces on the Rhodian Shore: Attitudes to Nature from Classical Times to 1800.* Berkeley: University of California Press.

Glantz, M. H. 1987. *Drought and Hunger in Africa: Denying Famine a Future.* Cambridge: Cambridge University Press.

Goodall, J. 1988. *My Life with the Chimpanzees.* New York: Pocket Books.

Gordon, R. J. 2003. "Fido: Dog Tales of Colonialism in Namibia." In *Social History and African Environments,* ed. W. Beinart and J. McGregor, 240–254. Oxford: James Currey.

Grove, A. T. 1978. "Geographical Introduction to the Sahel." *The Geographical Journal, 144*(3): 407–415.

———. 1993. "Africa's Climate in the Holocene." In *The Archaeology of Africa: Food, Metals and Towns,* ed. T. Shaw, P. Sinclair, and B. Andah, and A. Okpoko, 32–42. London: Routledge.

———. 1997. "Modern Climates and Vegetation Zones." In *Encyclopedia of Precolonial Africa: Archaeology, History, Languages, Cultures, and Environments,* ed. J. O. Vogel, 39–42. Walnut Creek: Alta Mira Press.

———. 1997. "Pleistocene and Holocene Climates and Vegetation Zones." In *Encyclopedia of Precolonial Africa: Archaeology, History, Languages, Cultures, and Environments,* ed. J. O. Vogel , 35–39. Walnut Creek: Alta Mira Press.

Grove, R. H. 1995. *Green Imperialism: Colonial Expansion, Tropical Island Edens and the Origins of Environmentalism, 1600–1866.* Cambridge: Cambridge University Press.

Grove, R. H., and Chappell, J., eds. 2000. *El Niño—History and Crisis: Studies from the Asia-Pacific Region.* Cambridge: White Horse Press.

Grove, R. H., and Chappell, J. 2000. "Introduction, Enso: A Brief Overview." In *El Niño—History and Crisis: Studies from the Asia-Pacific Region,* ed. R. H. Grove and J. Chappell, 1–4. Cambridge: White Horse Press.

———. 2000. "El Niño Chronology and the History of Global Crises During the Little Ice Age." In *El Niño—History and Crisis: Studies from the Asia-Pacific Region,* ed. R. H. Grove and J. Chappell, 5–35. Cambridge: White Horse Press.

Grzimek, B. 1970. *Among Animals of Africa.* Trans. J. M. Brownjohn. New York: Stein and Day.

Grzimek, B. and M. Grzimek.1961. *Serengeti Shall Not Die.* Trans. E. L. Rewald and D. Rewald. New York: Dutton.

Guy, J. 1987. "Analysing Pre-Capitalist Societies in Southern Africa." *Journal of Southern African Studies, 14*(1): 18–37.

Haaland, R. 1995. "Sedentism, Cultivation, and Plant Domestication in the Holocene Middle Nile Region." *Journal of Field Archaeology, 22*(2): 157–174.

———. 1996. "Dakawa: An Early Iron Age Site in the Tanzanian Hinterland." In J. Sutton, *The Growth of Farming Communities in Africa from the Equator Southwards,* 238–247. Nairobi: British Institute in Eastern Africa.

Hakansson, N. T. 1997. "Farming Societies in Sub-Saharan Africa." In *Encyclopedia of Precolonial Africa: Archaeology, History, Languages, Cultures, and Environments,* ed. J. O. Vogel, 215–222. Walnut Creek: Alta Mira Press.

Hall, M. 1987. *The Changing Past: Farmers, Kings, and Traders in Southern Africa.* Cape Town: D. Philip.

Hanotte, O., D. Bradley, J. Ochient, Y. Verjee, E. Hill, and E. Rege. 2002. "African Pastoralism: Genetic Imprints of Origins and Migrations." *Science, 296:* 336–339.

———. 1993. "The Tropical African Cereals." In *The Archaeology of Africa: Food, Metals and Towns,* ed. T. Shaw, P. Sinclair and B. Andah, and A. Okpoko, 53–60. London: Routledge.

Harlan, J. R. 1997. "Farming Methods." In *Encyclopedia of Precolonial Africa: Archaeology, History, Languages, Cultures, and Environments,* ed. J. O. Vogel, 215–222. Walnut Creek: Alta Mira Press.

———. 1997. "Food Crops." In *Encyclopedia of Precolonial Africa: Archaeology, History, Languages, Cultures, and Environments,* ed. J. O. Vogel, 222–225. Walnut Creek: Alta Mira Press.

Hart, K. 1982. *The Political Economy of West African Agriculture.* Cambridge: Cambridge University Press.

Hartwig, G. W., and K. D. Patterson, eds. 1978. *Disese in African History: An Introductory Survey and Case Studies.* Durham, NC: Duke University Press.

Harvey, L. D. D. 2000. *Climate and Global Environmental Change.* Harlow: Prentice Hall.

Hassan, F. A. 1984. "Environment and Subsistence in Predynastic Egypt." In *From Hunters to Farmers: The Causes and Consequences of Food Prodution in Africa,* ed. J. D. Clark and S. A. Brandt, 57–64. Berkeley: University of California Press.

———. 1997. "Egypt: Beginnings of Agriculture." In *Encyclopedia of Precolonial Africa: Archaeology, History, Languages, Cultures, and Environments,* ed. J. O. Vogel, 405–409. Walnut Creek: Alta Mira Press.

———. 1997. "Holocene Palaeoclimates of Africa." *African Archeaology Review, 144:* 213–230.

Hill, P. 1970. *Studies in Rural Capitalism in West Africa.* Cambridge: Cambridge University Press.

Hogendorn, J. S., and K. M. Scott. 1983. "Very Large-Scale Agricultural Projects: The Lessons of the East African Groundnut Scheme." In *Imperialism, Colonialism, and Hunger: East and Central Africa,* ed. R. I. Rotberg, 167–198. Lexington, MA: Lexington Books.

Holl, A. F. C. 1993. "Transition from Late Stone Age to Iron Age in the Sudano-Sahelian Zone: A Case Study from the Perichdian Plain." In *The Archaeology of Africa: Food, Metals and Towns,* ed. T. Shaw, P. Sinclair, and B. Andah, and A. Okpoko, 320–343. London: Routledge.

———. 1997. "Western Africa: The Prehistoric Sequence." In *Encyclopedia of Precolonial Africa: Archaeology, History, Languages, Cultures, and Environments,* ed. J. O. Vogel, 305–312. Walnut Creek: Alta Mira Press.

———. 1998. "The Dawn of African Pastoralisms: An Introductory Note." *Journal of Anthropological Archaeology, 17*(2): 81–96.

———. 1998. "Livestock Husbandry, Pastoralism, and Territoriality: The West African Record." *Journal of Anthropological Archaeology, 172*: 143–165.

Hoppe, K. A. 2003. *Lords of the Fly: Sleeping Sickness Control in British East Africa, 1900–1960.* Westport, CT: Praeger.

Horton, M. and N. Mudida. 1993. "Exploitation of Marine Resources: Evidence for the Origin of the Swahili Communities of East Africa." In *The Archaeology of Africa: Food, Metals and Towns,* ed. T. Shaw, P. Sinclair, and B. Andah, and A. Okpoko, 673–693. London: Routledge.

Huggett, R. J. B. 1991. *Climate, Earth Processes, and Earth History.* New York: Springer-Verlag.

Hulme, D., and Marshall Murphree, eds. 2001. *African Wildlife and Livelihoods: The Promise of Performance of Community Conservation.* Oxford: James Currey.

Hunter, S. 2003. *Black Death: Aids in Africa.* New York: Palgrave MacMillan.

Iliffe, J. 1979. *A Modern History of Tanganyika.* Cambridge: Cambridge University Press.

———. 1987. *The African Poor: A History.* Cambridge: Cambridge University Press.

———. 1995. *Africans: The History of a Continent.* Cambridge: Cambridge University Press.

Isaacman, A. 1996. *Cotton in the Mother of Poverty: Peasants, Work, and Rural Struggle in Colonial Mozambique, 1938–1961.* Portsmouth, NH: Heinemann.

Jacobs, N. J. 1996. "The Flowing Eye: Water Management in the Upper Kuruman Valley, South Africa, C. 1800–1962." *Journal of African History, 37*(2): 237–260.

———. 2003. *Environment, Power, and Injustice: A South African History.* Cambridge: Cambridge University Press.

Johnson, D. H., and D. M. Anderson, eds. 1988. *The Ecology of Survival: Case Studies from Northeast African History.* Boulder: Westview Press.

Johnson, T. C., C. A. Scholz, M. R. Talbot, K. Keits, R. D. Ricketts, G. Ngobi, et al. 1996. "Late Pleistocene Desiccation of Lake Victoria and Rapid Evloution of Cichlid Fishes." *Science, 273*: 1091–1093.

Jones, W. O. 1959. *Manioc in Africa.* Stanford: Stanford University Press.

Karega-Munene. 1996. "The East African Neolithic: An Alternative View." *African Archeaology Review, 13*(4): 247–254.

Kea, R. 1982. *Settlements, Trade, and Polities in the Seventeenth-Century Gold Coast.* Baltimore: Johns Hopkins University Press.

Kingdon, J. 1999. "Introduction: Geology, Ecology, and Biogeography." In *African Biogeography, Climate Change, and Human Evolution,* ed. T. G. Bromage and F. Schrenk, 91–93. Oxford: Oxford University Press.

Kiple, K. F. 1993. "Diseases of Sub-Saharan Africa to 1860." In *The Cambridge World History of Human Disease,* ed. K. F. Kiple, 293–298. Cambridge: Cambridge University Press.

Kiriama, H. O. 1993. "The Iron-Using Communities in Kenya." In *The Archaeology of Africa: Food, Metals and Towns,* ed. T. Shaw, P. Sinclair, and B. Andah, and A. Okpoko, 484–498. London: Routledge.

Kitching, G. 1980. *Class and Economic Change in Kenya: The Making of an African Petite Bourgeoisis, 1905–1970.* New Haven: Yale University Press.

Kjekshus, H. 1977. *Ecology Control and Economic Development in East African History: The Case of Tanganyika, 1850–1950.* London: Heinemann.

Klee, M., and B. Zach. 1999. "The Exploitation of Wild and Domesticated Food Plants at Settlment Mounds in North-East Nigeria 1800 CAL BC to Today." In *The Exploitation of Plant Resources in Ancient Africa,* ed. M. van der Veen, 81–88. New York: Kluwer Academic.

Klein, R. G. 1984. "The Prehistory of Stone Age Herders in South Africa." In *From Hunters to Farmers: The Causes and Consequences of Food Prodution in Africa,* ed. J. D. Clark and S. A. Brandt, 281–289. Berkeley: University of California Press.

Klein, R. G. 1999. *The Human Career.* Chicago: University of Chicago Press.

Klieman, K. A. 2003. *"The Pygmies Were Our Compass": Bantu and Batwa in the History of West Central Africa, Early Times to C. 1900 C.E.* Portsmouth, NH: Heinemann.

Koponen, J. 1986. "Population Growth in Historical Perspective: The Key Role of Changing Fertility." In *Tanzania: Crisis and Struggle for Survival,* ed. J. Boesen, K. J. Havnevid, J. Koponen, and R. Odgaard, 33–42. Uppsala: Scandinavian Institute for African Studies.

———. 1988. *People and Production in Late Precolonial Tanzania: History and Structures.* Uppsala, Sweden: Scandinavian Institute of African Studies.

———. 1995. *Development for Exploitation: German Colonial Policies in Mainland Tanzania, 1884–1914.* Helsinki: Finnish Historical Society.

Kopytoff, I. 1987. *The African Frontier: The Reproduction of Traditional African Societies.* Bloomington: Indiana University Press.

Kreike, E. 2003. "Hidden Fruits: A Social Ecology of Fruit Trees in Namibia and Angola." In *Social History and African Environments,* ed. W. Beinart and J. McGregor, 27–42. Oxford: James Currey.

Kusimba, C. M. 1999. *The Rise and Fall of Swahili States.* Walnut Creek: Alta Mira.

Kusimba, S. B. 1999. "Hunter-Gatherer Land Use Patterns in Later Stone Age East Africa." *Journal of Anthropological Archaeology, 18*(2): 165–200.

———. 2003. *African Foragers: Environment, Technology, Interactions.* Walnut Creek: Alta Mira.

Lamb, H. H. 1995. *Climate, History and the Modern World, Second Edition.* London: Routledge.

Lane, P. 1996. "The Use and Abuse of Ethnography in Iron Age Studies of Southern Africa." In J. Sutton, *The Growth of Farming Communities in Africa from the Equator Southwards,* 51–64. Nairobi: British Institute in Eastern Africa.

Lavachery, P. 2000. "The Holocene Archaeological Sequence of Shum Laka Rock Shelter Grassfields, Western Cameroon." *African Archaeological Review, 17*(4): 213–247.

Leach, M. 1992. "Women's Crops in Women's Spaces: Gender Relations in Mende Rice Farming." In *Bush Base: Forest Farm: Culture, Environment and Development,* ed. E. Croll and D. Parkin, 76–96. London: Routledge.

———. 1994. *Rainforest Relations: Gender and Resource Use Among the Mende of Gola, Sierra Leone.* Edinburgh: Edinburgh University Press.

Leach, M., and R. Mearns, eds. 1996. *The Lie of the Land: Challenging Received Wisdom on the African Environment.* Portsmouth, NH: Heinemann.

Leach, M., and R. Mearns. 1996. "Environmental Change and Policy: Challenging Received Wisdom in Africa." In *The Lie of the Land: Challenging Received Wisdom on the African Environment,* ed M. Leach and R. Mearns, 1–33. Portsmouth, NH: Heinemann.

Leach, M. and C. Green. 1997. "Gender and Environmental History: From Representations of Women and Nature to Gender Analysis of Ecology and Politics." *Environment and History, 3*(3): 343–370.

Le Houérou, H. N. 1989. *The Grazing Land Ecosystems of the African Sahel.* New York: Springer-Verlag.

Le Roy Ladurie, E. 1971. *Times of Feast, Times of Famine: A History of Climate Since the Year 1000.* Trans. B. Bray. Garden City, NY: Doubleday.

Lim, I. L. 1997. "Eastern African Rock Art." In *Encyclopedia of Precolonial Africa: Archaeology, History, Languages, Cultures, and Environments,* ed. J. O. Vogel, 362–368. Walnut Creek: Alta Mira Press.

Lindesay, J. A., and C. A. Vogel. 1990. "Historical Evidence for Southern Oscillation-Southern African Rainfall Relationships." *International Journal of Climatology, 10:* 679–689.

Livingston, D. A. 1984. "Interactions of Food Production and Changing Vegetation in Africa." In *From Hunters to Farmers: The Causes and Consequences of Food Production in Africa,* ed. J. D. Clark and S. A. Brandt, 22–25. Berkeley: University of California Press.

Lovejoy, P. E. 1986. *Salt of the Desert Sun: A History of Salt Production and Trade in the Central Sudan.* Cambridge: Cambridge University Press.

———. 2000. *Transformations in Slavery: A History of Slavery in Africa, Second Edition.* Cambridge: Cambridge University Press.

Lovejoy, P. E., and S. Baier. 1975. "The Desert-Side Economy of the Central Sudan." *The International Journal of African Historical Studies,* 8(4): 551–581.

Lyons, M. 1992. *A Social History of Sleeping Sickness in Northern Zaire, 1900–1940.* Cambridge: Cambridge University Press.

———. 1993. "Diseases of Sub-Saharan Africa Since 1860." In *The Cambridge World History of Human Disease,* ed. K. F. Kiple, 298–305. Cambridge: Cambridge University Press.

Maack, P. A. 1996. "'We Don't Want Terraces!' Protest and Identity Under the Uluguru Land Usage Scheme." In *Custodians of the Land: Ecology and Culture in the History of Tanzania,* ed. G. H. Maddox, J. L. Giblin, and I. Kimambo, 152–170. Athens: Ohio University Press.

MacDonald, K. 1997. "The Late Stone Age and Neolithic Cultures of West Africa and the Sahara." In *Encyclopedia of Precolonial Africa: Archaeology, History, Languages, Cultures, and Environments,* ed. J. O. Vogel, 394–398. Walnut Creek: Alta Mira Press.

———. 1997. "Western African and Southern Saharan Advanced Foragers." In *Encyclopedia of Precolonial Africa: Archaeology, History, Languages, Cultures, and Environments,* ed. J. O. Vogel, 330–335. Walnut Creek: Alta Mira Press.

MacKenzie, F. 1998. *Land, Ecology and Resistance in Kenya, 1880–1952.* Edinburgh: International African Institute.

MacKenzie, J. M. 1989. *The Empire of Nature: Hunting, Conservation and British Imperialism.* Manchester: Manchester University Press.

Maddock, L. 1979. "The 'Migration' and Grazing Succession." In *Serengeti: The Dynamics of an Ecosystem,* ed. A. R. E. Sinclair and M. Norton-Griffiths, 104–129. Chicago: University of Chicago Press.

Maddox, G. 1986. "*Njaa:* Food Shortages and Famines in Tanzania Between the Wars." *International Journal of African Historical Studies,* 19(1): 17–34.

Maddox, G. H. 1990. "*Mtunya:* Famine in Central Tanzania, 1917–1920." *Journal of African History,* 31(2): 181–198.

———. 1991. "Famine, Impoverishment and the Creation of a Labor Reserve in Central Tanzania." *Disasters,* 15(1): 35–41.

———. 1996. "Environment and Population Growth in Ugogo, Central Tanzania." In *Custodians of the Land: Ecology and Culture in the History of Tanzania,* ed. G. Maddox, J. Giblin, and I. N. Kimambo, 43–66. London: James Currey.

———. 1998. "Networks and Frontiers in Colonial Tanzania." *Environmental History,* 3(4): 436–459.

———. 1999. "Africa and Environmental History." *Environmental History*, 4(2): 162–167.

Maddox, G. H., J. L. Giblin, and I. N. Kimambo, eds. 1996. *Custodians of the Land: Ecology and Culture in Tanzanian History.* London: James Currey.

Maley, J. 1993. "The Climatic and Vegetational History of the Equatorial Regions of Africa During the Upper Quaternary." In *The Archaeology of Africa: Food, Metals and Towns*, ed. T. Shaw, P. Sinclair, B. Andah, and A. Okpoko, 43–52. London: Routledge.

———. 1996. "Middle to Late Holocene Changes in Tropical Africa and Other Continents: Paleomonsoon and Sea Surface Temperature Variations." In *Third Millennium BC Climate Change and Old World Collapse*, ed. H. N. Dalfes, G. Kukla, and H. Weiss, 611–640. Berlin: Springer.

Mamdani, M. 1996. *Citizen and Subject: Contemporary Africa and the Legacy of Late Colonialism.* Princeton: Princeton University Press.

Mandala, E. C. 1990. *Work and Control in a Peasant Economy: A History of the Lower Tchiri Valley in Malawi, 1859–1960.* Madison, Wisconsin: University of Wisconsin Press.

Manger, L. O. 1988. "Traders, Farmers and Pastoralists: Economic Adaptations and Environmental Problems in the Southern Nuba Mountains of the Sudan." In *The Ecology of Survival: Case Studies from Northeast African History*, ed. D. H. Johnson and D. M. Anderson, 155–172. Boulder: Westview Press.

Manning, P. 1990. *Slavery and African Life: Occidental, Oriental and African Slave Trades.* Cambridge: Cambridge University Press.

Marshall, F., and E. Hildebrand. 2002. "Cattle Before Crops: The Beginnings of Food Production in Africa." *Journal of World Prehistory*, 16: 99–143.

Mathiessen, P. 1991. *African Silences.* New York: Randon House.

Mbano, B. N. N., R. C. Malpas, M. K. S. Maige, P. A. K. Symonds, and D. M. Thompson. 1995. "The Serengeti Regional Conservation Strategy." In *Serengeti II: Dynamics, Management, and Conservation of an Ecosystem*, ed. A. R. E. Sinclair and P. Arcese, 605–616. Chicago: University of Chicago Press.

McBrearty, S., and A. S. Brooks. 2000. "The Revolution That Wasn't: A New Interpretation of the Origin of Modern Human Behavior." *Journal of Human Evolution*, 39(5): 453–563.

McCabe, J. T. 2002. "Giving Conservation a Human Face? Lessons from Forty Years of Combining Conservation and Development in the Ngorongoro Conservation Area, Tanzania." In *Conservation and Mobile Indigenous Peoples: Displacement, Forced Settlement and Sustainable Development*, ed. D. Chatty and M. Colchester, 61–76. New York: Berghahn Books.

———. 2003. "Summer, Sustainability and Livelihood Diversification Among the Masai of Northern Tanzania." *Human Organization*, 62(2): 100–111.

McCann, J. C. 1987a. "The Social Impact of Drought in Ethiopia: Oxen, Households, and Some Implications for Rehabilitation." In *Drought and Hunger in Africa: Denying Famine a Future*, ed. M. H. Glantz, 245–268. Cambridge: Cambridge University Press.

————. 1987b. *From Poverty to Famine in Northeast Ethiopia: A Rural History 1900–1935.* Philadelphia: University of Pennsylvania Press.

————. 1988. "History, Drought and Reproduction: Dynamics of Society and Ecology in Northeast Ethiopia." In *The Ecology of Survival: Case Studies from Northeast African History*, ed. D. H. Johnson and D. M. Anderson, 283–304. Boulder: Westview Press.

————. 1991. "Agriculture and African History" [Review article]. *Journal of African History, 32*(3): 507–513.

————. 1995. *People of the Plow: An Agricultural History of Ethiopia, 1800–1990.* Madison: University of Wisconsin Press.

————. 1997. "The Plow and the Forest: Narratives of Deforestation in Ethiopia, 1840–1992." *Environmental History, 2*(2): 138–159.

————. 1999a. "Climate and Causation in African History." *International Journal of African Historical Studies, 32*: 2–3, 261–279.

————. 1999b. *Green Land, Brown Land, Black Land: An Environmental History of Africa, 1800–1990.* Portsmouth, NH: Heinemann.

————. 2004. *Maize and Grace: Africa's Encounter with a New World Crop, 1500–2000.* Cambridge: Harvard University Press.

McCracken, J. 2003. "Conservation and Resistance in Colonial Malawi: The 'Dead North' Revisited." In *Social History and African Environments*, ed. W. Beinart and J. McGregor, 155–174. Oxford: James Currey.

McDonald, M. M. A. 1998. "Early African Pastoralism: View from Dakheleh Oasis South Central Egypt." *Journal of Anthropological Archaeology, 17*(2): 124–142.

McDougall, E. A. 1985. "The View from Awdaghust: War, Trade and Social Change in the Southwestern Sahara, from the Eighth to the Fifteenth Cenutry." *Journal of African History, 26*(1): 1–31.

McGregor, J. 2003. "Living with the River: Landscape and Memory in the Zambezi Valley, Northwest Zimbabwe." In *Social History and African Environments*, ed W. Beinart and J. McGregor, 87–106. Oxford: James Currey.

McIntosh, R. J. 1993. "The Pulse Model: Genesis and Accommodation of Specialization in the Middle Niger." *Journal of African History, 34*(2): 181–220.

————. 1997. "Agricultural Beginnings in Sub-Saharan Africa." In *Encyclopedia of Precolonial Africa: Archaeology, History, Languages, Cultures, and Environments*, ed. J. O. Vogel, 409–418. Walnut Creek: Alta Mira Press.

————. 1998. *The Peoples of the Middle Niger: The Island of Gold.* Oxford: Blackwell.

————. 1999. "Western Representations of Urbanism and Invisible African Towns." In *Beyond Chiefdoms: Pathways to Complexity in Africa*, ed. S. K. McIntosh, 56–65. Cambridge: Cambridge University Press.

————. 2000. "Social Memory in Mande." In *The Way the Wind Blows: Climate, History, and Human Action*, ed. R. J. McIntosh, J. A. Tainter, and S. K. McIntosh, 141–180. New York: Columbia University Press.

McIntosh, R. J., and S. K. McIntosh. 1984. "Early Iron Age Economy in the Inland Niger Delta (Mali)." In *From Hunters to Farmers: The Causes and Consequences of Food Production in Africa*, ed. J. D. Clark and S. A. Brandt, 158–172. Berkeley: University of California Press.

McIntosh, R. J., J. A. Tainter, and S. K. McIntosh. 2000. "Climate, History and Human Action." In *The Way the Wind Blows: Climate, History, and Human Action*, ed. R. J. McIntosh, J. A. Tainter, and S. K. McIntosh, 1–44. New York: Columbia University Press.

McIntosh, R. J., J. A. Tainter, and S. K. McIntosh, eds. 2000. *The Way the Wind Blows: Climate, History, and Human Action*. New York: Columbia University Press.

McIntosh, S. K., ed. 1999. *Beyond Chiefdoms: Pathways to Complexity in Africa*. Cambridge: Cambridge University Press.

———. 1999. "Modeling Political Organization in Large-Scale Settlement Clusters: A Case Study from the Inland Niger Delta." In *Beyond Chiefdoms: Pathways to Complexity in Africa*, ed. S. K. McIntosh, 56–65. Cambridge: Cambridge University Press.

———. 1999. "Pathways to Complexity: An African Perspective." In *Beyond Chiefdoms: Pathways to Complexity in Africa*, ed. S. K. Mcintosh, 1–30. Cambridge: Cambridge University Press.

McKee, J. K. 1999. "The Autocatalytic Nature of Hominid Evolution in African Plio-Pleistocene Environments." In *African Biogeography, Climate Change, and Human Evolution*, ed. T. G. Bromage and F. Schrenk, 57–67. Oxford: Oxford University Press.

McNeill, W. H. 1976. *Plagues and Peoples*. Garden City, NY: Anchor/Doubleday.

McNiell, J. R., and W. H. McNeill. 2003. *The Human Web: A Bird's-Eye View of World History*. New York: Norton.

Middleton, K. 2003. "The Ironies of Plant Transfer: The Case of Prickly Pear in Madagascar." In *Social History and African Environments*, ed. W. Beinart and J. McGregor, 43–59. Oxford: James Currey.

Miller, J. C. 1989. *Way of Death: Merchant Capitalism and the Angolan Slave Trade*. Madison: University of Wisconsin Press.

Miracle, M. P. 1966. *Maize in Tropical Africa*. Madison: University of Wisconsin Press.

Mitchell, P. J. 2000. "The Organization of Later Stone Age Lithic Technology in the Caledon Valley, Southern Africa." *African Archaeological Review*, 17(3), 141–176.

Moron, V., and M. N. Ward. 1999. "ENSO Teleconnections with Climate Variability in the European and African Sectors." *Monthly Weather Review: 127*, 287–295.

Mortimore, M. J. 1989. *Adapting to Drought: Farmers, Famines and Desertification in West Africa*. Cambridge: Cambridge University Press.

———. 1998. *Roots in the African Dust: Sustaining the Sub-Saharan Drylands*. Cambridge: Cambridge University Press.

Mortimore, M. J., and W. M. Adams. 1999. *Working the Sahel: Environment and Society in Northern Nigeria*. London: Routledge.

Mortimore, M. J., and M., Tiffen, 1995. "Population and Environment in Time Perspective: The Machkos Story." In *People and Environment in Africa*, ed. T. Binns, 69–89. Chichester: John Wiley and Sons.

Mutoro, H. W. 1996. "Tana Ware and the *Kaya* Settlements of the Coastal Hinterland of Kenya." In *The Growth of Farming Communities in Africa from the Equator Southwards*, ed. J. Sutton, 257–260. Nairobi: British Institute in Eastern Africa.

Muzzolini, A. 1993. "The Emergence of a Food-Producing Economy in the Sahara." In *The Archaeology of Africa: Food, Metals and Towns*, ed. T. Shaw, P. Sinclair, and B. Andah, and A. Okpoko, 227–239. London: Routledge.

Nana-Sinkam, S. C. 1995. *Land Environmental Degradation and Desertification in Africa*. Rome: FAO.

Nettle, D. 1998. "Explaining Global Language Diversity." *Journal of Anthropological Archaeology*, 18(4): 354–374.

Neumann, K. 1999. "Early Plant Food Production in the West African Sahel." In *The Exploitation of Plant Resources in Ancient Africa*, ed. M. van der Veen, 73–80. New York: Kluwer Academic.

Neumann, R. P. 1997. "Forest Rights, Privileges and Prohibitions: Contextualising State Forestry Policy in Tanganyika." *Environment and History*, 3(1): 45–68.

———. 1998. *Imposing Wilderness: Struggles over Livelihood and Nature Preservation in Africa*. Berkeley: University of California Press.

Nicholson, S. E. 1976. "A Climatic Chronology for Africa: A Synthesis of Geological, Historical and Meteorological Data." Ph.D. Dissertation. Madison: University of Wisconsin.

———. 1978. "Climatic Variations in the Sahel and Other African Regions During the Past Five Centuries." *Journal of Arid Environments*, 1: 3–24.

———. 1979. "The Methodology of Historical Climate Reconstruction and its Application to Africa. *Journal of African History*, 20: 31–49.

———. 1980. "Saharan Climates in Historic Times." In *The Sahara and the Nile*, ed. M. A. J. Williams and H. Faure. Rotterdam: Balkema.

———. 1981. "The Historical Climatology of Africa." In *Climate and History: Studies in Past Climates and Their Impact on Man*, ed. T. M. Wigley, M. J. Ingram and G. Farmer, 257–259. London.

Nicholson, S. E., and H. Flohn. 1980. "African Environmental and Climatic Changes and the General Atmospheric Circlation in Late Pleistocene and Holocene." *Climatic Change*, 2: 313–348.

Northrup, D. 1978. *Trade Without Rulers: Pre-Colonial Economic Development in South–Eastern Nigeria*. Oxford: Clarendon Press.

———. 1988. *Beyond the Bend in the River: African Labor in Eastern Zaire, 1865–1940*. (Monographs in International Studies Africa Series, no. 52). Athens, Ohio: Ohio University Center for International Studies.

Norton-Griffiths, M. 1979. "The Influence of Grazing, Browsing, and Fire on the Vegetation Dynamics of the Serengeti." In *Serengeti: The Dynamics of an Ecosystem*, ed. A. R. E. Sinclair and M. Norton-Griffiths, 310–352. Chicago: University of Chicago Press.

———. 1995. "Economic Incentives to Develop the Rangelands of the Serengeti: Implications for Wildlife Conservation." In *Serengeti II: Dynamics, Management, and Conservation of an Ecosystem*, ed. A. R. E. Sinclair and P. Arcese, 588–604. Chicago: University of Chicago Press.

Nurse, D. 1997a. "The Contributions of Linguistics to the Study of History in Africa." *Journal of African History*, 38(3): 359–391.

———. 1997b. "Languages of Eastern and Southern Africa in Historical Perspective." In *Encyclopedia of Precolonial Africa: Archaeology, History, Languages, Cultures, and Environments*, ed. J. O. Vogel, 159–166. Walnut Creek: Alta Mira Press.

Nyamweru, C. 1997. "Geography and Geology." In *Encyclopedia of Precolonial Africa: Archaeology, History, Languages, Cultures, and Environments*, ed. J. O. Vogel, 29–35. Walnut Creek: Alta Mira Press.

O'Brien, E. M., and C. R. Peters. 1999. "Landforms, Climate, Ecogeographic Mosaics, and the Potential for Hominid Diversity in Pliocene Africa." In *African Biogeography, Climate Change, and Human Evolution*, ed. T. G. Bromage and F. Schrenk, 115–137. Oxford: Oxford University Press.

Okafor, E. E. 1993. "New Evidence on Early Iron-Smelting from Southeastern Nigeria." In *The Archaeology of Africa: Food, Metals and Towns*, ed. T. Shaw, P. Sinclair, B. Andah, and A. Okpoko, 432–448. London: Routledge.

Olumwullah, O. A. 2002. *Dis-Ease in the Colonial State: Medicine, Society, and Social Change Among the Abanyole of Western Kenya*. Westport, CT: Greenwood Press.

Östberg, W. 1986. *The Kondoa Transformation: Coming to Grips with Soil Erosion in Central Tanzania*. (Scandinavian Institute of African Studies Research Report, no. 76). Uppsala: Scandinavian Institute of African Studies.

Owen-Smith, N. 1999. "Ecological Links Between African Savanna Environments, Climate Change, and Early Hominid Evolution." In *African Biogeography, Climate Change, and Human Evolution*, ed. T. G. Bromage and F. Schrenk, 138–149. Oxford: Oxford University Press.

Parker, I. S. C., and A. D. Graham. 1989. "Elephant Decline: Downward Trends in African Elephant Distribution and Numbers." *International Journal of Environmental Studies*, 34, 35: 13–26, 287–305.

Patterson, K. D. 1993. "Disease Ecologies of Sub-Saharan Africa." In *The Cambridge World History of Human Disease*, ed. K. F. Kiple, 447–452. Cambridge: Cambridge University Press.

Patterson, K. D., and G. W. Harwig. 1979. "The Disease Factor: An Introductory Overview." In *Disease in African History: An Introductory Survey and Case Studies*, ed. G. W. Hartwig and K. D. Patterson, 3–24. Durham, NC: Duke University Press.

Peires, J. B. 1989. *The Dead Will Arise: Nongqawuse and the Great Xhosa Cattle–Killing Movement of 1856–57.* Bloomington: Indiana University Press.

Pennington, R. L. 1997. "Disease as a Factor in African History." In *Encyclopedia of Pre-colonial Africa: Archaeology, History, Languages, Cultures, and Environments,* ed. J. O. Vogel, 45–48. Walnut Creek: Alta Mira Press.

Perkin, S. 1995. "Multiple Land Use in the Serengeti Region: The Ngorongoro Conservation Area." In *Serengeti II: Dynamics, Management, and Conservation of an Ecosystem,* ed. A. R. E. Sinclair and P. Arcese, 571–587. Chicago: University of Chicago Press.

Petite-Maire, N., L. Beufort, and N. Page. 1996. "Holocene Climate Change and Man in the Present Day Sahara Desert." In *Third Millennium BC Climate Change and Old World Collapse,* ed. H. N. Dalfes, G. Kukla, and H. Weiss, 297–308. Berlin: Springer.

Phillipson, D. W. 1984. "Early Food-Production in Central and Southern Africa." In *From Hunters to Farmers: The Causes and Consequences of Food Prodution in Africa,* ed. J. D. Clark and S. A. Brandt, 272–280. Berkeley: University of California Press.

———. 1993. "The Antiquity of Cultivation and Herding in Ethiopia." In *The Archaeology of Africa: Food, Metals and Towns,* ed. T. Shaw, P. Sinclair, and B. Andah, and A. Okpoko, 344–357. London: Routledge.

Phillippson, G., and S. Bahuchet. 1996. "Cultivated Crops and Bantu Migrations in Central and Eastern Africa: A Linguistic Approach." In *The Growth of Farming Communities in Africa from the Equator Southwards,* ed. J. Sutton, 103–120. Nairobi: British Institute in Eastern Africa.

Pikirayi, I. 2003. "Environmental Data and Historical Process: Historical Climatic Reconstruction and the Mutapa State 1450–1862." In *Social History and African Environments,* ed. W. Beinart and J. McGregor, 60–71. Oxford: James Currey.

Posnansky, M. 1984. "Early Agricultural Societies in Ghana." In *From Hunters to Farmers: The Causes and Consequences of Food Production in Africa,* ed. J. D. Clark and S. A. Brandt, 147–151. Berkeley: University of California Press.

Potts, R. 1996. *Humanity's Descent: The Consequences of Ecological Instability.* New York: Murrow.

Prins, F. E. 1996. "Climate Vegetation and Early Agriculturist Communities in Transkei and Kwazulu-Natal." In *The Growth of Farming Communities in Africa from the Equator Southwards,* ed. J. Sutton, 179–186. Nairobi: British Institute in Eastern Africa.

Pwiti, G. 1996. "Early Farming Communities of the Middle Zambezi Valley." In *The Growth of Farming Communities in Africa from the Equator Southwards,* ed. J. Sutton, 202–208. Nairobi: British Institute in Eastern Africa.

Quinn, W. H., and V. T. Neal. 1987. "El Niño Occurrences over the Past Four and a Half Centuries." *Journal of Geophysical Research,* 92: 14,449–14,461.

Ranger, T. 1999. *Voices from the Rocks: Nature, Culture and History in the Matopos Hills of Zimbabwe.* Bloomington, IN: Indiana University Press.

———. 2003. "Women and Environment in African Religion: The Case of Zimbabwe." In *Social History and African Environments*, ed. W. Beinart and J. McGregor, 72–86. Oxford: James Currey.

Raynaut, C., with E. Grégoire, P. Janin, J. Koechlin, and P. L. Delville. 1997. *Societies and Nature in the Sahel*. Trans. D. Simon and H. Kaziol. London: Routledge.

Reader, J. 1997. *Africa: The Biography of a Continent*. London: Hamish Hamilton.

Renfrew, C. 1992. "Archaeology, Genetics and Linguistic Diversity." *Man, 273*: 445–478.

Reynaud-Farrera, I. 1996. "Late Holocene Vegetational Changes in South-West Cameroon." In *Third Millennium BC Climate Change and Old World Collapse*, ed. H. N. Dalfes, G. Kukla, and H. Weiss, 641–652. Berlin: Springer.

Richards, P. 1985. *Indigenous Agricultural Revolution: Ecology and Food Production in West Africa*. London: Hutchinson.

———. 1996. *Fighting for the Rain Forest: War, Youth and Resources in Sierra Leone*. Portsmouth: Heinemann.

Robbins, L. H. 1984. "Late Prehistoric Aquatic and Pastoral Adaptations West of Lake Turkana, Kenya." In *From Hunters to Farmers: The Causes and Consequences of Food Production in Africa*, ed. J. D. Clark and S. A. Brandt, 206–211. Berkeley: University of California Press.

Robertshaw, P. 1993. "The Beginnings of Food Production in Southwestern Kenya." In *The Archaeology of Africa: Food, Metals and Towns*, ed. T. Shaw, P. Sinclair, and B. Andah, and A. Okpoko, 358–371. London: Routledge.

———. 1999. "Seeking and Keeping Power in Bunyoro-Kitara, Uganda." In *Beyond Chiefdoms: Pathways to Complexity in Africa*, ed. S. K. McIntosh, 124–135. Cambridge: Cambridge University Press.

Rodney, W. 1982. *How Europe Underdeveloped Africa*. V. Harding, Introduction, A. M. Babu. Washington: Howard University Press.

Roosevelt, T. 1988. *African Game Trails: An Account of the African Wanderings of an American Hunter-Naturalist*. New York: St. Martin's Press. Original work published 1910.

Rossel, G. 1996. "*Musa* and *Ensete* in Africa: Taxonomy, Nomenclature and Uses." In *The Growth of Farming Communities in Africa from the Equator Southwards*, ed. J. Sutton, 130–146. Nairobi: British Insitute in Eastern Africa.

Rowley-Conway, P., W. Deakin, and C. H. Shaw. 1999. "Ancient DNA from Sorghum: The Evidence from Qasr Ibrim, Egyptian Nubia." In *The Exploitation of Plant Resources in Ancient Africa*, ed. M. van der Veen, 55–62. New York: Kluwer Academic.

Runyoro, B. A., H. Hofer, E. B. Chausi, and P. D. Moehlman. 1995. "Long-Term Trends in the Herbivore Populations of the Ngorongoro Crater, Tanzania." In *Serengeti II: Dynamics, Management, and Conservation of an Ecosystem*, ed. A. R. E. Sinclair and P. Arcese, 146–168. Chicago: University of Chicago Press.

Sadr, K. 2003. "The Neolithic of Southern Africa." *Journal of African History, 44*(2): 195–210.

Saro-Wiwaand Ken. 1995. *A Month and a Day: A Detention Diary.* New York: Penguin.

Schmidt, P. R., ed. 1996. "The Agricultural Hinterland Settlement Trends in Tanzania." In *The Growth of Farming Communities in Africa from the Equator Southwards,* ed. J. Sutton, 261–262. Nairobi: British Institute in Eastern Africa.

———. 1996a. *The Culture and Technology of African Iron Production.* Gainesville: University Press of Florida.

———. 1996b. "Cultural Representations of African Iron Production." In *The Culture and Technology of African Iron Production,* ed. P. R. Schmidt, 1–28. Gainesville: University Press of Florida.

———. 1997a. "Archaeological Views on a History of Landscape Change in East Africa." *Journal of African History, 383*: 393–421.

———. 1997b. *Iron Technology in East Africa: Symbolism, Science and Archaeology.* Bloomington: Indiana University Press.

Schoenbrun, D. L. 1993. "We Are What We Eat: Ancient Agriculture Between the Great Lakes." *Journal of African History, 34*(1): 1–31.

———. 1996. "Social Aspects of Agricultural Change Between the Great Lakes, AD 500–1000." In *The Growth of Farming Communities in Africa from the Equator Southwards,* ed. J. Sutton, 270–282. Nairobi: British Institute in Eastern Africa.

———. 1998. *A Green Place, a Good Place: Agrarian Change, Gender, and Social Identity in the Great Lakes Region to the 15th Century.* Portsmouth, NH: Heinemann.

———. 1999. "The (In)Visible Roots of Bunyoro-Kitara and Buganda in the Lakes Region: AD 800–1300." In *Beyond Chiefdoms: Pathways to Complexity in Africa,* ed. S. K. McIntosh, 136–150. Cambridge: Cambridge University Press.

Schroeder, R. A. 1999. *Shady Practices: Agroforestry and Gender Politics in the Gambia.* Berkeley: University of California Press.

Schroeder, W. A., E. S. Munger, and Darleeen R. Powars. 1990. "Sickle Cell Anaemia, Genetic Variations and the Slave Trade to the United States." *Journal of African History, 31*(2): 163–180.

Scott, J. 1998. *Seeing Like a State: How Certain Schemes to Improve the Human Condition Have Failed.* New Haven: Yale University Press.

Scott, L. 1997. "Quaternary Environment." In *Encyclopedia of Precolonial Africa: Archaeology, History, Languages, Cultures, and Environments,* ed. J. O. Vogel, 42–46. Walnut Creek: Alta Mira Press.

Seavoy, R. E. 1989. *Famine in East Africa: Food Production and Food Policies.* New York: Greenwood Press.

Shaw, T. 1984. "Archaeological Evidence and Effects of Food-Producing in Nigeria." In *From Hunters to Farmers: The Causes and Consequences of Food Prodution in Africa,* ed. J. D. Clark and S. A. Brandt, 152–157. Berkeley: University of California Press.

Shaw, T., P. Sinclair, B. Andah, and A. Okpoko, eds. 1993. *The Archaeology of Africa: Food, Metals and Towns.* London: Routledge.

Sheriff, A. 1987. *Slaves, Spices and Ivory in Zanzibar.* Athens, Ohio: Ohio University Press.

Shinnie, P. L. 1996. *Ancient Nubia.* London: Kegan Paul International.

Sikes, N. E. 1999. "Plio-Pleistocene Floral Context and Habitat Preferences of Sympatric Hominid Species in East Africa." In *African Biogeography, Climate Change, and Human Evolution,* ed. T. G. Bromage and F. Schrenk, 301–315. Oxford: Oxford University Press.

Sinclair, A. R. E. 1979a. "Dynamics of the Serengeti Ecogystem: Process and Pattern." In *Serengeti: The Dynamics of an Ecosystem,* ed. A. R. E. Sinclair and M. Norton-Griffiths, 1–30. Chicago: University of Chicago Press.

———. 1979b. "The Serengeti Environment." In *Serengeti: The Dynamics of an Ecosystem,* ed. A. R. E. Sinclair and M. Norton-Griffiths, 31–45. Chicago: University of Chicago Press.

———. 1995a. "Population Limitation of Resident Herbivores." In *Serengeti II: Dynamics, Management, and Conservation of an Ecosystem,* ed. A. R. E. Sinclair and P. Arcese, 194–221. Chicago: University of Chicago Press.

———. 1995b. "Serengeti Past and Present." In *Serengeti II: Dynamics, Management, and Conservation of an Ecosystem,* ed. A. R. E. Sinclair and P. Arcese, 3–30. Chicago: University of Chicago Press.

Sinclair, A. R. E., and M. Norton-Griffiths, eds. 1979. *Serengeti: The Dynamics of an Ecosystem.* Chicago: University of Chicago Press.

Sinclair, A. R. E., and P. Arcese, eds. 1995. *Serengeti II: Dynamics, Management, and Conservation of an Ecosystem.* Chicago: University of Chicago Press.

Sinclair, A. R. E., and P. Arcese. 1995. "Serengeti in the Context of Worldwide Conservation Efforts." In *Serengeti II: Dynamics, Management, and Conservation of an Ecosystem,* ed. A. R. E. Sinclair and P. Arcese, 31–46. Chicago: University of Chicago Press.

Sinclair, P. J. J., T. Shaw, and B. Andah. 1993. "Introduction." In *The Archaeology of Africa: Food, Metals and Towns,* ed. T. Shaw, P. Sinclair, B. Andah, and A. Okpoko, 1–32. London: Routledge.

Smith, A. B. 1984. "Origins of the Neolithic in the Sahara." In *From Hunters to Farmers: The Causes and Consequences of Food Prodution in Africa,* ed. J. D. Clark and S. A. Brandt, 84–92. Berkeley: University of California Press.

Smith, F. 1997a. "Hominid Origins." In *Encyclopedia of Precolonial Africa: Archaeology, History, Languages, Cultures, and Environments,* ed. J. O. Vogel, 247–257. Walnut Creek: Alta Mira Press.

———. 1997b. "Modern Human Origins." In *Encyclopedia of Precolonial Africa: Archaeology, History, Languages, Cultures, and Environments,* ed. J. O. Vogel,

257–266. Walnut Creek: Alta Mira Press.

Soper, R. 1997. "Eastern African Terraced-Irrigation Systems." In *Encyclopedia of Precolonial Africa: Archaeology, History, Languages, Cultures, and Environments*, ed. J. O. Vogel, 227–231. Walnut Creek: Alta Mira Press.

Spear, T. 1997. *Mountain Farmers: Moral Economies of Land and Agricultural Development in Arusha and Meru.* Berkeley: University of California Press.

Spear, T., and R. Waller. 1992. *Being Masai.* Athens, OH: Ohio University Press.

Stahl, A. B. 1984. "A History and Critique of Investigations into Early African Agriculture." In *From Hunters to Farmers: The Causes and Consequences of Food Production in Africa*, ed. J. D. Clark and S. A. Brandt, 9–21. Berkeley: University of California Press.

———. 1993. "Intensification in the West African Late Stone Age: A View from Central Ghana." In *The Archaeology of Africa: Food, Metals and Towns*, ed. T. Shaw, P. Sinclair, and B. Andah, and A. Okpoko, 261–273. London: Routledge.

———. 1999. "Perceiving Variability in Time and Space: The Evolutionary Mapping of African Societies." In *Beyond Chiefdoms: Pathways to complexity in Africa*, ed. S. K. McIntosh, 39–55. Cambridge: Cambridge University Press.

Stanley, S. M. 1992. "An Ecological Theory for the Origin of Homo." *Paleobiology, 183*: 237–257.

Steinhart, E. I. 1989. "Hunters, Poachers and Gamekeepers: Towards a Social History of Hunting in Colonial Kenya." *Journal of African History, 30:* 247.

———. 2005. *Black Poachers, White Hunters: A Social History of Hunting in Colonial Kenya.* Athens: Ohio University Press.

Street-Perrott, F. A., and R. A. Perrott. 1993. "Holocene Vegetation, Lake Levels, and Climate of Africa." In *Global Climates Since the Last Glacial Maximum*, ed. H. E. J. Wright, J. E. Kutzback, T. I. Webb, W. F. Ruddiman, F. A. Street-Perrott, and P. J. Bartlein, 318–356. Minneapolis: University of Minnesota Press.

Stringer, C. 2003. "Human Evolution: Out of Ethiopia." *Nature, 423:* 692–695.

Stringer, C., and R. McKie. 1996. *African Exodus: The Origins of Modern Humanity.* London: Cape.

Sutton, J. E. G. 1984. "Irrigation and Soil Conservation in African Agricultural History: With a Reconsideration of the Inyanga Terracing (Zimbabwe) and Engaruka Irrigation Works (Tanzania)." *Journal of African History, 25*(1), 25–41.

———. 1989a. "Editor's Introduction: Fields, Farming and History in Africa." *Azania, 24:* 6–11.

———. 1989b. "Towards a History of Cultivating the Fields." *Azania, 24:* 98–112.

———. 1990. *A Thousand Years of East Africa.* British Institute in East Africa: Nairobi.

———. 1993. "The Antecedents of the Interlacustrine Kingdoms." *Journal of African History, 34*(1): 33–64.

———. 1996a. *The Growth of Farming Communities in Africa from the Equator South-*

wards. Nairobi: British Institute in Eastern Africa.

———. 1996b. "The Growth of Farming and the Bantu Settlement on and South of the Equator: Editor's Introduction." In *The Growth of Farming Communities in Africa from the Equator Southwards*, ed. J. Sutton, Nairobi: British Institute in Eastern Africa.

Swart, S. 2003. "The Ant of the White Soul: Popular Natural History, the Politics of Afrikaner Identity, and the Entomological Writings of Eugène Marais." In *Social History and African Environments*, ed. W. Beinart and J. McGregor, 219–239. Oxford: James Currey.

Syfert, D. N. 1977. "The Liberian Coasting Trade, 1822–1900." *Journal of African History, 18*(2): 217–235.

Sykes, B. 2001. *The Seven Daughters of Eve: The Science That Reveals Our Genetic Ancestry.* New York: Norton.

Szalay, F. S. 1999. "Paleontology and Macroevolution: On the Theoretical Conflict Between an Expanded Synthesis and Hierarchic Punctuationism." In *African Biogeography, Climate Change, and Human Evolution*, ed. T. G. Bromage and F. Schrenk, 35–56. Oxford: Oxford University Press.

Tainter, J. A. 2000. "Global Change, History, and Sustainability." In *The Way the Wind Blows: Climate, History, and Human Action*, ed. R. J. McIntosh, J. A. Tainter, and S. K. McIntosh, 331–356. New York: Columbia University Press.

Taylor, D. and R. Marchant, 1996. "Human Impact in the Interlacustrine Region: Long-Term Pollen Records from the Rukiga Highlands." In *The Growth of Farming Communities in Africa from the Equator Southwards*, ed. J. Sutton, 283–295. Nairobi: British Institute in Eastern Africa.

Thomas, D. S. G., and T. Middleton. 1994. *Desertification: Exploding the Myth.* Chichester: John Wiley and Sons.

Thompson, G., and R. Young. 1999. "Fuels for the Furnace: Recent and Prehistoric Ironworking in Uganda and Beyond." In *The Exploitation of Plant Resources in Ancient Africa*, ed. M. van der Veen, 221–239. New York: Kluwer Academic.

Thornton, R. 1998. *Africa and Africans in the Making of the Atlantic World.* Cambridge: Cambridge University Press.

Tiffen, M. 1996. "Land and Capital: Blind Spots in the Study of the 'Resource-Poor' Farmer." In *The Lie of the Land: Challenging Received Wisdom on the African Environment*, ed. M. Leach and R. Mearns, 168–185. Portsmouth, NH: Heinemann.

Tilley, H. 2003. "African Environments and Environmental Sciences: The African Research Survey, Ecological Paradigms and British Colonial Development 1920–1940." In *Social History and African Environments*, ed. W. Beinart and J. McGregor, 109–130. Oxford: James Currey.

Togola, T. 2000. "Memories, Abstractions, and Conceptualizations of Ecological Crisis in the Mande World." In *The Way the Wind Blows: Climate, History, and Human Action*, ed. R. J. McIntosh, J. A. Tainter, and S. K. McIntosh, 181–192. New York:

Columbia University Press.

Turner, A. 1999. "Evolution in the African Plio-Pleistocene." In *African Biogeography, Climate Change, and Human Evolution*, ed. T. G. Bromage and F. Schrenk, 76–87. Oxford: Oxford University Press.

Turshen, M. 1984. *The Political Economy of Disease in Tanzania*. New Brunswick, New Jersey: Rutgers University Press.

van Beusekom, M. M. 2002. *Negotiating Development: African Farmers and Colonial Experts at the* Office Du Niger, *1920–1960*. Portsmouth, NH: Heinemann.

van der Veen, M. 1999a. *The Exploitation of Plant Resources in Ancient Africa*. New York: Kluwer Academic.

———. 1999b. "Introduction." In *The Exploitation of Plant Resources in Ancient Africa*, ed. M. van der Veen, 1–10. New York: Kluwer Academic.

Vandervort, B. 1998. *Wars of Imperial Conquest in Africa, 1830–1914*. London: UCL Press.

Vansina, J. 1990. *Paths in the Rainforests: Toward a History of Political Tradition in Equatorial Africa*. Madison: University of Wisconsin Press.

———. 1995. "New Linguistic Evidence and 'the Bantu Expansion.'" *Journal of African History*, 36(2): 173–195.

———. 1996. "A Slow Revolution: Farming in Subequatorial Africa." In *The Growth of Farming Communities in Africa from the Equator Southwards*, ed. J. Sutton, 15–26. Nairobi: British Institute in Eastern Africa.

———. 1999. "Pathways of Political Development in Equatorial Africa and Neo-Evolutionary Theory." In *Beyond Chiefdoms: Pathways to Complexity in Africa*, ed. S. K. McIntosh, 166–172. Cambridge: Cambridge University Press.

Van Sittert, L. 2004. "The Nature of Power: Cape Environmental History, the History of Ideas and Neoliberal History." *Journal of African History*, 43(2): 305–313.

Vaughan, M. 1987. *The Story of an African Famine: Gender and Famine in Twentieth-Century Malawi*. Cambridge: Cambridge University Press.

Vincens, A., D. Schwartz, H. Elenga, I. Reynaud-Ferrera, A. Alexandre, J. Bertaux, et al. 1999. "Forest Response to Climate Change in Atlantic Equatorial Africa During the Last 4,000 Years BP and Inheritance on the Modern Landscapes." *Journal of Biogeography*, 26(4): 879–885.

Vogel, J. O., ed. 1997a. *Encyclopedia of Precolonial Africa: Archaeology, History, Languages, Cultures, and Environments*. Walnut Creek: Alta Mira Press.

———. 1997b. "Earliest African Cultures." In *Encyclopedia of Precolonial Africa: Archaeology, History, Languages, Cultures, and Environments*, ed. J. O. Vogel, 297–305. Walnut Creek: Alta Mira Press.

Vrba, E. S. 1999. "Habitat Theory in Relation to the Evolution in African Neogene Biota and Hominids." In *African Biogeography, Climate Change, and Human Evolution*, ed. T. G. Bromage and F. Schrenk, 19–34. Oxford: Oxford University Press.

Waller, R. 1985. "Ecology, Migration and Expansion in East Africa." *African*

Affairs, 84: 347.

———. 1988. "Emutai: Crisis and Response in Masailand 1883–1902." In *The Ecology of Survival: Case Studies from Northeast African History,* ed. D. H. Johnson and D. M. Anderson, 73–114. Boulder: Westview Press.

———. 1990. "Tsetse Fly in Western Narok, Kenya." *Journal of African History, 31*(1): 81–102.

Wasylikowa, K., J. R. Harlan, J. Evans, F. Wendorf, R. Schild, A. E. Close, et al. 1993. "Examination of Botanical Remains from Early Neolithic Houses at Nabta Playa, Western Desert, Egypt, with Special Reference to Sorghum Grains." In *The Archaeology of Africa: Food, Metals and Towns,* ed. T. Shaw, P. Sinclair, and B. Andah, and A. Okpoko, 154–164. London: Routledge.

Wasylikowa, K., and J. Dahlberg. 1999. "Sorghum in the Economy of the Early Neolithic Nomadic Tribes at Nabta Playa, Southern Egypt." In *The Exploitation of Plant Resources in Ancient Africa,* ed. M. van der Veen, 11–32. New York: Kluwer Academic.

Watts, M. 1983. *Silent Violence: Food, Famine and Peasantry in Northern Nigeria.* Berkeley: University of California Press.

Watts, S. 1997. *Epidemics and History: Disease, Power and Imperialism.* New Haven: Yale University Press.

Webb, J. L., Jr. 1995. *Desert Frontier: Ecological and Economic Change Along the Western Sahel, 1600–1850.* Madison: University of Wisconsin Press.

Wendorf, F., and R. Schild. 1998. "Nabta Playa and Its Role in Northeastern African Prehistory." *Journal of Anthropological Archaeology, 17*(2): 97–123.

Wetterstrom, W. 1993. "Foraging and Farming in Egypt: The Transitions from Hunting and Gathering to Horticulture in the Nile Valley." In *The Archaeology of Africa: Food, Metals and Towns,* ed. T. Shaw, P. Sinclair, and B. Andah, and A. Okpoko, 165–226. London: Routledge.

———. 1997. "Nile Valley: Pre-Agricultural Cultures." In *Encyclopedia of Precolonial Africa: Archaeology, History, Languages, Cultures, and Environments,* ed. J. O. Vogel, 386–389. Walnut Creek: Alta Mira Press.

Whitelaw, G. 1996. "Towards an Early Iron Age Worldview: Some Ideas from Kwazulu-Natal." In *The Growth of Farming Communities in Africa from the Equator Southwards,* ed. J. Sutton, 37–50. Nairobi: British Institute in Eastern Africa.

Widgren, M., and J. E. G. Sutton, eds. 2004. *Islands of Intensive Agriculture in Eastern Africa.* Oxford: James Currey.

Wigboldus, J. S. 1996. "The Spread of Crops into Sub-Equatorial Africa During the Early Iron Age." In *The Growth of Farming Communities in Africa from the Equator Southwards,* ed. J. Sutton, 121–129. Nairobi: British Institute in Eastern Africa.

Wilks, I. 1977. "Land, Labour, Capital and the Forest Kingdom of Asante: A Model of Early Change." In *The Evolution of Social Systems,* ed. J. Friedman and M. J. Rowlands, Pittsburgh: University of Pittsburgh Press.

Williams, M. A. J. 1984. "Late Quaternary Prehistoric Environments in the Sahara." In *From Hunters to Farmers: The Causes and Consequences of Food Production in Africa*, ed. J. D. Clark and S. A. Brandt, 74–83. Berkeley: University of California Press.

Williamson, K. 1997. "Western African Languages in Historical Perspective." In *Encyclopedia of Precolonial Africa: Archaeology, History, Languages, Cultures, and Environments*, ed. J. O. Vogel , 166–170. Walnut Creek: Alta Mira Press.

Wilmsen, E. 1989. *Land Filled with Flies: A Political Economy of the Kalahari*. Chicago: University of Chicago Press.

Wolpoff, M., and R. Caspari. 1997. *Race and Human Evolution*. New York: Simon and Schuster.

Wright, H. T. 1993. "Trade and Politics on the Eastern Littoral of Africa, AD 800–1300." In *The Archaeology of Africa: Food, Metals and Towns*, ed. T. Shaw, P. Sinclair, and B. Andah, and A. Okpoko, 484–498. London: Routledge.

Yngstrom, I. 2003. "Representations of Custom, Social Identity and Environmental Relations in Central Tanzania 1926–1950." In *Social History and African Environments*, ed. W. Beinart and J. McGregor, 175–196. Oxford: James Currey.

Young, R. and G. Thompson. 1999. "Missing Plant Foods? Where Is the Archaeobotanical Evidence for Sorghum and Finger Millet in East Africa?" In *The Exploitation of Plant Resources in Ancient Africa*, ed. M. van der Veen, 63–72. New York: Kluwer Academic.

INDEX

Page numbers in **boldface** indicate major discussions in the text.

Abacha, Sani, 164
Acheulian culture, 17–18
Adams, J.S., 186
Africa
  chronology of events, 257–266
  climate. *See* Climate
  eco-physiognomic zones and physiographic regions, 6
  geography of, 5(map), **5–9**. *See also* Geography of Africa; Maps
  popular images of, 139
  regional divisions, 4, 6
  seasons, **11**, 36, 186, 285, 294
  See also specific regions and countries
African peoples
  and Apartheid system in South Africa, 124–125, 153–154
  colonial policies continued by African elites, 138
  colonial powers' use of African troops, 120–121
  colonialism and African farmers, 131–133, 151
  conflicts between settlers and African livestock holders, 130
  displaced by colonial settlers in South Africa, 124, **294**
  displaced by conservation programs, 138, 146, 148, 151, 154, 190, 193–194
  fertility control, 150–151
  genetic diversity of, 19

personal narratives of. *See* Kongola, Musa
  resistance to colonial rule, 120(map), 138, 150, 153–160, 264–265
  resistance to conservation programs that deny their rights to resources and heritage, 141, 191–192, 194–195, 272
  resistance to destocking policies, 157–159, 162, 270
  See also Civilizations and settlements, pre-Columbian; Diseases; Farmers; Fishing; Foraging; Humans, early; Hunting; Labor force; Nomadism; Pastoralism; Population declines; Population growth; Slave trade; Slave trade, and Islamic world; Slave trade, trans-Atlantic; Slavery, inter-African; Social organization of African societies; *specific peoples*
African states, post-colonial, 160–166
  civil strife in, 138, 162
  and climate change, 162, 164
  conservation as unaffordable luxury, 139
  continuation of colonial conservation policies, 138, 162
  decline of state legitimacy, 138
  weak state institutions in, 139
An African Survey (Hailey), 152
Afrikaans, 99
Afro-Asiatic language speakers, 34, **267**, 274
Age of Exploration, 80–84

325

Agriculture, **23–47**, 64–65
  chronology, 258–265
  and climate, 32, 36, 75
  colonial development schemes,
      137–138, 147–148, 151, 158–159,
      277, **287**
  crops. *See* Crops
  diffusion of agricultural practices and
      crops, **43–46**
  and disease, 40–42, 51
  drying of the Sahara and expansion of
      agricultural societies, **39–43**, 60,
      170, 173–174, 290
  and early societies, **49–74**, 197–205,
      258–260
  forest/savannah frontier, 38–39, 51
  intensification and early urbanization,
      59–74
  mixed farming/herding cultures, 58
  origins of, **31–39**
  and rainforests, 8–9, 55, 197–198
  and river systems, 7–8, 51, 54–55
  techniques. *See* Agriculture
      techniques
  and transformation of landscapes. *See*
      Landscape alteration
  and variable vegetation bands of
      western Africa, 64–65
  and wildlife populations, 32, 142, 204
  See also Farmers; *specific crops and*
      regions
Agriculture techniques
  assumptions about superiority of
      Western science, 147–148
  citimene/visoso system of burning, 156
  and colonialism, 132–134, 152,
      154–155, 157–159, 287
  and commerce revolution of the
      nineteenth century, 106
  flood recession cultivation, 36, 60, 287
  "green revolution" technologies, 166,
      254–256
  intensification of
      production/reduction of fallow
      following reduced access to land,
      151, 292
  and iron working, 53–55, 204–205,
      293

ox-drawn plowing, 62, 63, 111, 259
  recognition of efficacy of African
      techniques, 152
  swidden agriculture (long fallow
      periods following clearance by
      burning), 51, 90, 124, 204, **292**
  terracing, 63, 156, 157, 158–159, 259
  tool development, 33
  urban farming, 135
  and "village" slavery, 106
Agroforestry, 146, 159, 198, **276**
Agropastoralism, 198, 258
Ahaggar Mountains, 172, **267**
Ahaggar Plateau, 12
Akan peoples, 91, 199, **268**
Albert National Park, 146, 295
Albert Nyanza, 227–233
Algeria, 178
Almohad movement, 177
Almoravid movement, 177
Alodia, 62
Altitude, 4, 6, 8, 62
Amazon Basin, 86
Amhara language, 63
Angola
  and cassava, 83
  civil war, 164
  colonial settlers, 129
  early mixed farming/herding cultures,
      58
  European farming operations, 155
  expansion of early agricultural
      communities, 56
  Imbangala raiders, 94
  Kalundu culture, 58
  and maize, 83
  oil production, 155, 164
  and slave trade, 90, 91, 94, 108
Animal husbandry
  African resistance to destocking
      policies, 157–159, 162, 270
  chronology, 258–-265
  conflicts between settlers and African
      livestock holders, 130
  development of, 23, 170, 173–174
  diffusion of, 41, **43–46**, 50, 56, 203
  and disease, 40–41, 198. *See also*
      Cattle diseases

and drought, 70

drying of the Sahara and expansion of agricultural societies, 40, 170, 173–174

earliest domestication of cattle, 34–36, 170, **270**

and early societies, **49–74**, 198

importation of sheep and goats, 36, 43–44

mixed farming/herding cultures, 58

origins of, **31–39**

risk-reduction strategies, 130, 157, 270

and variable vegetation bands of western Africa, 64–65

See also Cattle; Chickens; Goats; Guinea fowl; Horses; Pastoralism; Sheep

Animal populations. *See* Wildlife; Wildlife conservation; *specific animals*

Ankole state, 71

Anthropogenic change and fire, 18, 23, 28

and iron working. *See* Iron working

Apartheid, 124–125, 153–154. *See also* Strangled peasantry

Aquatic civilization, 32, 258, **268**, 287

Arab people in the Sahara Desert, 177, 178, 290

Arabia, 45, 63, 71

Archaeological evidence and cotton cultivation, 37

and food production, 36, 37

Kintampo culture, 38, 54, **281**

for population increases, 50

Wilton culture, **295–296**

See also Art; Iron working; Pottery; Tool use by early humans

Ardipithecus ramidus, 14

Argentina, 87

Art, 29, 30(fig), 31, 40

Asante, kingdom of, 92, 93, 108, 119

Asia crops introduced from, 45–46, 78, 260

crops introduced to, 78

domesticated animals imported from, 39, 198

and historical linguistics, 34

and hominid evolution, 14–20

Askumite state, 63

Atlas Mountains, **268**

Auroch, domestication of, 35

Australopithecines, 15, 188, **268**, **282–283**

Awdaghust, 175

Baaka foragers, 55, 142, 275

Baastards, 99

Baker, Florence von Sass, 227

Baker, Samuel W., 227–233

Bambara state of Karatu, 105

Bananas, 70, **268**

and climate, 32

and coffee production, 133

introduced from Asia, 45–46, 78, 260

large populations sustained by, 58, 70–71, 205, 261

and societal change in Great Lakes region, 69–71

Banda, Hastings, 157

Bantu language speakers, **268–269**

and disease, 42

in eastern Africa, 56–57, 199–202

and iron working, 54, 201–202, 260

languages descended from proto-Bantu, 52–53

other populations absorbed by, 203

social organization, 199, 204

in southern Africa, 98, 202–203, 260

and spread of agricultural practices, 46, 50, 197–203, 260

in Upemba Depression, 73

Barley, 36, 62, 63, 174, 287

Batwa foragers, 55, 275

Belgian colonialism, 120(map), 121, 125, 145, 146, 271–272, 295

Benguela current, **269**

Benin, 67, 91

Berber language speakers, 45, 65–66, 175, 283

Biafra, 162

"The Big Dry," 65, 175

Bight of Benin, **269**

Bight of Biafra (Bight of Bonny), 107, **269**

Bilma, 175
Black rhino, 187, 195
Black wattle, 146, 159, 276
Blue Nile, 286
Boer Republics, 123
Boer War, 118, 120(map), 124, 131
Boers, 104, 112–118, 281
    armed resistance to colonial rule, 120(map)
    British–Boer conflict, 116–118, 124. *See also* Boer War
    expansion of settlements, 123
    "Great Trek," 116, 142
    and hunting, 142, 143
    and linguistics, 99–100
    and property ownership, 115
    and slavery, 116
    and Zulus, 116
Botswana
    early mixed farming/herding cultures, 58
    early pastoralism, 56
    HIV/AIDS, 166
    Kalundu culture, 58
    Khoisan speakers, 58
Brazil, 89, 108, 111
Bridewealth, 123
British colonialism, 118–135, 120(map), 145
    British–Boer conflict, 116–118, 124. *See also* Boer War
    cash crop policies, 134
    chartering companies, 125
    and Chelimbwe uprising, 150
    Colonial Development Act of 1940, 152
    Colonial Welfare and Development Fund, 154
    and commerce revolution of the nineteenth century, 106
    and conservation, 146, 151–152, 155–156
    development plans, 151, 154–155, 158. *See also* groundnut scheme *under this heading*
    and diamond mines, 117–118
    and East Africa, 122, 127, 190
    game controls, 142–146

    groundnut scheme, 158, **277**
    oil palm imports, 106–107
    and railroads, 127
    slave trade outlawed, 105
    Society for the Preservation of [Wild] Fauna of the Empire, 145–146
    soil conservation, 155–157
    and South Africa, 104, 112–118, 123–124
    use of African troops, 121
    use of quinine, 119
    and West Africa, 119, 145
    and Xhosa disaster of 1857, 116–117
    and Zulu War of 1879, 121
British Imperial College of Tropical Agriculture, 147
British South Africa Company, 125, 144
Brockingham, Daniel, 160
Buffalo, 40, 190, 273
Buganda state, 71, 157
Bukoba region, 70, 133
Bunyoro state, 71
Burroughs, Edgar Rice, 139
Burudi, 71, 130
"Bush meat," 141, 166

Camels, 170, 175, 260, **269**
    introduction into Africa, 45
    military advantage of, 178
    and trans-Sahara trade route, 77, 80
    and variable vegetation bands of western Africa, 64–65
Cameroon, 52, 67
Cameroon Highlands, **269**
CAMPFIRE, 271
Camwood, 107
Cape Colony, 116–117, 131, 269, 281
Cape of Good Hope, **269**
Cape Town, 97, 98, 262, 269
Cape Verde Islands, 82–83
Caribbean Islands, 86, 87
Carter, Jimmy, 254
Cash crops, 132–134. *See also* Commodity production
Cassava (manoic), **269–270**
    drought tolerance of, 84, 135
    introduced from New World, 76, 82–84, 261

population increases made possible
by, 96
as staple crop, 134
Caste and ethnic specialization, 55–56
Catarrhal fever, 40
Cattle, **270**
African resistance to destocking
policies, 157–159, 162, 270
bred to live in forest–savannah
borderlands, 39
central role in southern African
societies, 58, 72–73
conflicts between settlers and African
livestock holders, 130
and disease. *See* Cattle diseases
domestication of, 35–36, 77, 170, 258
and drought, 70
emergence of pastoralism, 32, 35
expansion into southern Africa, 203
Great Lakes region, 69
knowledge of cattle management
brought to New World by West
Africans, 87
risk-reduction strategies, 130, 157,
270
and Serengeti, 188
stone enclosures, 72
use of milk and blood of, 35
and variable vegetation bands of
western Africa, 175
and Xhosa disaster of 1857, 116–117
See also Animal husbandry; Maasai
herders; Pastoralism
Cattle diseases, 33, 39, 273
cattle range limited by tsetse fly, 39,
51, 55, 56, 68, 198
East Coast fever, 40, 41, **273**
list of, 40
lung sickness, 116, 190
rinderpest epidemic. *See* Rinderpest
and smallpox, 40
Cavalli-Sforza, L. Luca, 34
Cazengo, 108
Central Africa
and cassava, 84, 96
and colonialism, 125
and commerce revolution of the
nineteenth century, 103

declining wildlife populations, 141
demand for "bush meat," 141, 166
early agricultural communities, 56
early foragers, 198
early urbanization, 73
forest conservation, 146
iron-working sites, 54
ivory trade, 103
mining, 153
miombo woodlands, 285
rainforests, 8, 11, 274
region defined, 6
rubber boom, 263
variability of rainfall, 8
See also Rainforests; *specific regions
and countries*
Cheetahs, 187, 292
Chelimbwe uprising, 150
Chickens, 39, 46, 68, 78
Chilies, 85
Chimpanzees, 165
Cholera, 122
Christianity, 62, 119–120, 150
Chronology of events, 257–266
CITES. *See* Convention on
International Trade in Endangered
Species of Wild Flora and Fauna
Citimene system of agriculture, 156
Civilizations and settlements, pre-
Columbian, **49–74**
aquatic civilization, 32, 258, **268**
caste and ethnic specialization in
early West African societies, 55–56
central Africa, 56, 73
and drying of the Sahara, 9, 12
early urbanization, **59–65**, 73
eastern Africa, 56–58, 62–64, 68–72
Great Zimbabwe, 72–73, 261, **296**
and iron working, 49–50, 53–55, 57–58
Kintampo culture, 38
Luba Empire, 295
mixed farming/herding cultures, 58
northeastern Africa, 12, 59–62
southern Africa, 58–59, 72–73, 296
symbiotic relationship between
farmers and foragers, 55–56
waterways and navigation, 6–7
western Africa, 51–56, 64–68

Civilizations and settlements,
    pre-Columbian *(cont.)*
  See also Population declines;
    Population growth; Social
    organization of African societies
Climate, **10–13**
  and agriculture, 32, 36, 75
  altitude as most important factor in
    local climate, 6, 8
  change and variability. *See* Climate
    change; Climate variability
  and disease, 41–42
  Holocene Optimum, 173, 258, **278**
  and Intertropical Convergence Zone,
    10–11, 170, 285
  microclimates, 11
  Serengeti, 184, 188
  See also Rainfall; Temperature;
    specific regions
Climate change, 10–13, 169–174
  chronology, 257–265
  dry phases, 15, 65, 174(table), 175
  drying of the Sahara and expansion of
    agricultural societies, **39–43**, 60,
    170, 173–174, 290
  and European assumptions about
    "desertification," 156, 164, 171,
    179–181, 291
  glaciation, 12, 24, 26–27, 173
  global warming, 10, 13
  and human evolution, 15, 19–20
  long-term changes, 12
  Milankovitch cycles, 25
  post-colonial era, 162, 164
  and pressure on resources, 50
  and Serengeti, 188
  and transition to the Holocene, 24–27
  See also Drought; Rainfall
Climate variability, 3
  annual variations, 11–12, 290–291
  and Bantu language speakers, 204
  and El Niño-Southern Oscillation, 3,
    11–12
  and North Atlantic Oscillation, 3, 11
  Sahel and Sudan as areas of high
    variability, 173, 290
  and urbanization of Inner Niger Delta,
    65–68

  and variable vegetation bands of
    western Africa, 64–65, 175
  See also Rainfall
Cloves, 103, 111
Club du Sahel, 180
Coal mines, 131
Cocoa, 133
Coffee, 133, **270–271**
  and acid soils, 152
  Africans prohibited from growing, 129
  and commerce revolution of the
    nineteenth century, 108
  early cultivation of, 38
  expatriate-owned plantations, 130
  origins of, 62
  spread beyond Africa, 45
Colonial Development Act of 1940, 152
Colonial Development Fund, 151
Colonial Welfare and Development
  Fund, 154
Colonialism, **103–135**, 120(map), **271**
  and agriculture, 132–134, 137–138,
    147–148, 151, 158–159, **277**, **287**
  Berlin Conference (1884–1885), 121
  chartering companies, 125, 127
  chronology, 264–265
  colonial powers' use of African troops,
    120–121
  commerce revolution of the
    nineteenth century, 105–111
  conflicts between settlers and African
    livestock holders, 130
  and conservation, 105, 137–160, 191,
    276. *See also* Conservation
  development plans, 137, 147–148,
    151, 158–159, **277**, **287**
  differential commodity prices for
    whites vs. African sellers, 130
  eastern Africa, 108–111, 122–123
  ecological stresses caused by, 118,
    121–123, 151, 179
  European assumptions about Africans'
    use of land, 151, 156, 178–181
  European conquests after the 1870s,
    118–123
  Europeans' improved abilities to
    operate in African environments,
    118–121

and famine, 122–123, 132
fiscal policies and cash crops, 134
and gold trade, 130–131
hunting by European settlers, 141
and labor force. *See* Labor force
landscape alteration/reorganization of
  space under colonial rule, 123–135
population decline due to, 104, 118,
  122
population growth fueled by demand
  for labor, 137, 151, 160
post-1870s European scramble for
  control over African resources,
  118–123
and quinine, 119
and railroads, 125, 127–128
resistance to colonial conservation
  programs, 138
resistance to colonial rule, 138, 150,
  153–160
resource mining replaced by
  commodity production, 128–129
and rinderpest epidemic, 104, 121–122
and Sahara Desert, 178
and sanitation practices, 119
settler colonies, 129–130, 264
and slavery, 103–104, 108
southern Africa, 104, 111–118,
  123–125
and taxation, 132
and urbanization, 153
western Africa, 103, 105–108, 122
women's rights repressed under
  colonially sanctioned "native law,"
  132
Columbian exchange era, 75–77, **79–87**,
  261–262
  Age of Exploration, 80–84
  Dutch settlements in southern Africa,
    96–100
  New World crops, 76, 82–85
  precursors to, 76–79
  slave trade, 76, 86–96
Commodity production, 262
  colonialism and African farmers,
    131–133
  and colonialist taxation systems, 132
  differential commodity prices for

whites vs. African sellers, 130
increases in response to urbanization,
  135
population growth fueled by demand
  for labor, 137, 151, 160
and railroads, 128–129
resource mining replaced by, 128–129
rise in demand after WWII, 155
Congo
  and cassava, 83
  and colonialism, 104, 129
  copper production, 155
  early agricultural communities, 56
  early mixed farming/herding cultures,
    58
  early urbanization, 71
  expatriate-owned plantations, 130
  and Portuguese traders, 81
  and slave trade, 90, 91, 93–94
  Virungu National Park, 146, **295**
  See also Kongo, kingdom of; Upemba
    Depression
Congo Free State, 122, 125
Congo River, 7–8, **271–272**
Congo River Basin
  equatorial rainforest described, **274**
  plantain cultivation, 56, 70
  symbiotic relationship between
    farmers and foragers, 55–56
  See also Conrad, Joseph
Conrad, Joseph, 238–242
Conservation, **137–168**
  colonial policies continued in post-
    colonial era, 138, 162
  and colonialism, 105, 137–160, 191,
    292–293
  community-based conservation, 271
  and culture of expertise, 152, 162
  debt for conservation process, **272**
  and "development" policies, 147,
    154–156
  and economic problems of the 1970s,
    138, 195
  "fortress conservation," 141, 160
  game laws enforced only in reserves
    and parks, 148
  hunting and wildlife conservation,
    139–153

Conservation *(cont.)*
  international agencies' direction of,
    138
  international conventions on, 145,
    148, 195, 242–251, 272
  opposition from European settlers,
    145, 146
  opposition to destocking policies,
    157–158, 162, 270
  opposition to labor-intensive colonial
    programs, 138, 157, 159, 293
  opposition to programs that deny
    Africans' rights to resources and
    heritage, 141, 191–192, 194–195
  origins of colonial effort, 141–148
  and poor states, 139
  and resistance to colonial rule, 138,
    153–160
  and rinderpest epidemic, 141–142, 144
  segregation of human population from
    reserves, 136, 138, 146, 148, 151,
    154, 190–191, 193–194
  and warfare, 141
  See also Environmental degradation;
    Forest conservation; Pasture
    conservation; Soil conservation;
    Wildlife conservation
Convention for the Protection of
  African Fauna and Flora (1932), 148
Convention on International Trade in
  Endangered Species of Wild Flora
  and Fauna (CITES), 195, **242–251**,
  272
Copper, 53, 74, 131, 155
Corn. *See* Maize
Corruption, 157, 164–165
Côte d'Ivoire
  early agricultural communities, 52
  early urbanization, 67
  expatriate-owned plantations, 130
  French colonialism, 129
  Kintampo culture, 38, 54, **281**
Cotton, 133
  and colonialism, 132
  differential commodity prices for
    whites vs. African sellers, 130
  early cultivation of, 37
  spread beyond Africa, 45

Cowry shells, 81
Crater Highlands, 185, 187, 193
Crops
  African crops introduced abroad, 45, 78
  Asian crops, 45–46, 78
  and climate, 32, 36, 75
  Congo River Basin, 70
  cultivation of cash crops compelled by
    colonial taxation system, 132–134
  eaten by weaver birds, 289
  forest/savannah frontier, 51
  Great Lakes region, 69, 70–71, 200,
    205
  highlands, 62–63, 79, 83
  New World crops, 76, 82–85, 261
  Nile Valley, 174
  rainforests, 79, 84, 198
  resource mining replaced by
    commodity production, 129
  savannahs, 83, 85
  See also Commodity production;
    specific crops
Crosby, Alfred, 75, 100
Cuba, 89
Cummings, R.G.G., 139
Currency, 81, 90
Curtin, Philip, 1, 76, 86
Cushitic language speakers, 34, 53, 258
  and animal husbandry, 41
  and Bantu language speakers, 201
  domestication of plants in Ethiopian
    Highlands, 62
  in Kenya, 57
  movement south from Ethiopia, 56
  in Tanzania, 57

Dahomey, 108
Dahomey Gap, **272**
DDT, 284
de Brazza, Savorgana, 121
Debt for conservation, **272**
Deforestation, 3, **272**
  and elephant population, 194
  Ethiopian Highlands, 64, 156
  and iron working, 54, 197–199,
    202–203, 260
  land clearance by burning banned,
    156–157

See also Landscape alteration

Democratic Republic of the Congo (DRC), 7

Demographic transition theory, 159, 166

Demography, 6

Desertification, 265
  "degradation" vs., 164, 181, **273**, 291
  European perceptions about, 156, 164, 171, 179–181, 273

Development
  assumptions about superiority of Western science, 147–148, 151, 152
  colonial policies, 137–138, 147, 151, 154–156, 158–159, **277**, **287**
  and resistance to colonial rule, 138
  and transportation infrastructure, 155
  and WWII, 152

Diamond, Jared, 75–76

Diamonds
  and bridewealth, 123
  discovery of, 104, 112, 117, 263
  effects of discovery, 123, 131, 263
  labor force for, 117–118, 123, 131

Dias, Bartholomeu, 80

Digging sticks, 33, 296

Dingiswayo, 112

Diseases
  and cattle. *See* Cattle diseases
  contact-era disease epidemics, 76
  Europeans' improved survival rates following discovery of quinine and sanitation practices, 119
  Europeans' lack of disease tolerance, 81, 86
  and expansion of agricultural societies, 40, 41–42
  and floodplains, 8
  and forests, 67–68
  HIV/AIDS, 139, 160, 161, 165–166
  New World depopulated by, 85–86
  plague, 261
  population decline due to, 66
  post–WWII medical advances, 159–160
  and slave trade, 76, 87, 88, 96
  and small populations of foragers/pastoralists, 98

smallpox among Khoisan speakers, 99

smallpox among Maasai pastoralists, 190

smallpox and cholera in colonial era, 122

and transportation infrastructure, 150

tuberculosis, 150, 166

water-borne diseases as inhibitors of agriculture, 51

wildlife protected in environments hostile to humans due to diseases, 29, 39, 41, 144

See also Malaria; Rinderpest; River blindness; Schistosomiasis; Tsetse fly; Yellow fever

DNA, 19, 36

Dodomo region (Tanzania), 207–211

Dog, wild, 187, 292

Donkey, 80, **273**

Dorobo people, 142, 275

Drakensberg Mountains, 98, 112, **273**

Drought
  "The Big Dry" (300 B.C.E. to 300 C.E.), 65, 175
  chronology, 257–265
  and conservation programs, 138
  drought of 1890s, 179
  drought of 1910–1915, 179, 264, 291
  drought of the 1970s, 138, 162, 180, 273, 291
  drought-tolerant crops. *See* Cassava; Millet; Teff
  dry century (950 C.E.), 69–70
  and famine, 162, 179, 264, 274
  and intensification of livestock keeping, 70
  and Little Ice Age, 188
  maize's vulnerability to, 134
  post-colonial era, 162, 164
  western Africa, 66–67
  See also Sahara Desert; Sahel

Dry Veld, 97

Durban, 131

Dutch
  early settlers in southern Africa, 96–97, 97(map), 262
  and slave trade, 86–87
  See also Boers

Dutch East India Company, 98, 99, 142, 269

Early Stone Age, 17
East Coast fever, 40, 41, **273**
East London, 131
Eastern Africa
  and banana cultivation. *See* Bananas
  colonial game reserves, 146
  and colonialism, 108–111, 125, 127, 129, 132, 190–191
  and commerce revolution of the nineteenth century, 103, 108–111
  and cotton, 132
  early agricultural communities, 56–58, 198–202
  early urbanization, 62–64, 68–72
  famines, 122–123, 150
  geographical characteristics, 6, 62
  and human origins, 14–15
  iron-working sites, 54. *See also* Iron working
  and Islam, 79
  ivory trade, 103, 108–111
  land clearance as check on animal populations, 142
  and maize, 83
  Maji Maji Revolt, 132, 150
  miombo woodlands, 285
  "monsoon exchange" across Indian Ocean, 43–46, **76–79**
  and peanuts, 85
  railroads, 127
  rainfall, 8, 11, 62
  region defined, 6
  resettlement schemes, 158
  and rinderpest epidemic, 122, 190–191
  settler colonies, 129
  and slave trade, 79, 94–95, 108
  South Arabian immigrants, 63
  urbanization and outside contact (Swahili civilization), 71–72
  wildlife, 9, 142
  Wilton culture, **295–296**
  See also British colonialism; Cushitic language speakers; German colonialism; Great Lakes; Portuguese colonialism; Rift Valley;

Serengeti; *specific regions and countries*
Eastern Arc Mountains, 71, **273**
Eco-physiognomic zones, 6
Ecological stresses
  and Apartheid system, 154
  climate change vs. landscape degradation through overexploitation, 162, 164, 171, 179–181, 273
  and colonial conquest, 118, 121–123, 151, 179
  degradation of the Sahel, 173, 273
  and intensification of production, 137, 151
  and population growth, 137
  See also Environmental crises; Natural resources, pressures on
Economics
  crisis of 1970s and 1980s, 165
  and demographic transition theory, 159
  economic crisis of the 1970s, and conservation, 138, 195
  and globalization, 164
  See also Colonialism; Labor force; Trade
Ecosystems, 13(map), 26(map). *See also* specific regions and types of ecosystems, such as Rainforests
Ecotourism, **273–274**
Egypt, 7
  cattle iconography, 40
  and commerce revolution of the 19th century, 111
  cultivation of sorghum and millet, 36
  domestication of livestock, 35–36
  and drying of the Sahara, 40, 170
  recessional cultivation, 60
  urbanization, 60–62, 170
  See also Nile Valley
Ehret, Christopher, 34, 37, 45, 197, 202
El Niño, 3, 11–12, 164
Elephants
  decline in population, 109, 128, 142, 193
  poaching, 165, 195
  recovery of population, 165, 194, 196

and Serengeti, 187, 189–190
as threatened species, 141
See also Ivory trade
Elmina, 81
Empakaai Crater, 191
Emutai, 190
Endemism, 6, 9, 38, 62, 273
England, and slave trade, 86–87, 93. *See also* British colonialism
Enset, 38, 62, **274**
Environmental change. *See* Ecological stresses; Environmental degradation; Landscape alteration
Environmental crises
  and Apartheid system, 154
  and collapse of Great Zimbabwe, 73
  and colonialism, 104, 121–123, 132
  crisis of 300–500 C.E. (Great Lakes region), 57–58
  crisis of 950 C.E. (Great Lakes region), 69–71
  and early urban societies, 60, 64, 66–67, 261
  Ethiopian Highlands, 64
  subsistence crises under colonialism, 132
  and variable vegetation bands of western Africa, 64–65
  western Africa, 66–67
  See also Natural resources, pressures on
Environmental degradation, 3
  and climate change, 162, 164, 181, 273, 291
  European perceptions about, 138, 151, 156, 171, 179–181, 273
  intensification of production/reduction of fallow following reduced access to land, 151
  and oil industry, 164
  post-colonial era, 138, 162, 164, 181, 273
  and Sahel, 173, 273
  and soil, 179–181
  and top-down colonial policies, 156
Environmental variability, 3
  eco-physiognomic zones and

physiographic regions, 6
  and human evolution, 4, 9
  and nomadism, 64–65, 175, 178–179
  and Rift Valley, 9
  and urbanization in western Africa, 66
  variable vegetation bands of western Africa, 64–65, 175
  See also Climate variability
Environments, colonial era. *See* Colonialism
Environments, Columbian era. *See* Columbian exchange era
Environments, post-colonial era, **160–166**
Environments, pre-Columbian era. *See* Civilizations and settlements, pre-Columbian; Humans, early
Ergs, 172
Eritrean Highlands, 62–64
Erosion, 3. *See also* Environmental degradation; Soil conservation
Ethiopia
  Askumite state, 63
  and commerce revolution of the 19th century, 111
  famine and warfare, 162
  Ge'ez daughter languages, 63
  Italian invasion defeated (1896), 121
  and rinderpest epidemic, 122
  and slave trade, 108
Ethiopian Highlands, 63(photo)
  crops and climate, 62–63, 294
  defined/described, **274**
  early agriculture, 38, 259
  early urbanization, 62–64
  environmental zones, 62
  European perceptions about lack of tree cover, 156
  plant endemism, 62
  soil fertility, 62
  vulnerability to crises in production, 64
  wildlife, 9
  See also Rift Valley
Eucalyptus, 146, 159, 276
Europe
  Age of Exploration, 80–84
  and hominid evolution, 14–20

Europeans
  contact-era disease epidemics, 76
  improved ability to operate in African
    environments, 118–121
  lack of disease tolerance, 81, 86
  See also Colonialism; Columbian
    exchange era; Missionaries
Extractive industries. *See* Coal mines;
  Copper; Diamonds; Gold trade; Iron
  working; Oil industry

Fachi, 175
Fagan, Brian, 12, 169
Famine
  and colonialism, 122–123, 132
  and drought, 162, 179, 264, 274
  post-colonial era, 160–162
  and Saskawa/Global 2000, 254–256
  and transportation infrastructure, 150
  and WWI, 150
  and WWII, 152–153
Farmers
  and Apartheid system, 124–125, 154
  and changes in women's status, 132
  and colonialism, 124, 131–133, 151
  symbiotic relationship between
    farmers and foragers, 55–56, 142
  trade with pastoralists, 66
  See also Agriculture; Agriculture
    techniques; Commodity production;
    Labor force
Fauna Preservation Society of London,
  192
Fertility control, 150–151
Finger millet, 63, **274–275**
Fire, 18
  and alteration of landscapes for
    agriculture/grazing, 18, 23, 28,
    40–41, 55, 198
  citimene/visoso system of burning, 156
  and European misperceptions about
    "desertification," 179
  intense fires in lands no longer burned
    regularly, 194
  land clearance by burning banned,
    156–157, 191, 194
  and reduction of tsetse-bearing bush,
    41

  and Serengeti, 188, 191
  See also Swidden agriculture
Fish River, **275**
Fishing
  and early agricultural societies, 39, 173
  Middle Stone Age, 29
  and Nile Valley, 33
  in western Africa, 68
  See also Aquatic civilization
Floodplains, 7, 8, 36, 60, 287
Fonio, **275**
Food production, **23–47**
  cash crops vs. subsistence crops, 132,
    134
  and changes in women's status under
    colonialism, 132
  chronology, 258–265
  climate change and transition to the
    Holocene, **24–27**
  diffusion of agricultural practices,
    crops, and livestock, **43–46**
  and disease, 39–42
  drying of the Sahara and expansion of
    agricultural societies, **39–43**, 60,
    170, 173–174, 290
  early food-producing societies, **49–74**
  early urbanization, **59–65**
  and linguistics. *See* Linguistics;
    specific languages
  and Saskawa/Global 2000, 254–256
  and standard of living, 50
  and transformation of landscapes,
    197–205. *See also* Landscape
    alteration
  transition to agriculture and animal
    husbandry, **31–39**
  See also Agriculture; Animal
    husbandry; Foraging
Foot and mouth disease, 40
Foraging, 23–24, **27–31**, 198, **275–276**
  and disease, 42
  foraging Khoisan speakers, 58–59, 98
  relative freedom from disease of small
    mobile populations, 98
  and Serengeti, 188
  small number of sites compared to
    agricultural sites, 50
  symbiotic relationship between

farmers and foragers, 55–56, 142
symbiotic relationship between
    pastoralists and foragers, 56, 142
Forest conservation
  and colonialism, 137, 146–147, **276**
  land clearance by burning banned,
    156–157, 191, 194
  Maathai's contributions, 168
  segregation of human population from
    reserves, 138, 146, 148
Forest–savannah border
  agricultural zones of western Africa,
    52(map)
  crops suitable for, 51
  early agricultural communities, 51
  early urbanization, 67
"Fortress conservation," 141, 160
Fossey, Diane, 140, **277**
Frankfort Zoo, 139, 192
French colonialism, 118–135, 120(map),
    122, 125, 127, 129, 134
  cash crop policies, 134
  chartering companies, 125
  and commerce revolution of the
    nineteenth century, 106
  development plans, 151, 154–155,
    158, 179, **287**
  game controls, 145
  game reserves, 146
  lack of settler colonies, 129
  and peanuts, 106
  and railroads, 127, 128
  and Sahel/Sahara Desert, 178
  and slave trade, 86–87, 92–93, 178
  sugar plantations on Indian Ocean
    islands, 95
  use of African troops, 120–121
  use of quinine, 119
Frontier concept, and African social
    organization, 50–51
Fulani language speakers, 65–66, 270

Gabon, 155, 164
Gambia Valley, 105–106
Gao, 66
Gaza, 114
Ge'ez language, 63
Genetic diversity of African

populations, 19
Geography of Africa, 5(map), **5–9**
  deserts, 9, 171–172
  Great Lakes region, 69, 200–201
  Rift Valley, 8, 9
  Serengeti, 181–182, 184–187
German colonialism, 118–135,
    120(map), 129, 132
  chartering companies, 125
  and cotton cultivation, 132
  and game reserves, 144, 146
  and Herero Revolt, 150, 264
  and hunting, 144
  and railroads, 127
  scorched earth policy in East Africa,
    122
  settlers, 129
  and spread of tsetse-bearing bush
    following rinderpest epidemic,
    190–191
  use of African troops, 121
Ghana
  early urbanization, 67
  Kintampo culture, 38, 54, **281**
  and Saskawa/Global 2000, 254
  and slave trade, 92
Ghana, Empire of, 66, 177
Giraffe, 188, 190
Glaciation, 12, 173, 257–258
  and African climates, 26–27, 188
  and fluctuation in human population,
    19, 33
  and Holocene, 24
  and human evolution, 19
Global warming, 10, 13, 25–26, 164
Globalization, 5, 164
  and commerce revolution of the
    nineteenth century, 105
  See also Slave trade, trans-Atlantic
Goats, 39, **277**
  expansion into southern Africa, 56
  and forest–savannah borderlands, 51
  importation into Africa, 36, 43–44,
    198
  resistance to disease, 44, 199
  in western Africa, 68
Gold Coast
  and British colonialism, 119

Gold Coast (cont.)
  and commerce revolution of the
    nineteenth century, 108
  and gold trade, 83
  and maize, 83
  and oil palm, 108
  population increase on, 83
  and slave trade, 91
Gold trade
  and colonialism, 130–131
  early trade systems, 66
  effects of, 123, 130–131, 263
  location of mining activity, 72–73
  and Portuguese traders, 81, 83, 91, 95
  and slave trade, 91
  and South Africa, 104, 112, 118, 123,
    282
  and Swahili civilization, 72
  and trans-Sahara trade route, 177
  and Zimbabwe, 79, 95, 296
Goodall, Jane, 140, **277**
Gorillas, 32, 146
Grazing lands. See Pasture conservation
Great Lakes (eastern Africa), 8, **277**
  banana cultivation, 69–71, 205, 261
  crisis of 300–500 C.E., 57–58
  crisis of 950 C.E., 69–71
  crops and climate, 69, 200–201
  decline in tree cover following advent
    of iron working, 54, 204, 205, 260
  early agricultural communities, 198,
    200
  early Bantu-speaking communities,
    199–202
  environmental costs of agricultural
    expansion and iron working, 57–58
  geographical characteristics, 69,
    200–201
  iron-working sites, 69
  landscapes altered by agriculture, 71,
    204
  and Rift Valley, 289
  trade near, 109
  and transition to the Holocene, 27
Great Limpopo Transfrontier Park, 281
Great Rift Valley. See Rift Valley
Great Trek, 116, 142
Great Zimbabwe, 72–73, 261, 270, **296**

Grindstones, 33, 35, 36
Griqua, 99, 115, 142
Groundnuts. See Peanuts
Grumeti Conservation Area, 182,
  183(map)
Grzimek, Bernhard, 139, 192, **277–278**
Guinea fowl, 39, 46, **278**
Guinea rice, 51–52
Guy, Jeff, 114

Hadley Circulation, 170
Hadza people, 58
Haggard, Ryder, 139
Hailey, Lord William, 152
Haitian Revolution, 88–89
Harlon, Jack, 37
Hausa people, 65–66
The Heart of Darkness (Conrad), 238–242
Herero Rebellion, 149(photo), 150, 264
Hippo, 187
HIV/AIDS, 139, 160, 161, 165–166
Holocene
  chronology, 258
  climate changes during, 10–13
  Holocene Optimum, 173, 258, **278**
  "optimum" end, 60, 69–70
  and Sahara Desert, 170
  transition to, 24–27
  as warm interglacial period, 25
Hominids, 14–21
Homo antecessor/maritanicus, 17
Homo erectus, 14, 15, 17, 19, 20, 169,
  188, 257, **278**
Homo ergaster, 17, 257, **278**
Homo habilis, 15, 17, 188, 257, **278–279**
Homo heidelbergensis/rhodesiensis, 19
Homo neanderthalensis, 18–19, 20
Homo rudolfensis, 15
Homo sapiens, 19–20, 257–258, **279**.
    See also African peoples; Humans,
    early
Horn of Africa, **279**
Horses
  and hunting, 142
  introduction into Africa, 44
  military advantage of, 178
  and trade in the Sudan, 80
  and trans-Saharan contact, 65

Human evolution, **14–20, 278–279**
chronology, 257–258
and climate change, 15, 19–20,
169–170
co-evolution of humans and wildlife,
29
and disease, 41–42
and environmental variability, 4, 9
movement out of Africa, 12, 17, 20
and Serengeti, 188
Humans, early, **14–20**
adaptability of, 18, **27–31**
alteration of landscapes by, 13–14, 18,
23–24, 28, **197–205**
art, 29, 30(fig), 31
chronology, 257–260
control of fire, 18, 198
Early Stone Age, 17
emergence of *Homo sapiens*, 19–20
fluctuation in population, 19, 33
food production. *See* Food production
foraging, 23–24, **27–31**
languages, 34, 37
lifeways and social organization,
17–18
and malaria, 198
metalworking technology, 24
Middle Stone Age, 17, 29
origins of, 14–21
origins of agriculture and animal
husbandry, **31–39**
and Sahara Desert, 169–177
and Serengeti, 188
tool use, 17–18, 28–29
Wilton culture, **295–296**
See also Iron working
Hunting
colonial game laws, 142–145
and Convention of 1900, 145
demand for "bush meat," 141, 166
and early agricultural societies, 32, 39,
51
and early forest communities, 68
extermination of springbok, 143–144
extinction of species caused by,
115–116
game laws enforced only in reserves
and parks, 148

hunting by Africans banned, 143, 144,
191, 288–289
and improvements in firearms, 142
indigenous methods, 142
poaching, 165, 194–195, **288–289**
safaris/tourist hunting, 196, 264, 288,
289–290
and Serengeti, 194–195
and settler colonies, 129, 144
Stone Age techniques, 29
and threatened species, 193
and "vermin," 145
and wildlife conservation, 139–148
See also Ivory trade; Wildlife
conservation
Hyena, 187, 292

Ibn Khaldun, 65, 177
Ice ages. *See* Glaciation
Ikoma WMA, 183(map)
Ikorongo Game Reserve, 182, 183(map)
Ile-Ife, 67
Iliffe, John, 153
Imbangala people, 94
Indian Ocean
"monsoon exchange," 43–46, **76–79**
slave trade, 108
sugar plantations on Indian Ocean
islands, 95
Industrialization, 131, 153
Inner Delta (Niger River), 8, 54–55,
64–68, **279**, 286
International Conference for the
Protection of Nature (1931; Paris),
148
International Monetary Fund, 138
International organizations, 138, 162,
193, 272
Intertropical Convergence Zone (ITCZ),
10–11, 39, 170, 285
Iron working, **279–280**
and Bantu language speakers, 201–202
and clay soils, 54
and decline in forest cover, 54,
197–199, 202–203, 205, 260
early evidence for, 49–50, 53–55, 69,
279, 287
and environmental crises, 57–58

Iron working *(cont.)*
 iron trade networks, 204
 Lake Victoria region, 68(map)
 origins of, 202, 259
 and sorghum, 53, 293
Irrigation
 colonial development schemes, 137,
  151, 158, 179
 in Ethiopian Highlands, 63
 and Nile Valley, 60
Islam
 and agricultural labor, 106
 and East Africa, 79
 effect on African societies, 79
 and North Africa, 79, 178, 260
 and resistance to colonial rule,
  120(map)
 and Swahili civilization, 72
 See also Slave trade, and Islamic
  world
Italian colonialism, 118, 120(map)
 game controls, 145
 and spread of rinderpest in African
  cattle, 121–122, 263, 289
Ivory Coast, 133. *See also* Côte d'Ivoire
Ivory trade, 79, 110(photo), 262, **280**
 and commerce revolution of the
  nineteenth century, 103, 108–111
 and decline in elephant populations,
  109, 128, 142
 international ban on, 165, 195,
  242–251
 limited by Convention of 1900, 145
 and railroads, 128
 and Serengeti, 189–190

Jenné-jeno, 65, 66, **280**
Johannesburg, 131
Johnston, Sir Harry, 145
Jonglei Canal, 180

Kalahari Desert, 9, 27, **281**
 personal narrative of David
  Livingstone, 233–238
Kalundu culture, 58
Kamba people, 109
Karatu, 105
Karoo, **281**

Kenya
 Bantu language speakers, 54, 57
 and colonialism, 104, 125, 129, 190
 Cushitic language speakers, 57
 destocking policies, 157–158
 European farming operations, 155
 expansion of cattle from, 56
 famines, 152–153
 herders' use of land reserved for white
  farms, 154
 HIV/AIDS, 166
 independence, 193
 iron-working sites, 54
 ivory trade, 109
 Maathai's contribution to
  conservation efforts, 168
 MauMau uprising, 157
 nature reserves, 146, 182, 183(map),
  185. *See also* Serengeti
 and railroads, 127–128
 settler colonies, 129, 130
 terrace farming, 158
 See also Rift Valley; Serengeti
Kerma, 61
Khami, 73
Khoi people, 59, 98, 99, 275, **281**. *See
 also* Khoisan speakers
Khoisan speakers, 296
 and animal husbandry, 41
 popular views of, 58–59
 relative freedom from disease of small
  mobile populations, 98
 symbiotic relationship between
  pastoralists and foragers, 56
Kilimanjaro. *See* Mount Kilimanjaro
Kilwa, 72, 95, 109, 296
Kimberley, diamond mines, 117, 123,
 131
Kin groups, and "frontier" concept, 51.
 See also Social organization of
 African societies
Kintampo culture, 38, 54, **281**
Klieman, Kairn, 197
Kob, white-eared, 186
Kola nuts, 108
Kondoa Highlands, 157, 162
Kongo, kingdom of, 81, 93–94, 271
Kongola, Musa, 207–211

Kopytoff, Igor, 50
Kruger National Park, 125, **281–282**
Kumasi, 92
Kumbi Saleh, 66
Kush civilization, 61(photo)
Kwale culture, 58

Labor force, 123–127, 130, 131, 292
  and colonial conservation programs,
    138, 156–157, 159, 293
  and colonial development schemes,
    151
  and colonialist taxation systems, 132
  and commerce revolution of the
    nineteenth century, 103–104, 108,
    118
  and commodity production, 129, 130
  and diamond mines, 117–118, 123
  and inter-African slavery, 103–104,
    106, 108
  labor shortages caused by population
    declines, 148
  population growth fueled by demand
    for labor, 137, 151, 160
  and property ownership, 118, 124
  and railroad placement, 127–128
  recruiting systems, 124
  and spread of infectious disease, 150
  and subsistence crises under
    colonialism, 132, 134
  and Xhosa disaster of 1857, 116–117
Lake Albert, 227–233, 277
Lake Chad, 8, **270**
  contraction of, 39, 54
  expansion of, 27, 33, 37, 173
  and glaciation cycles, 12
  iron working and clay soils, 54
  and transition to the Holocene, 27
Lake Malawi, 8, 27, 58, 277, 289
Lake Manyara National Park, 185
Lake sediment studies, 50, 54, 204
Lake Tanganyika, 8, 58, 277, 289
Lake Turkana, 41
Lake Victoria Nyanza, 8, 27, 68(map),
    277, 289
Lamu, 72
Land ownership
  Africans displaced from traditional

lands by conservation programs,
    138, 146, 148, 151, 154, 190–191,
    193–194
  conflicts between settlers and African
    livestock holders, 130
  and settler colonies, 129–130
  South Africa, 115–118, 124, **294**
Landscape alteration, 3, 71, **197–205**,
    295
  and colonialism, 115–118, 123–135
  and early human populations, 13–14,
    18, 23–24, 28
  and early urban societies, 55, 71
  and fire, 18, 23, 28, 55
  intense fires in lands no longer burned
    regularly, 194
  and iron working, 54, 197–199,
    202–203, 260
  land clearance by burning banned,
    156–157, 191, 194
  landscapes managed as nature
    preserves, 138. *See also* Serengeti
  and mining, 130–131
  and railroads, 127–128
  and rinderpest epidemic, 122,
    141–142, 190, 263, 285, 295
  western African mosaic, 8–9
  and wildlife populations, 32, 142, 204
  See also Deforestation; Environmental
    degradation
Languages. *See* Linguistics
Late Stone Age tool development, 29,
    33
Laterite soils, 8, 90
Leakey, Mary, Louis, and Richard, **282**
Leopard, 187
Leopold, King of Belgium, 121, 125,
    126(fig), 271
Lesotho, 114, 166
Liberia, 107–108, 130
Libya, 120(map)
Libyan Desert, 172, 290
Limpopo National Park, 281
Limpopo River, 8, **282**
Lineage slavery, 90
Linguistics, 56–57
  and agriculture, 34, 37, 38, 46
  and animal husbandry, 41

Linguistics *(cont.)*
    and caste/ethnic specialization in
        western Africa, 65–66
    and foragers, 275–276
    languages descended from proto-
        Bantu, 52–53
    and Madagascar, 46, 78–79
    and multicultural society of southern
        Africa, 99
    and spread of early food-producing
        communities, 50, 52–53, 56–57,
        197–203
    and trans–Indian Ocean trade, 78–79
    See also specific language families
Lion, 187, 193, 295
Little Ice Age, 188
Livestock. *See* Animal husbandry;
    Cattle; Goats; Horses; Pastoralism;
    Sheep
Livingstone, David, 139, 233–238
Locust invasions, 64
Logging. *See* Timber trade
Loniondo Game Controlled Area, 182,
    183(map)
Lualaba River, 7, 73–74
Luanda, 94
Luba Empire, 295
Luba Empire, state ancestral to, 73–74
Lucy, **282–283**
Lugard, Alfred, 121
Lunda Empire, 94
Luso-African people, 89, 94, 108

Maasai herders, 188, 190–196, 262, 270
    confined to reserves, 190, 194
    and land emptied by rinderpest
        epidemic, 144, 190, 192
    and Ngorongoro Crater, 286
    resistance to conservation programs
        that deny their rights to resources
        and heritage, 191–192, 194–195
    and smallpox, 190
Maathai, Wangari, 167(photo), 168, 266,
    283
Madagascar, 46, 78–79, 260, **283**
Maghreb, **283**
Maize, **283–284**
    and diamond mines, 123

introduced from New World, 76,
    82–83, 261
population increases made possible
    by, 96
rapid spread of, 82–83
and Saskawa/Global 2000, 254–256
spread in southern Africa, 114
as staple crop, 134
vulnerability to drought, 134
Maji Maji Revolt, 132, 150
Malagasy language speakers, 78–79,
    260, 283
Malaria, **284–285**
    and early human populations, 198
    Europeans' improved survival rates
        following discovery of quinine,
        119
    lack of European immunity to, 86
    limited range of, 75
    pathology of, 284
    prevalence in floodplains, 8
    prevalence in forests, 67–68
    and quinine, 119, 159, 284
    resistant strains, 160
    and spread of agriculture, 41–42
    spread to New World, 85–86
Malawi, 58, 104, 156. *See also*
    Nyasaland
Malawi kingdom, 73
Malebo Pool, 7
Mali, 128, 151, 294
Mali, Empire of, 81, 91, 177
Malignant catarrhal fever, 40
Mamdani, Mahmood, 112
Mande language speakers, **285**
    caste and ethnic specialization, 55,
        65–66
    and commerce revolution of the
        nineteenth century, 105
    and Portuguese traders, 81, 91
    and slave trade, 92
Mangrove swamp, 52(map), 64, 79
Manning, Patrick, 87, 95
Manioc. *See* Cassava
Maps
    agricultural zones of western Africa,
        52(map)
    colonial powers, 120(map)

Dutch settlements in southern Africa, 97(map)
early industries in the Lake Victoria region, 68(map)
relief map, 5(map)
Serengeti reserve systems, 183(map)
slave trade, 94(map)
southern Africa in the nineteenth century, 114(map)
vegetation belts, 13(map), 26(map)
West African trade states, 92(map)
Mapungubwe, 73
Mara Ranches, 182, 183(map), 185
Masai Mara National Reserve, 182, 183(map)
Maswa Game Reserve, 182, 183(map)
Mauritania, 66
Mbulu resettlement plan, 158
McCann, James, 156
McIntosh, Roderick, 65, 174
McIntosh, Susan Keech, 174
McNeill, John and William, 43
McShane, T.O., 186
Men
  and cash crops, 132
  sex ratio in slave populations, 96
Meroe, 54
Meroe (Sudan), 61(photo)
Mexico, 87
Mfecane, 114–115
Microclimates, 11
Middle Stone Age, 17, 29
Migrations
  springbok, 143–144, 186, **293**
  white-eared kob, 186
  wildebeest, 194, **291–292**
Milankovitch cycles, 25
Miller, Joseph, 91
Millet, 62
  and climate, 32
  early cultivation of, 35, 36, 37, 63, 200
  finger millet, 62, 63, **274–275**
  Great Lakes region, 69, 200
  pearl millet (bulrush millet), **288**
  spread to India, 45
Miombo woodlands, **285**
Missionaries, 119–120, 150
  personal narrative of David

Livingstone, 233–238
Mogadishu, 71
Mombassa, 72, 83, 95, 127
Monsoon exchange, 43–46, **76–79**
Monsoons, 6, 8, 36, **285**, 290
Montane environments, 52(map), 70
Morocco, 177
MOSOP. *See* Movement for the Survival of the Ogoni People
Mosquito, 41–42, 85. *See also* Malaria; Yellow fever
Mount Kenya, 70
Mount Kilimanjaro, 70, 127, 129–130, 133, **281**
Mount Meru, 70, 133
Mountain gorillas, 146
Movement for the Survival of the Ogoni People (MOSOP), 165
Mozambique
  and colonialism, 125, 129–130, 132
  and cotton cultivation, 132
  differential commodity prices for whites vs. African sellers, 130
  early mixed farming/herding cultures, 58
  European farming operations, 155
  and Portuguese traders, 95
  and railroads, 127
  settler colonies, 129
Mwene Mutapa, 73
The Myth of Wild Africa (Adams and Mcshane), 186

Namibia, 58, 166
Napata, 61–62
Natal, 112–114, 116–117, 143
National parks
  and colonialism, 137
  first park, 146, 295
  games laws enforced in, 148
  segregation of human population from reserves, 136, 138, 146, 148, 151, 154, 190–191, 193–194
  South Africa, 125, 281
  success of, 140–141
  Tanzania, 150, 182. *See also* Serengeti
  See also Conservation; Forest conservation; Kruger National Park;

National parks *(cont.)*
    Limpopo National Park; Pasture
    conservation; Selous National Park;
    Virungu National Park; Wildlife
    conservation
Nationalism, and populist critiques of
    conservation programs, 138
Native Americans, 85–86, 98
"Native law," 132
Natural resources, pressures on
    and climate change, 50, 64
    and collapse of Great Zimbabwe, 73
    and commerce revolution of the
        nineteenth century, 103–111
    Ethiopian Highlands, 64
    and population growth, 50, 64
    resource mining in the post-colonial
        era, 165–166
    and variable vegetation bands of
        western Africa, 64–65
    See also Ecological stresses;
        Environmental crises
Nature Conservancy, 272
Nature reserves. *See* Forest
    conservation; Serengeti; Wildlife
    conservation
Navigation, 6–7, 80
Nbata Playa, 36, 39–40, **285–286**
Ndebele, 114, 116
Near East, livestock acquired from,
    43–44
New World
    crops introduced from, 76, 82–85, 261
    depopulated due to post-contact
        diseases, 85–86
    environmental knowledge of Africans,
        87
    slave trade's role in repopulation of, 85
Ngorongoro Conservation Area, 182,
    183(map), 185, 193, 194, 271, 286
Ngorongoro Crater, 185, 187, 191,
    195–196, **286**
Nguni-speaking peoples, 112
Ngwenya iron ore mine, 201(photo)
Niger, 54
Niger Bend, 7–8, 66, 174
Niger River, 6–7, **286**
    and agriculture, 7–8, 259

colonial development schemes, 151,
    158, 179, **287**
and commerce revolution of the
    nineteenth century, 105
environmental destruction from oil
    industry, 164
and glaciation, 27
personal narrative of Mungo Park,
    211–225
and trade, 80
See also Inner Delta (Niger River)
Nigeria
    and colonialism, 125
    early agricultural communities, 52, 53
    early urbanization, 66, 67
    famine and warfare, 162
    iron-working sites, 54
    Nok culture, 53, **287**
    oil production, 155, 164
    Sokoto Caliphate, 178
Nile River, 6–7, **286–287**
    and agriculture, 7–8, 32
    and commerce revolution of the
        nineteenth century, 111
    flood level, 8, 11
    flood level estimation, 69, 70(photo)
    and glaciation, 27
    and Great Lakes, 8
    and ivory trade, 111
    and navigation, 7
Nile Valley
    crops, 37, 174
    and drying of the Sahara, 40, 170
    early sedentary communities, 32,
        33–34, 36, 258
    introduction of sheep and goats, 43
    urbanization of, 12, 60–62, 170
Nilotic language speakers, 53, 178
Nkope culture, 58
Nobadia, 62
Nobel Peace Prize, 168, 266, 283
Nok culture, 53, **287**
Nomadism
    and slavery, 175, 178
    trade and conflict with sedentary
        peoples, 175, 177, 178
    and variable climate/vegetation bands,
        64–65, 175, 178–179

See also Pastoralism
North Africa
  contact with sub-Saharan Africa lost
    during "Big Dry," 175
  early urbanization, 59–62
  and Islam, 79, 260
  Italian Libya, 120(map)
  region defined, 6
  See also Egypt; Nile River; Nile Valley
North Atlantic Oscillation, 3, 11
Northern Rhodesia, 125, 155
Northrup, David, 106
Nubia, 60–62, 287
Nyamwezi people, 109
Nyasaland
  Chelimbwe uprising, 150
  European farming operations, 155
  expatriate-owned plantations, 130
  land clearance by burning banned, 156
Nyerere, Julius, 160–161

Oases, 175
Office du Niger, **287**
Ogoni people, 164–165
Oil industry, 131, 155, 164–165
Oil palm, **287**
  and commerce revolution of the
    nineteenth century, 106–108
  early cultivation of, 38, 51, 198, 199
  expansion of agricultural system
    based on, 52–53, 199
  Great Lakes region, 69
  and Kintampo culture, 54
Olduvai Gorge, 188, **287–288**
Olduwan industry, 17
Oman, 109, 111
Orange Free State, 116, 142
Organization for Economic Co-
  operation and Development, 180
Ox-drawn plowing, 62, 63, 111, 259
Oyo federation, 67

Palm oil. *See* Oil palm
Park, Mungo, 211–225
Pastoralism, **288**
  African resistance to destocking
    policies, 157–158, 162
  conflicts between nomadic and

sedentary people, 65
  and drying of the Sahara, 39–40
  earliest evidence for, 35
  early pastoralism in southern Africa,
    56
  origins of, 32, 36
  pastoral Khoisan speakers, 58–59, 98
  relative freedom from disease of small
    mobile populations, 98
  risk-reduction strategies, 130, 157,
    270
  trade with farmers, 66
  trade with foragers, 56, 142
  and variable vegetation bands of
    western Africa, 64–65, 175
  and wildlife conservation, 191–193
  See also Cattle; Maasai herders;
    Nomadism
Pasture conservation, 147, 157–158, 179
Patronage, 162
Peanuts (groundnuts)
  British "groundnut scheme," 158, **277**
  and commerce revolution of the
    nineteenth century, 106
  introduced from New World, 85, 261
  labor force for, 106
  origins of, 87
  and railroads, 128
Pearl millet, **288**
Pearsall, W.H., 192
Pemba, 83
Peters, Karl, 121
Physiographic regions, 6
Plague, 66, 261
Plant populations, species endemism, 6,
  9, 38, 62
Plantains, 45–46, 56, 70, **268**
Plantations, 130, 262, 263
Pleistocene, 24
Poaching, 165, 194–195, **288–289**
Pollution, 3
Population declines
  and colonialism, 104, 118, 122, 148
  contact-era disease epidemics, 76
  and demographic transition theory,
    166
  eastern Africa, 58, 64, 70, 150, 190
  Ethiopian Highlands, 64

Population declines (cont.)
  and famine during WWI, 150
  Great Lakes region, 58, 70
  and HIV/AIDS, 166
  labor shortages, 148
  reversed by introduction of banana
    cultivation, 58
  and slave trade, 95
  western Africa, 66–67
  western Sahel, 261
  See also Diseases; Famine
Population growth
  and banana cultivation, 46, 70–71,
    205, 261
  demographic transition theory, 159
  dense populations in forest regions,
    55, 67
  desertification blamed on, 179
  fertility control, 150–151
  fueled by demand for labor, 137, 151,
    160
  and intensification of production, 51,
    137
  and medical advances, 159–160
  and New World crops, 96
  and Portuguese trade on the Gold
    Coast, 83
  reasons for post–WWI growth,
    150–151, 159
  recovery of population in the 1930s,
    148, 150, 265
  societal responses to, 50
  and spread of early food-producing
    communities, 50
Population statistics, 6
Port Elizabeth, 131
Portugal
  and Age of Exploration, 80–83
  and gold trade, 81, 83
  use of maize, 83
Portuguese colonialism, 118, 120(map),
  121, 125
  and East Africa, 95, 125
  game controls, 145
  and gold trade, 91, 95
  and slave trade, 86, 91, 93–94, 108
  use of African troops, 121
Pottery, 36, 50, 58

Public health, 150, 159–160. See also
  Sanitation

Quagga, 115–116
Quaternary Era, 24
Quattara Depression, 172
Quelea birds, **289**
Quinine, 119, 159, 284

"Race," as historical construct, 20
Railroads, 125, 127–128, 150
Rainfall
  crises due to variability of, 64, 69–70
  and dense populations in forest
    regions, 67
  distribution of, 11
  environmental zones of Eritrean and
    Ethiopian Highlands, 62
  estimation by Rodah Nilometer, 69,
    70(photo)
  and Great Lakes, 8
  increase after 1985, 181
  rainfall catchments and shadows, 11
  Sahara Desert, 172
  and seasons, 11
  Serengeti, 184, 188
  variability of, 3, 4, 8, **11–12**, 64, 69–70
  See also Monsoons
Rainforests
  agricultural zones of western Africa,
    52(map)
  and agriculture, 8–9, 38–39, 84, 86,
    106–107, 197–199
  botanical climax, 8
  and cassava, 84, 96
  and climate change, 27, 39
  and commerce revolution of the
    nineteenth century, 106–107
  corruption spurred by wood-cutting
    rules, 157
  and disease, 67–68
  early farmers, 38–39, 197–199
  early foragers, 198, 199
  early settlement patterns, 55, 197–198
  early urbanization, 67
  equatorial rainforest described, **274**
  goats' ability to survive in, 199
  high population density in, 55, 96

and logging, 9, 79, 272
loss of high canopy, 8
and oil palm, 106–107, 198
population increases in and resulting
 clearing of, 83
and rainfall, 11
and rubber, 263
soil characteristics, 67
wildlife, 9
See also Deforestation; Forest
 conservation; Forest–savannah
 border
Regs, 172
Republic of South Africa (Transvaal),
 116
Rhinoceros, 141, 165, 187, 195–196, 286
Rhodes, Cecil, 121, 125
Rhodesia, 125, 130, 144, 155, 291. *See
 also* Zimbabwe
Rice
 African rice, **267**
 Asian rice introduced into eastern
  Africa, 46, 78
 domestication of, 8
 early cultivation of, 38, 51, 259
 Guinea rice, 51–52
 spread in West African coastal
  regions, 55
Rift Valley, 8, **289**
 and Bantu language speakers, 57
 ecological diversity of, 9, 56
 and human origins, 14–15
 Olduvai Gorge, 188, **287–288**
 spread of early food-producing
  societies, 56–57
Rift Valley fever, 40
Rinderpest, 104, 121–122, 263, **289**, 295
 and Maasai pastoralists, 190
 nature reserve idea originating from
  lands emptied by epidemic,
  141–142, 144, 292
 reestablishment of the tsetse-bearing
  bush following epidemic, 122,
  141–142, 190, 263, 285, 295
 and Sahel/Sahara Desert, 179
 and southern Africa, 282
 vaccine for, 192, 194, 289
 and wildlife populations, 190, 192, 293

Risk-reduction strategies among
 livestock holders, 130, 157, 270
River blindness, 8, 29, 42, 51
Rivers
 and early agricultural communities,
  32
 and early urban societies. *See* Inner
  Delta; Nile Valley
 and navigation, 6–7
 water-borne diseases as inhibitors of
  agriculture, 51
 See also specific rivers
Roads, 150, 155
Rodah Nilometer, 70(photo)
Rodney, Walter, 90
Roosevelt, Theodore, 139, 140(photo),
 264, 290
Ruaha game reserve, 150
Rubber, 128–129, 130, 146, 263
Rufiji Delta, 8
Ruwenzori mountain range, 200
Rwanda
 and cattle, 270
 coffee production, 130
 crisis of 300–500 C.E., 58
 Tutsi-dominated state, 71

Safaris, 264, 288, **289–290**
Sahara Desert, **169–181**, 290
 and Arab people, 177, 178, 290
 as barrier/filter, 4, 9, 77, 169, 170
 "The Big Dry" (300 B.C.E. to 300 C.E.),
  65, 175
 and camels, 45, 80, 170, 175, 260
 chronology, 257–265
 and climate change, 26–27, 169–174,
  290
 and colonialism, 178
 dry phases, 4, 9, 12, 27, 65, 174(table),
  174–175, 177–178
 drying of the Sahara and expansion of
  agricultural societies, **39–43**, 60,
  170, 173–174, 290
 geographical characteristics, 171–172
 and glaciation, 12, 173
 "green" phase, 33, 44, 173, 290
 and human civilization, 9
 and human evolution, 169–170

Sahara Desert *(cont.)*
  as pump, 12, 64–65, 169–170
  rainfall, 172
  salt trade, 66, 80, 175, 176(photo)
  and trans-Atlantic trade, 177–178
  trans-Sahara trade route, 77, 175, 177,
    260, 290, 294
  See also Berber language speakers;
    Desertification; Drought
Sahel, **290–291**
  banded ethnic distribution, 175
  collapse of population (1400), 261
  and colonialism, 178
  defined, 172
  drought of 1890s, 179
  drought of 1910–1915, 179, 264
  drought of the 1600s to 1800s,
    177–178
  drought of the 1940s, 265
  drought of the 1970s, 138, 162, 180, 273
  environmental degradation of, 173
  and famine, 179
  and Holocene Optimum, 173
  and irrigation, 180
  and millet, 288
  moist phase, 260
  personal narrative of Mungo Park,
    211–225
  rinderpest epidemic, 179
Sahelian language speakers, 41
Salt trade, 60, 66, 80, 175, 176(photo)
  Lake Victoria region, 68(map)
  and slavery, 175, 178
San people, 59, 98, 275, 281, **291**
  and Boer expansion, 115
  decimated by smallpox, 99
  trade with agricultural and pastoral
    people, 142
  See also Khoisan speakers;
    Livingstone, David
Sandawe people, 58
Sanhaja people, 177
Sanitation, 119, 150
Saro-wiwa, Ken, 164–165
Saskawa/Global 2000, 254–256
Savannah
  and banding effect in western Africa,
    52(map), 64–65

and Bantu language speakers, 202–203
  crops, 83, 85
  early urbanization, 65
  land clearance as check on animal
    populations, 142
  and rinderpest epidemic, 122
  See also Forest–savannah border;
    Serengeti
Schistosomiasis, 29, 42, 51
Schmidt, Peter, 53, 57–58, 69, 202
Schoenbrun, David, 197
Scott, James, 138
Seasons, **11**, 186, 285, 294
Selous, F.C., 139, 146, **291**
Selous National Park, 150, 291
Semitic language speakers, 34, 63
Senegal, 178
Senegal River, 6, 27, 80, 81, 178
Senegal Valley, 66, 105, 106
Serengeti, **181–196**, **291–292**
  and Bantu language speakers, 57
  cattle in, 188, 191–192
  climate, 184, 188
  conflicts over land use, 191–192, 194
  conflicts over local peoples' hunting
    for meat, 194–195, 196
  early human populations, 188
  and economic problems of the 1970s,
    195
  farmland surrounding, 196
  geographical characteristics, 181–182,
    184–187
  human communities in, 184, 188,
    192
  and ivory trade, 189–190
  landscape managed through burning,
    188, 191
  and Little Ice Age, 188
  and Maasai pastoralists, 144, 188,
    190–196
  recovery of elephant population, 194,
    196
  reserves and parks listed, 182,
    183(map)
  and rinderpest epidemic, 144, 190, 192
  and tourism, 184, 193–194, 195, 196
  and tsetse fly, 188
  wildlife, 182, 184–187, 194, 291–292

Serengeti Must Not Die (film), 139–140, 192, 278
Serengeti National Park, 182, 183(map)
Sesame, 45
Shaka, 104, 112, 113
Shari River, 27
Sheep, **292**
    expansion into southern Africa, 56, 203
    importation into Africa, 36, 43–44, 198
    resistance to disease, 41, 44
    and South Africa, 117
Shell Oil, 165
Shifting cultivation, **292**. *See also* Swidden agriculture
Sickle cell trait, 67, 85
Sierra Leone, 106
Sisel, 127, 130
Slash and burn agriculture. *See* Swidden agriculture
Slave trade, 94(map)
    and African social organization, 90
    Africa's ability to supply numerous slaves, 89–92
    and commerce revolution of the nineteenth century, 111
    decline of, 108
    eastern Africa, 108
    and gold trade, 83
    and Islam. *See* Slave trade, and Islamic world
    and Portuguese traders, 81, 91
    slave raids, 89, 92
    trans-Atlantic. *See* Slave trade, trans-Atlantic
    trans–Indian Ocean, 109
    and underdevelopment, 90
Slave trade, and Islamic world, 79, 94(map)
    early trade systems, 66
    formalization of slavery, 90
    intensification after 800 C.E., 72
    intensification during drought, 66–67
    Islamic reform movement of early 1600s, 92
Slave trade, trans-Atlantic, 76, 82, 94(map), 108, 262
    and commerce revolution of the nineteenth century, 105, 108, 111
    and cycle of violence, 91–92, 95
    decline in demand, 88–89
    and disease environments, 76, 85–86, 87, 96
    end of, 105–106
    environmental impact on Africa, **86–96**
    and environmental knowledge of Africans, 87
    and French presence in West Africa, 178
    high mortality/high imports, 88
    mortality during process of enslavement, 88, 95
    negative population growth rate in New World, 87–88
    origins of, 85–87
    and population decline in western Africa, 95
    price of slaves, 91–92
    sex ratio of slaves, 96
    "slaving frontier," 91, 92, 94
    statistics, 88, 95
    See also Park, Mungo
Slavery, inter-African
    and commerce revolution of the nineteenth century, 103–106, 108, 115
    "lineage" slavery, 90
    and salt mines, 175, 178
    "slaving frontier," 108
    and South Africa, 115, 116
    "village" slavery, 106
Smallpox, 40, 122, 190
Soba, 62
Social cooperation among early humans, 17–18
Social organization of African societies
    and banana cultivation in Great Lakes region, 69–71
    caste and ethnic specialization in West African societies, 55–56, 65–66, 175
    and colonial taxation systems, 132
    compared to other parts of the world, 51

Social organization of
     African societies *(cont.)*
  diamond mines and bridewealth, 123
  and early Bantu speakers, 199, 204
  and fertility control, 150–151
  and "frontier" concept, 50–51
  and Islam, 79
  kin groups, 51
  and livestock, 58, 72–73, 130, 157,
     204, 270
  and property ownership, 115
  risk-reduction strategies among
     livestock holders, 130, 157, 270
  and slave trade, 90
  status of women, 132
  See also Kongola, Musa
Society for the Preservation of [Wild]
     Fauna of the Empire, 145–146
Soil
  and cassava cultivation, 135
  and coffee production, 152
  conservation of. *See* Soil conservation
  degradation of, 179–181. *See also*
     Environmental degradation
  fragility of, 90
  infertility of, 8, 90
  laterite soils, 8, 90
  maintaining fertility of, 51, 90, 124,
     204
  volcanic soils, 8, 62
Soil conservation, 264, **292–293**
  and colonialism, 137, 155–157, 158,
     292–293
  extensive labor required for, 138,
     156–157, 293
  resettlement schemes, 158
Sokoto Caliphate, 178
Somalia, 71
Songhai, Empire of, 177
Songhai people, 65–66, 92
Sonnike-speaking people, 106
Sorghum, **293**
  early cultivation of, 35, 36, 37, 51, 200
  gathering techniques, 37, 53
  Great Lakes region, 69, 200
  and iron tools, 53, 293
  spread to Arabia, 45
  spread to Asia, 78

Sotho people, 100, 115, 123
South Africa, 293
  agricultural science in, 147
  Apartheid, 124–125, 153–154. *See also*
     Strangled peasantry
  bifurcated landscape, 131, 294
  borders (1910), 120(map)
  British–Boer conflict, 116–118, 124
  coal mines, 131
  and commerce revolution of the
     nineteenth century, 104
  conflicts over grazing lands, 157
  diamond mines, 104, 117–118, 131
  Dutch settlers, 96–97
  geographical characteristics, 97–98
  gold mines, 104, 118
  HIV/AIDS, 166
  industrialization, 131, 153
  Khoisan speakers, 58
  Kwale culture, 58
  labor force, 124–125
  and land ownership, 115–118, 124
  landscape alteration/reorganization of
     space under colonial rule, 115–118,
     131
  national parks, 125
  pasture conservation, 147
  racial division of wealth and power,
     104, 124–125, **294**
  railroads, 127
  rise of South African state, 112–118
  and slave trade, 108
  soil conservation, 147
  springbok exterminated, 143–144,
     186, **293**
  "squatters," 154
  wool industry, 117
  See also Boers; Transvaal; Zulu
     kingdom
Southern Africa, 114(map)
  and Bantu language speakers, 260
  cattle and social organization, 72–73
  climate, 97–98
  commerce in the nineteenth century,
     111–118
  decline in wildlife population, 141,
     142, 263
  diamond mines, 112

early Dutch settlers, 86–87, 96–97, 97(map), 262
early mixed farming/herding cultures, 58
early pastoral cultures, 56
early urbanization, 72–73, 261
and European contact, **96–100**
geographical characteristics, 6
gold mines, 112, 282
miombo woodlands, 285
multicultural society of, 99
and peanuts, 85
personal narrative of David Livingstone, 233–238
region defined, 6
and rinderpest epidemic, 282
and slave trade, 94–95, 115
spread of iron working and agricultural communities, 58–59
variability of rainfall, 8
wildlife, 9
Wilton culture, **295–296**
and Xhosa disaster of 1857, 116–117
See also Boers; Khoisan speakers; Limpopo River; Zambezi River; Zulu kingdom; specific countries and regions
Southern Oscillation, 3, 11–12
Southern Rhodesia, 125, 129, 154, 291. See also Zimbabwe
Spain, 86, 118, 120(map)
Speke, John H., 225–227
Springbok, 143–144, 186, **293**
Stanley, Henry, 121, 139, 233
Steppe, 52(map), 64
Stone Age, 17, 29, 30(fig), **31–39**
Strangled peasantry, **294**
Sudan, **294**
  and agriculture, 32–38
  Arabization of, 178
  and colonialism, 122, 178
  donkeys and horses used for trade in, 80
  drought of 1890s, 179
  drying of the Sahara and expansion of agricultural societies, 40, 60, 174
  famines, 152–153, 162
  and gold trade, 177

iron working, 54
Kush civilization, 61(photo)
land clearance as check on animal populations, 142
lineage slavery, 90
and rinderpest epidemic, 122
white-eared kob migration, 186
See also Inner Delta (Niger River)
Sudanic language speakers, 201, 258
Sudd, **294**
Sugar cane, 46, 78, 103
Sugar plantations, 86, 88, 111, 262
Swahili language speakers
  and commerce revolution of the nineteenth century, 109, 111
  and Islam, 72
  and ivory trade, 109
  urbanization and outside contact, 71–72
Swazi state, 114
Swaziland, 166
Swidden agriculture (long fallow), 51, 90, 124, 204, **292**

Ta-Seti state, 60
Table Bay, 97
Tanganyika
  coffee production, 129–130, 133
  colonial forest conservation efforts, 146
  destocking policies, 157–158
  expatriate-owned plantations, 130
  failure of British "groundnut scheme," 158
  failure of terrace farming scheme, 159
  famines, 152–153
  independence, 160, 193
  mechanized wheat farming, 191
  resettlement schemes, 158
  and Serengeti, 191
  shooting culture, 148
  See also Tanzania
Tangire National Park, 185
Tanzania
  Bantu language speakers, 56
  border closed, 195
  and colonialism, 104
  crisis of 300–500 C.E. (Great Lakes region), 58

Tanzania *(cont.)*
  Cushitic language speakers, 57
  differential commodity prices for
    whites vs. African sellers, 130
  early urbanization, 71
  economic crisis of the 1970s, 195
  expansion of cattle into, 56
  Hadza people, 58
  HIV/AIDS, 166
  Khoisan speakers, 58
  nature reserves, 182, 183(map), 185
  Olduvai Gorge, 188, **287–288**
  personal narrative of Musa Kongola,
    207–211
  population decline, 70
  and resistance to destocking policies,
    157, 162
  Sandawe people, 58
  Selous National Park, 150
  settler colonies, 130
  and tourism, 196
  wildlife policy, 251–254
  See also Ngorongoro Crater; Rift
    Valley; Serengeti; Tanganyika;
    Zanzibar
Taro, 79
Taxation, 104, 132
Tea, 130
Tectonic activity, 9
Teff, 38, 62, 63, **294**
Temperature, 4, 6, 12. *See also* Global
    warming
Terrace farming, 63, 156, 157, 158–159,
    259
Theileriosis. *See* East Coast fever
Thomson's gazelles, 186
Thornton, John, 90–91
Tibesti Massif, 12, 172, 177, 290
Tigré language, 63
Tigrinya language, 63
Timber trade, 9, 79, 272
Timbuktu, 66, **294**
Tobacco, 76, 85
Tomatoes, 85
Tool use by early humans, 17–18,
    28–29, 33, 36
Torwa state, 73
Tourism, 193–194, 288

  ecotourism, **273–274**
  and Serengeti, 184, 195, 196
  tourist hunting, 196, 264, **289–290**
Trade
  and collapse of Great Zimbabwe, 73
  Columbian exchange era, **75–101**
  commerce between East Africa and
    the Arabian Peninsula, 71
  commerce limited by transportation
    issues, 80, 108–109
  commerce revolution of the
    nineteenth century, 103, 105–111
  early Common Era, 59–60, 66–67
  and European Age of Exploration,
    80–83
  "monsoon exchange" across Indian
    Ocean, 43–46, **76–79**
  and New World crops, 76
  trans-Sahara trade, 77, 80, 177, 260,
    290, 294
  urbanization and outside contact
    (Swahili civilization), 71–72
  See also Copper; Diamonds; Gold
    trade; Ivory trade; Salt trade; Slave
    trade, and Islamic world; Slave
    trade, trans-Atlantic; Timber trade
Trade winds, 80–81
Transportation
  and cash crops, 133–134
  colonial development schemes, 155
  difficulties of water transport, 127
  and famine, 152–153
  infrastructure and extension of
    colonial markets, 150
  railroads, 125, 127–128, 150
  and spread of infectious disease, 150
Transvaal
  gold mines, 118, 123, 131
  Kruger National Park, 125, **281–282**
  laborers attracted by meat, 142
  wildlife conservation, 143
Trypanosomiasis. *See* Tsetse fly
Tsetse fly, 29, 40–41, **294–295**
  and agriculture/animal husbandry,
    39–43
  cattle range limited by, 39, 51, 55, 56,
    68, 198
  commerce limited by, 80

game animals as reservoir for tsetse, 40, 145, 198, 204
habitat for, 285
horse range limited by, 144
and land clearance, 40–41
pathology of trypanosomiasis, 40–41, **294–295**
and preservation of game populations, 141, 144
reduced incidence of, 41, 159
and reestablishment of tsetse-bearing bush following rinderpest epidemic, 122, 141–142, 190, 295
and Serengeti, 188
and settlement/land-use patterns, 41, 141
Tswana peoples, 100, 233
Tuberculosis, 150, 166
Tutsis, 71
Tutu, Osei, 92, 93
Twa peoples, 199

Ubangi River, 7
Uganda
coffee production, 130
early urbanization, 71
HIV/AIDS, 165, 166
and railroads, 127
resistance to soil conservation rules, 157
Uhuru Park, 168
Uluguru Land Usage scheme, 159
Underdevelopment, and slave trade, 90
United Nations, 162, 180
United States, 106, 292. *See also* Slave trade, trans-Atlantic
Upare (mountain), 70
Upemba Depression, 73–74, **295**
Upper Nile, 7–8
Urbanization
and caste/ethnic specialization in western Africa, 65–66
central Africa, 73
and commodity production, 135
and crises of production and survival, 60, 66–67
earliest urban centers, **59–65**, 170, 280
eastern Africa, 62–64, 68–72

and late colonial era, 153
northeastern Africa, 59–62, 170
southern Africa, 72–73, 131, 261
urban farming, 135
walled towns, 67
western Africa, 64–68
Usambara (mountain), 70

Vaccine for rinderpest, 192, 194, 289
Vansina, Jan, 197
Vegetation, 11, 13(map), 26(map). *See also* Forest–savannah border; Landscape alteration; Miombo woodlands; Rainforests; Savannah
Virungu National Park, 146, **295**
Vivax malaria, 284
Volcanism, 9, 19–20, 24–25, 62
von Wissmann, Hermann, 144

Warfare
Boer War, 118, 120(map), 124, 131
civil strife in independent post-colonial states, 138
colonial powers' use of African troops, 120–121
and colonialism, 104, 120(map), 122, 124, 264–265
European advantages, 118–119
and famine, 162
impact on conservation programs, 141
Zulu wars, 116, 121
Weaver birds, **289**
Webb, James L.A., 177
Wendorf, F., 34, 35
Western Africa, 94–95
agricultural zones, 52(map), 64–65
and cassava, 83, 96
caste and ethnic specialization, 55, 65–66, 175
climate, 8
and colonialism, 103, 105–108, 119
and commerce revolution of the nineteenth century, 103, 105–108
contact-era disease epidemics, 76
contact with North Africa lost during "Big Dry," 175
corruption spurred by wood-cutting rules, 157

Western Africa *(cont.)*
  declining wildlife populations, 141, 142
  dense populations in forest regions, 55, 96
  early agricultural communities/settlement patterns, 51–56, 174, 197–198, 259
  early urbanization, 64–68, 280
  expansion of iron-working sites, 54
  famines, 152–153
  forest conservation, 146
  game controls, 145
  game reserves, 146
  geographical characteristics, 6
  and gold trade, 81. *See also* Gold trade
  lineage slavery, 90
  and maize, 82, 83
  mosaic landscape, 8–9
  and oil palm, 106–107, **287**
  and peanuts, 85, 106
  and Portuguese traders, 81
  proto-Bantu language, 52–53
  "pulse" effect of changing conditions, 65, 66
  railroads, 127, 128
  rainforests, 8, 55. *See also* Rainforests
  region defined, 6
  rubber boom, 263
  and slave trade, 81, 87–95, 178
  trade, 66–67. *See also* Gold trade
  trade states (slave trade), 92(map)
  and yams, **296**
  See also British colonialism; French colonialism; Niger River; Sahel; Senegal River; *specific countries and regions*
Wheat
  and climate, 32, 75
  early cultivation of, 36, 62, 63, 174, 287
  mechanized wheat farming in Tanganyika, 191
White Nile, 286
Wildebeest, 32, 186–187, 190, 194, 196, 291–292
Wildlife
  animals best adapted to landscapes

  altered for agriculture, 32
  co-evolution of humans and wildlife, 29
  decimated by rinderpest, 190, 192
  decline in southern Africa, 141, 263
  and early urban societies, 51
  extermination of springbok, 143–144, 186
  extinction of quagga, 115–116
  game animals as reservoir for tsetse, 40, 145, 198, 204
  game populations in regions unsuitable for human habitation due to disease, 29, 39, 41, 144
  game reserves. *See* Serengeti; Wildlife conservation
  land clearance as check on animal populations, 142, 204
  and miombo woodlands, 285
  and Ngorongoro Crater, 286
  poaching, 165
  species endemism, 6, 9, 273
  See also specific animals
Wildlife conservation, 137, **139–153**
  chronology, 264–266
  colonial game laws, 142–145
  colonial reserve systems, 145–146, 282
  "fortress conservation," 32, 141
  and land emptied by rinderpest epidemic, 141–142, 144
  and pastoralism, 191–193
  recovery of elephant population, 165, 194, 196
  segregation of human population from reserves, 136, 146, 148, 151, 190, 194
  success of, 140–141
  Tanzania's wildlife policy, 251–254
  and tourism, 184
  See also Serengeti
Wilks, Ivor, 91
Wilton culture, **295–296**
Wolof language speakers, 65–66
Women
  food production and changes in women's status under colonialism, 132

sex ratio in slave populations, 96
women's rights promoted by Wangari
  Maathai, 167, 168
women's rights repressed under
  colonially sanctioned "native law,"
  132
See also Kongola, Musa; Park, Mungo
World Bank, 138, 162
World War I, 127, 146, 150, 191
World War II, 152–153
World Wide Fund for Nature (WWF),
  272

Xhosa people, 100, 275
  and Boer expansion, 115
  and British colonialism, 123
  disaster of 1857, 116–117

Yams, **296**
  agricultural zones of western Africa,
    52(map)
  early cultivation of, 51, 198, 199
  expansion of agricultural system
    based on, 52–53
  Great Lakes region, 69, 201
  and Kintampo culture, 54
  as staple crop, 52
Yellow fever, 67–68, 85–86
Yoruba cities, 67

Zaire River. *See* Congo River
Zambezi River, 8, 277, **296**
Zambia
  and colonialism, 104
  early mixed farming/herding cultures,
    58
  expansion of early agricultural
    communities, 56
  HIV/AIDS, 166
Zanga people, 177
Zanzibar, 72, 83, 95. *See also* Tanzania
Zebra, 32, 186, 190, 292
Zimbabwe
  CAMPFIRE conservation program,
    271
  and colonialism, 104
  early mixed farming/herding cultures,
    58
  and gold trade, 72, 79, 296
  Great Zimbabwe state's rise and fall,
    72–73, 261, **296**
  HIV/AIDS, 166
  Kalundu culture, 58
  Nkope culture, 58
  See also Southern Rhodesia
Zimbabwe Plateau, 73
Zulu kingdom, 104, 262
  formation of, 112–114
  Zulu wars, 115, 121